AF676010

Yeast Biotechnology 6.0

Yeast Biotechnology 6.0

Editor

Ronnie Willaert

MDPI

Basel • Beijing • Wuhan • Barcelona • Belgrade • Novi Sad • Cluj • Manchester

Editor
Ronnie Willaert
Department of
Bioengineering Sciences
Vrije Universiteit Brussel
Brussels
Belgium

Editorial Office
MDPI AG
Grosspeteranlage 5
4052 Basel, Switzerland

This is a reprint of articles from the Special Issue published online in the open access journal *Fermentation* (ISSN 2311-5637) (available at: https://www.mdpi.com/journal/fermentation/special_issues/yeast_6).

For citation purposes, cite each article independently as indicated on the article page online and as indicated below:

Lastname, A.A.; Lastname, B.B. Article Title. *Journal Name* **Year**, *Volume Number*, Page Range.

ISBN 978-3-7258-2293-5 (Hbk)
ISBN 978-3-7258-2294-2 (PDF)
doi.org/10.3390/books978-3-7258-2294-2

© 2024 by the authors. Articles in this book are Open Access and distributed under the Creative Commons Attribution (CC BY) license. The book as a whole is distributed by MDPI under the terms and conditions of the Creative Commons Attribution-NonCommercial-NoDerivs (CC BY-NC-ND) license.

Contents

About the Editor

Ronnie Willaert

Ronnie Willaert (Vrije Universiteit Brussel, Brussels, Belgium) has extensive research expertise in yeast (*Saccharomyces cerevisiae, Candida* species), yeast space biology, yeast adhesins), cell (yeast) immobilisation biotechnology, fermentation technology, brewing science and technology, single-molecule biophysics (atomic force microscopy), micro(bio)fabrication, and bone cell mechanobiology.

fermentation

Editorial

Yeast Biotechnology 6.0

Ronnie G. Willaert [1,2]

1 Research Group Structural Biology Brussels, Vrije Universiteit Brussel, 1050 Brussels, Belgium; ronnie.willaert@vub.be; Tel.: +32-2-629-18-46
2 International Joint Research Group VUB-EPFL BioNanotechnology & NanoMedicine (NANO), 1015 Lausanne, Switzerland

1. Yeast Biotechnology 6.0

This Special Issue continues the "Yeast Biotechnology" Special Issue series of the MDPI journal Fermentation. This issue compiles the current state-of-the-art research and technology around "yeast biotechnology". This issue highlights prominent current research directions in the fields of yeasts as a cell factory, yeast nanobiotechnology, wine yeasts and wine fermentation, yeasts and food fermentation, and biocontainment for yeast biotechnology. We very much hope that you enjoy reading the articles contained herein and are looking forward to continuing this Special Issue series in the Topical Collection "Yeast Biotechnology".

2. Yeasts as Cell Factory

Park et al. [1] reviewed the role of CRISPR-Cas engineered nonconventional yeasts (NCYs) as emerging microbial cell factories. Conventional widely used yeasts such as *Saccharomyces cerevisiae*, which have been widely used as a microbial cell factory [2,3], have some disadvantages. Product profiles are often restricted due to the Crabtree-positive nature of *S. cerevisiae*, and ethanol production from lignocellulose is possibly enhanced by developing alternative stress-resistant microbial platforms. Alternatively, NCYs may be considered an alternative microbial platform for industrial fermentations since they have desirable metabolic pathways and regulation, and they have a strong resistance to diverse stress factors [4,5]. This review describes the current status of and recent advances in promising NCYs in terms of industrial and biotechnological applications, highlighting CRISPR-Cas9-system-based metabolic engineering strategies.

Noseda et al. [6] developed a cost-effective process for the heterologous production of SARS-CoV-2 spike receptor binding domain using *Pichia pastoris* (*Komagataella phaffii*) in a stirred-tank bioreactor. The spike protein of SARS-CoV-2 is one of the most exposed proteins [7]. The receptor-binding domain (RBD), which is a fragment of the Spike protein, interacts with the ACE2 receptors of human cells, allowing the entrance of viruses. The RBD has been proposed to be an interesting protein for the development of diagnosis tools and treatments for and the prevention of this disease [8]. A method for recombinant RBD production using *P. pastoris* as a cell factory in a stirred-tank bioreactor was developed by the cited authors. The proposed method represents a feasible, simple, scalable, and inexpensive procedure for the production of RBD.

Carneiro et al. [9] reviewed advances in *K. phaffii* engineering for the production of renewable chemicals and proteins. *K. phaffii* has been extensively used in the production of heterologous proteins [10–12] and, recently, as a cell factory to produce various chemicals through new metabolic engineering and synthetic biology tools [13]. This review summarizes *Komagataella* taxonomy, diversity, and recent approaches in cell engineering to producing renewable chemicals and proteins. Finally, strategies for optimizing and developing new fermentative processes using *K. phaffii* are discussed.

Citation: Willaert, R.G. Yeast Biotechnology 6.0. *Fermentation* **2024**, *10*, 172. https://doi.org/10.3390/fermentation10030172

Received: 26 February 2024
Accepted: 18 March 2024
Published: 19 March 2024

Copyright: © 2024 by the author. Licensee MDPI, Basel, Switzerland. This article is an open access article distributed under the terms and conditions of the Creative Commons Attribution (CC BY) license (https://creativecommons.org/licenses/by/4.0/).

3. Yeast Nanobiotechnology

Radonicic et al. [14] developed a rapid antifungal susceptibility testing (AFST) method based on optical nanomotion detection (ONMD) and applied it to the opportunistic fungal pathogen *Candida albicans*. In ONMD, an optical microscope is used to record the nanometric-scale movements of living cells. These recorded cellular nanomotions cease upon cell death, allowing the susceptibility of a cell or specific strain to be easily assessed. Other cellular nanomotion methods have been developed, such as methods based on AFM-cantilever sensors [15–17], plasmonic imaging of the z motion of attached bacteria [18], sensing of the attached bacteria's vibrations using the phase noise of a resonant crystal [19], tracking the x-y motion of attached uropathogenic *Escherichia coli* [20], and subcellular fluctuation imaging, which is based on total internal reflection microscopy (TIRM) [21], as well as optically tracking bacterial responses on micropillar architectures using intrinsic phase-shift spectroscopy [22]. The fast emergence of multi-resistant pathogenic *Candida* species is caused by the extensive and sometimes unnecessary use of broad-spectrum antifungals [23]. Hence, the development of rapid antifungal sensitivity tests would allow for the identification and use of more-selective antifungals, thereby reducing the spread of pathogenic fungi and their evolution toward a multi-resistant phenotype [24,25]. In this study, a microfluidic chip containing an array of microwells that were designed to trap yeast cells was developed. Yeast cell entrapment in a microwell allowed for a very rapid exchange of growth medium with the antifungal, which enabled the performance of single-cell ONMD measurements on the same cell before and after antifungal treatment. The chip was used to quantify the real-time response of individual *C. albicans* cells to the antifungal treatment in as fast as 10 min.

Villalba et al. [26] developed a new method for measuring the adhesion of the opportunistic pathogenic yeast *C. albicans* since classical methods have some drawbacks; for example, the corresponding measurement is relatively complex, requires sophisticated equipment, and, in most cases, cannot be carried out without breaking the links between the studied cell and its target [27–30]. The applied force in the new method is generated by the cell itself, whereas cellular movements are detected via optical microscopy and developed dedicated software. The authors demonstrated that the measurement was non-destructive and single-cell-sensitive and permitted observation of the evolution of adhesion as a function of time. The new cellular nanomotion-based technique was applied for different *C. albicans* strains adhering to a fibronectin-coated surface. This novel approach could significantly simplify, accelerate, and make more affordable living cell–substrate adhesion measurements.

Dekhtyar et al. [31] studied the possible influences of differently electrically charged diamond nanoparticles [32,33] on the physiological characteristics of the yeast *S. cerevisiae*. They revealed that the adverse impact of these nanoparticles can manifest not only against prokaryotes but also against eukaryotic yeast cells. The results also indicated that it is possible to reduce and, most likely, eliminate the dangerous effects of nanoparticles on cells by using special physical approaches. A comparison of non-arylated and arylated nanoparticles showed that in terms of changes in the physiological activity of cells, the selection of certain nanoparticles (non-arylated or arylated) may be necessary in each specific case, depending on the purpose of their use.

4. Wine Yeasts and Wine Fermentation

Akan et al. [34] explored the potential of NCYs in wine fermentation [35,36] with a focus on *Saccharomycopsis fermentans*. Mutant strains resistant to the toxic compound trifluoroleucine (TFL) were selected, mutations in the SfLEU4 gene were verified, and the ability of the resulting strains to contribute to fermentation bouquets was characterized. Resistance to TFL relieved feedback inhibition in the leucine biosynthesis pathway and resulted in increased leucine biosynthesis. The *S. fermentans* TFL-resistant mutants generated increased amounts of isoamyl alcohol and isovalerate during wine fermentation. The selection of TFL-resistant strains provided a generally applicable strategy for the improvement of NCYs and their utilization in co-fermentation processes for different grape must varieties.

Fernández-Fernández et al. [37] used immobilized yeasts to improve the production of sparkling wines. Verdejo sparkling wines were elaborated according to the "champenoise" method, and the second fermentation was developed with the same free or alginate-immobilized [38–41] *Saccharomyces cerevisiae bayanus* yeast strain. These sparkling wines showed no significant differences among the two typologies in terms of enological parameters (pH, total acidity, volatile acidity, reducing sugars, and alcoholic strength), effervescence, or spectrophotometric measurements. The free amino nitrogen content was significantly higher in the sparkling wines obtained from immobilized yeasts; the levels of neutral polysaccharides and total proteins were lower. No significant differences in the volatile composition were found, except for only two volatile compounds (isobutyric acid and benzyl alcohol) that were present at levels below their respective olfactory thresholds. The sensory analysis conducted by consumers showed identical preferences for both types of sparkling wines, except in terms of color acceptability. The descriptive analysis carried out by a tasting panel revealed that sensorial differences between both sparkling wines were only found regarding the smell of dough. The authors showed that using immobilized yeasts for the second fermentation of sparkling wines can reduce and simplify some enological practices such as riddling and disgorging, with no impact on quality parameters.

5. Yeasts and Food Fermentation

Bencresciuto et al. [42] evaluated starter cultures of lactic acid bacteria (*Lactobacillus plantarum* strains) and killer yeasts (*Wickerhamomyces anomalus* and *S. cerevisiae)* for the fermentation of table olives to debitter olives [43] and improve their organoleptic quality and safety [44,45]. This study aimed to assess their potential to avoid pretreatments and the use of excessive salt in the brines and preservatives. The final oleuropein levels in the olives were unaffected by the treatments, but the use of these starters did not improve the LABs' growth nor prevent the growth of *Enterobacteriaceae* and molds. The nutraceutical value of the olives could be improved due to the higher production of hydroxytyrosol.

6. Biocontainment for Yeast Biotechnology

Pavão et al. [46] reviewed biocontainment techniques and applications for yeast biotechnology. Biocontainment techniques for genetically modified yeasts (GMYs) [47] are pivotal due to the importance of these organisms in biotechnological processes and due to the design of new yeast strains using synthetic biology tools and technologies. The different biocontainment technologies currently available for genetically modified yeasts (GMYs) were evaluated. Uniplex-type biocontainment approaches (UTBAs), which rely on nutrient auxotrophies induced by gene mutation or deletion or the expression of the simple kill switches apparatus, are still the major biocontainment approaches in use for GMY. While bacteria such as *E. coli* account for advanced biocontainment technologies based on synthetic biology and multiplex-type biocontainment approaches (MTBAs), GMYs are distant from this scenario for many reasons. Therefore, a comparison of different UTBAs and MTBAs applied for GMYs and genetically engineered microorganisms (GEMs) was made, indicating the major advances in biocontainment techniques for GMYs.

Funding: This work was supported by the Belgian Federal Science Policy Office (Belspo) and the European Space Agency, grant number PRODEX project Flumias Nanomotion; The Research Foundation—Flanders (FWO), grant number I002620; and FWO-SNSF, grant number 310030L_197946.

Conflicts of Interest: The author declares no conflicts of interest.

References

1. Park, J.; Kim, I.J.; Kim, S.R. Nonconventional Yeasts Engineered Using the CRISPR-Cas System as Emerging Microbial Cell Factories. *Fermentation* **2022**, *8*, 656. [CrossRef]
2. Parapouli, M.; Vasileiadis, A.; Afendra, A.-S.; Hatziloukas, E. Saccharomyces Cerevisiae and Its Industrial Applications. *AIMS Microbiol.* **2020**, *6*, 1–31. [CrossRef] [PubMed]
3. Rainha, J.; Gomes, D.; Rodrigues, L.R.; Rodrigues, J.L. Synthetic Biology Approaches to Engineer Saccharomyces Cerevisiae towards the Industrial Production of Valuable Polyphenolic Compounds. *Life* **2020**, *10*, 56. [CrossRef]

4. Spencer, J.; Ragout de Spencer, A.; Laluce, C. Non-Conventional Yeasts. *Appl. Microbiol. Biotechnol.* **2002**, *58*, 147–156. [CrossRef] [PubMed]
5. Binati, R.L.; Salvetti, E.; Bzducha-Wróbel, A.; Bašinskienė, L.; Čižeikienė, D.; Bolzonella, D.; Felis, G.E. Non-Conventional Yeasts for Food and Additives Production in a Circular Economy Perspective. *FEMS Yeast Res.* **2021**, *21*, foab052. [CrossRef] [PubMed]
6. Noseda, D.G.; D'Alessio, C.; Santos, J.; Idrovo-Hidalgo, T.; Pignataro, F.; Wetzler, D.E.; Gentili, H.; Nadra, A.D.; Roman, E.; Paván, C.; et al. Development of a Cost-Effective Process for the Heterologous Production of SARS-CoV-2 Spike Receptor Binding Domain Using *Pichia pastoris* in Stirred-Tank Bioreactor. *Fermentation* **2023**, *9*, 497. [CrossRef]
7. Perico, L.; Benigni, A.; Remuzzi, G. SARS-CoV-2 and the Spike Protein in Endotheliopathy. *Trends Microbiol.* **2024**, *32*, 53–67. [CrossRef] [PubMed]
8. Jackson, C.B.; Farzan, M.; Chen, B.; Choe, H. Mechanisms of SARS-CoV-2 Entry into Cells. *Nat. Rev. Mol. Cell Biol.* **2022**, *23*, 3–20. [CrossRef]
9. Carneiro, C.V.G.C.; Serra, L.A.; Pacheco, T.F.; Ferreira, L.M.M.; Brandão, L.T.D.; Freitas, M.N.D.M.; Trichez, D.; Almeida, J.R.M.D. Advances in *Komagataella phaffii* Engineering for the Production of Renewable Chemicals and Proteins. *Fermentation* **2022**, *8*, 575. [CrossRef]
10. Mastropietro, G.; Aw, R.; Polizzi, K.M. Expression of Proteins in *Pichia pastoris*. In *Methods in Enzymology*; O'Dell, W.B., Kelman, Z., Eds.; Recombinant Protein Expression: Eukaryotic Hosts; Academic Press: Cambridge, MA, USA, 2021; Volume 660, pp. 53–80.
11. Khlebodarova, T.M.; Bogacheva, N.V.; Zadorozhny, A.V.; Bryanskaya, A.V.; Vasilieva, A.R.; Chesnokov, D.O.; Pavlova, E.I.; Peltek, S.E. *Komagataella phaffii* as a Platform for Heterologous Expression of Enzymes Used for Industry. *Microorganisms* **2024**, *12*, 346. [CrossRef]
12. Barone, G.D.; Emmerstorfer-Augustin, A.; Biundo, A.; Pisano, I.; Coccetti, P.; Mapelli, V.; Camattari, A. Industrial Production of Proteins with *Pichia pastoris—Komagataella phaffii*. *Biomolecules* **2023**, *13*, 441. [CrossRef] [PubMed]
13. Bernauer, L.; Emmerstorfer-Augustin, A. *Komagataella phaffii* as Emerging Model Organism in Fundamental Research. *Front. Microbiol.* **2021**, *11*, 607028. [CrossRef] [PubMed]
14. Radonicic, V.; Yvanoff, C.; Villalba, M.I.; Devreese, B.; Kasas, S.; Willaert, R.G. Single-Cell Optical Nanomotion of *Candida albicans* in Microwells for Rapid Antifungal Susceptibility Testing. *Fermentation* **2023**, *9*, 365. [CrossRef]
15. Longo, G.; Alonso-Sarduy, L.; Rio, L.M.; Bizzini, A.; Trampuz, A.; Notz, J.; Dietler, G.; Kasas, S. Rapid Detection of Bacterial Resistance to Antibiotics Using AFM Cantilevers as Nanomechanical Sensors. *Nat. Nanotechnol.* **2013**, *8*, 522–526. [CrossRef] [PubMed]
16. Caruana, G.; Kritikos, A.; Vocat, A.; Luraschi, A.; Delarze, E.; Sturm, A.; Verge, M.P.; Jozwiak, G.; Kushwaha, S.; Delaloye, J.; et al. Investigating Nanomotion-Based Technology (Resistell AST) for Rapid Antibiotic Susceptibility Testing among Adult Patients Admitted to a Tertiary-Care Hospital with Gram-Negative Bacteraemia: Protocol for a Prospective, Observational, Cross-Sectional, Single-Arm Study. *BMJ Open* **2022**, *12*, e064016. [CrossRef] [PubMed]
17. Vocat, A.; Sturm, A.; Jóźwiak, G.; Cathomen, G.; Świątkowski, M.; Buga, R.; Wielgoszewski, G.; Cichocka, D.; Greub, G.; Opota, O. Nanomotion Technology in Combination with Machine Learning: A New Approach for a Rapid Antibiotic Susceptibility Test for *Mycobacterium tuberculosis*. *Microbes Infect.* **2023**, *25*, 105151. [CrossRef] [PubMed]
18. Syal, K.; Iriya, R.; Yang, Y.; Yu, H.; Wang, S.; Haydel, S.E.; Chen, H.-Y.; Tao, N. Antimicrobial Susceptibility Test with Plasmonic Imaging and Tracking of Single Bacterial Motions on Nanometer Scale. *ACS Nano* **2016**, *10*, 845–852. [CrossRef]
19. Johnson, W.L.; France, D.C.; Rentz, N.S.; Cordell, W.T.; Walls, F.L. Sensing Bacterial Vibrations and Early Response to Antibiotics with Phase Noise of a Resonant Crystal. *Sci. Rep.* **2017**, *7*, 12138. [CrossRef]
20. Syal, K.; Shen, S.; Yang, Y.; Wang, S.; Haydel, S.E.; Tao, N. Rapid Antibiotic Susceptibility Testing of Uropathogenic *E. coli* by Tracking Submicron Scale Motion of Single Bacterial Cells. *ACS Sens.* **2017**, *2*, 1231–1239. [CrossRef]
21. Bermingham, C.R.; Murillo, I.; Payot, A.D.J.; Balram, K.C.; Kloucek, M.B.; Hanna, S.; Redmond, N.M.; Baxter, H.; Oulton, R.; Avison, M.B.; et al. Imaging of Sub-Cellular Fluctuations Provides a Rapid Way to Observe Bacterial Viability and Response to Antibiotics. *Biorxiv* **2018**, 460139. [CrossRef]
22. Leonard, H.; Halachmi, S.; Ben-Dov, N.; Nativ, O.; Segal, E. Unraveling Antimicrobial Susceptibility of Bacterial Networks on Micropillar Architectures Using Intrinsic Phase-Shift Spectroscopy. *ACS Nano* **2017**, *11*, 6167–6177. [CrossRef] [PubMed]
23. Dadar, M.; Tiwari, R.; Karthik, K.; Chakraborty, S.; Shahali, Y.; Dhama, K. *Candida albicans*—Biology, Molecular Characterization, Pathogenicity, and Advances in Diagnosis and Control—An Update. *Microb. Pathog.* **2018**, *117*, 128–138. [CrossRef] [PubMed]
24. Durand, C.; Maubon, D.; Cornet, M.; Wang, Y.; Aldebert, D.; Garnaud, C. Can We Improve Antifungal Susceptibility Testing? *Front. Cell. Infect. Microbiol.* **2021**, *11*, 720609. [CrossRef] [PubMed]
25. Knabl, L.; Lass-Flörl, C. Antifungal Susceptibility Testing in Candida Species: Current Methods and Promising New Tools for Shortening the Turnaround Time. *Expert Rev. Anti-Infect. Ther.* **2020**, *18*, 779–787. [CrossRef] [PubMed]
26. Villalba, M.I.; LeibundGut-Landmann, S.; Bougnoux, M.-E.; d'Enfert, C.; Willaert, R.G.; Kasas, S. *Candida albicans* Adhesion Measured by Optical Nanomotion Detection. *Fermentation* **2023**, *9*, 991. [CrossRef]
27. Hu, L.; Zhang, X.; Miller, P.; Ozkan, M.; Ozkan, C.; Wang, J. Cell Adhesion Measurement by Laser-Induced Stress Waves. *J. Appl. Phys.* **2006**, *100*, 084701. [CrossRef]
28. Khalili, A.A.; Ahmad, M.R. A Review of Cell Adhesion Studies for Biomedical and Biological Applications. *Int. J. Mol. Sci.* **2015**, *16*, 18149–18184. [CrossRef]
29. Zhou, D.W.; García, A.J. Measurement Systems for Cell Adhesive Forces. *J. Biomech. Eng.* **2015**, *137*, 020908. [CrossRef]

30. Gunaratnam, G.; Dudek, J.; Jung, P.; Becker, S.L.; Jacobs, K.; Bischoff, M.; Hannig, M. Quantification of the Adhesion Strength of *Candida albicans* to Tooth Enamel. *Microorganisms* **2021**, *9*, 2213. [CrossRef]
31. Dekhtyar, Y.; Abols, D.; Avotina, L.; Stoppel, A.; Balakin, S.; Khroustalyova, G.; Opitz, J.; Sorokins, H.; Beshchasna, N.; Tamane, P.; et al. Effects of Diamond Nanoparticles Immobilisation on the Surface of Yeast Cells: A Phenomenological Study. *Fermentation* **2023**, *9*, 162. [CrossRef]
32. Hemelaar, S.R.; van der Laan, K.J.; Hinterding, S.R.; Koot, M.V.; Ellermann, E.; Perona-Martinez, F.P.; Roig, D.; Hommelet, S.; Novarina, D.; Takahashi, H.; et al. Generally Applicable Transformation Protocols for Fluorescent Nanodiamond Internalization into Cells. *Sci. Rep.* **2017**, *7*, 5862. [CrossRef] [PubMed]
33. Morita, A.; Hamoh, T.; Sigaeva, A.; Norouzi, N.; Nagl, A.; van der Laan, K.J.; Evans, E.P.P.; Schirhagl, R. Targeting Nanodiamonds to the Nucleus in Yeast Cells. *Nanomaterials* **2020**, *10*, 1962. [CrossRef] [PubMed]
34. Akan, M.; Gudiksen, A.; Baran, Y.; Semmler, H.; Brezina, S.; Fritsch, S.; Rauhut, D.; Wendland, J. Exploring the Potential of Non-Conventional Yeasts in Wine Fermentation with a Focus on Saccharomycopsis Fermentans. *Fermentation* **2023**, *9*, 786. [CrossRef]
35. Fleet, G.H. Yeast Interactions and Wine Flavour. *Int. J. Food Microbiol.* **2003**, *86*, 11–22. [CrossRef] [PubMed]
36. Jolly, N.P.; Varela, C.; Pretorius, I.S. Not Your Ordinary Yeast: Non-Saccharomyces Yeasts in Wine Production Uncovered. *FEMS Yeast Res.* **2014**, *14*, 215–237. [CrossRef] [PubMed]
37. Fernández-Fernández, E.; Rodríguez-Nogales, J.M.; Vila-Crespo, J.; Falqué-López, E. Application of Immobilized Yeasts for Improved Production of Sparkling Wines. *Fermentation* **2022**, *8*, 559. [CrossRef]
38. Kourkoutas, Y.; Bekatorou, A.; Banat, I.M.; Marchant, R.; Koutinas, A.A. Immobilization Technologies and Support Materials Suitable in Alcohol Beverages Production: A Review. *Food Microbiol.* **2004**, *21*, 377–397. [CrossRef]
39. López de Lerma, N.; Peinado, R.A.; Puig-Pujol, A.; Mauricio, J.C.; Moreno, J.; García-Martínez, T. Influence of Two Yeast Strains in Free, Bioimmobilized or Immobilized with Alginate Forms on the Aromatic Profile of Long Aged Sparkling Wines. *Food Chem.* **2018**, *250*, 22–29. [CrossRef]
40. Benucci, I.; Cerreti, M.; Maresca, D.; Mauriello, G.; Esti, M. Yeast Cells in Double Layer Calcium Alginate–Chitosan Microcapsules for Sparkling Wine Production. *Food Chem.* **2019**, *300*, 125174. [CrossRef]
41. Prokes, K.; Baron, M.; Mlcek, J.; Jurikova, T.; Adamkova, A.; Ercisli, S.; Sochor, J. The Influence of Traditional and Immobilized Yeast on the Amino-Acid Content of Sparkling Wine. *Fermentation* **2022**, *8*, 36. [CrossRef]
42. Bencresciuto, G.F.; Mandalà, C.; Migliori, C.A.; Cortellino, G.; Vanoli, M.; Bardi, L. Assessment of Starters of Lactic Acid Bacteria and Killer Yeasts: Selected Strains in Lab-Scale Fermentations of Table Olives (*Olea europaea* L.) Cv. Leccino. *Fermentation* **2023**, *9*, 182. [CrossRef]
43. Marsilio, V.; Lanza, B.; Pozzi, N. Progress in Table Olive Debittering: Degradationin Vitro of Oleuropein and Its Derivatives byLactobacillus Plantarum. *J. Am. Oil Chem. Soc.* **1996**, *73*, 593–597. [CrossRef]
44. Ambra, R.; Natella, F.; Bello, C.; Lucchetti, S.; Forte, V.; Pastore, G. Phenolics Fate in Table Olives (*Olea europaea* L. Cv. Nocellara Del Belice) Debittered Using the Spanish and Castelvetrano Methods. *Food Res. Int.* **2017**, *100*, 369–376. [CrossRef] [PubMed]
45. Xu, F.; Li, Y.; Zheng, M.; Xi, X.; Zhang, X.; Han, C. Structure Properties, Acquisition Protocols, and Biological Activities of Oleuropein Aglycone. *Front. Chem.* **2018**, *6*, 00239. [CrossRef]
46. Pavão, G.; Sfalcin, I.; Bonatto, D. Biocontainment Techniques and Applications for Yeast Biotechnology. *Fermentation* **2023**, *9*, 341. [CrossRef]
47. Hanlon, P.; Sewalt, V. GEMs: Genetically Engineered Microorganisms and the Regulatory Oversight of Their Uses in Modern Food Production. *Crit. Rev. Food Sci. Nutr.* **2021**, *61*, 959–970. [CrossRef] [PubMed]

Disclaimer/Publisher's Note: The statements, opinions and data contained in all publications are solely those of the individual author(s) and contributor(s) and not of MDPI and/or the editor(s). MDPI and/or the editor(s) disclaim responsibility for any injury to people or property resulting from any ideas, methods, instructions or products referred to in the content.

fermentation

MDPI

Article

Application of Immobilized Yeasts for Improved Production of Sparkling Wines

Encarnación Fernández-Fernández [1,*], José Manuel Rodríguez-Nogales [1], Josefina Vila-Crespo [2] and Elena Falqué-López [3]

[1] Área de Tecnología de Alimentos, Escuela Técnica Superior de Ingenierías Agrarias, Universidad de Valladolid, Av. Madrid 50, 34004 Palencia, Spain
[2] Área de Microbiología, Escuela Técnica Superior de Ingenierías Agrarias, Universidad de Valladolid, Av. Madrid 50, 34004 Palencia, Spain
[3] Área de Química Analítica, Facultad de Ciencias, Universidade de Vigo, As Lagoas s/n, 32004 Ourense, Spain
* Correspondence: encarnacion.fernandez@uva.es

Abstract: Verdejo sparkling wines from two consecutive vintages were elaborated following the "champenoise" method. The second fermentation was developed with the same free or immobilized *Saccharomyces cerevisiae bayanus* yeast strain, carrying out four batch replicates each year. The sparkling wines were analyzed after 9 months of aging, showing no significant differences among the two typologies in the enological parameters (pH, total acidity, volatile acidity, reducing sugars, and alcoholic strength), the effervescence, or the spectrophotometric measurements. The free amino nitrogen content was significantly higher in the sparkling wines obtained from immobilized yeasts, nevertheless, the levels of neutral polysaccharides and total proteins were lower. No significant differences in the volatile composition were found, except for only two volatile compounds (isobutyric acid and benzyl alcohol); however, these two substances were present at levels below their respective olfactory thresholds. The sensory analysis by consumers showed identical preferences for both types of sparkling wines, except for the color acceptability. The descriptive analysis by a tasting panel revealed that sensorial differences between both sparkling wines were only found for the smell of dough. Therefore, the use of immobilized yeasts for the second fermentation of sparkling wines can reduce and simplify some enological practices such as the procedure of riddling and disgorging, with no impact on the so-mentioned quality parameters.

Keywords: sparkling wine; free yeasts; immobilized yeasts; sensory characteristics; chemical characteristics

Citation: Fernández-Fernández, E.; Rodríguez-Nogales, J.M.; Vila-Crespo, J.; Falqué-López, E. Application of Immobilized Yeasts for Improved Production of Sparkling Wines. *Fermentation* **2022**, *8*, 559. https://doi.org/10.3390/fermentation8100559

Academic Editor: Ronnie Willaert

Received: 26 September 2022
Accepted: 17 October 2022
Published: 20 October 2022

Publisher's Note: MDPI stays neutral with regard to jurisdictional claims in published maps and institutional affiliations.

Copyright: © 2022 by the authors. Licensee MDPI, Basel, Switzerland. This article is an open access article distributed under the terms and conditions of the Creative Commons Attribution (CC BY) license (https://creativecommons.org/licenses/by/4.0/).

1. Introduction

The "champenoise" method of sparkling wine production involves the second fermentation in a bottle of wine to which sugar and yeasts have been added. During this second fermentation, the yeast metabolism and autolysis participate in the typical effervescence, aroma, and flavor of the sparkling wine [1]. After the second fermentation and aging, the yeast sediment (lees) is settled on the stopper and "disgorged" in a process known as riddling. Traditionally, this phase takes place manually over the course of several weeks, twisting and inclining the bottle gradually until all the yeast cells settle down to the bottleneck. This process is both long and arduous and needs a significant percentage of the cellars to be devoted to it. One alternative is the use of an automatic machine called "gyropalette", which greatly reduces the times of this process (at 4–5 days); however, a large investment is required [2].

To improve this process, over the last decades some alternatives have been studied [3], such as the use of chemical additives, flocculent yeast strains, or immobilized yeasts [2,4–9]. The employ of immobilized yeasts is an attractive strategy to reduce the time and the cost of manufacturing sparkling wine since they can be removed just in a few seconds instead of the long-lasting traditional process [1]. After the patent of Bidan et al. [10], most efforts have focused on the application of alginate as organic support for the immobilization of yeasts for the second fermentation of sparkling wines [4,7,8,11,12]. The use of calcium alginate gels is advantageous since they are easily prepared and allow the incorporation of yeasts under mild conditions, which almost completely rules out the death of the cells [13]. Moreover, calcium alginate has been approved by the International Organization of Vine and Wine (OIV) as a fining agent in the production of sparkling wine and as a matrix for yeast and bacteria encapsulation [14]. In certain cases, a release of yeasts from the matrix of alginate gel has been described [15,16]. However, this issue has not been addressed in many papers dealing with the use of alginate-encapsulated yeasts in sparkling wine production [4,7,8,11]. Some authors have proposed coating finished biocatalysts (prepared using calcium alginate) to solve this problem [13].

Currently, a commercial immobilized *Saccharomyces cerevisiae bayanus* yeast (Cremanti®, Erbslöh Cavis, Mainz, Germany) designed for the second fermentation of bottle-fermented sparkling wine is available for winemakers. In this commercial preparation, the yeasts are immobilized on double-layer beads made of calcium alginate in order to prevent the release of cells from the beads to the wine, thus avoiding turbidity. Then, the immobilized yeasts are dried in order to increase their shelf-life.

There is a lack of scientific information about the effect of the application of these immobilized yeasts (Cremanti®) on the chemical and sensorial characteristics of sparkling wines. Rather than small cellar experiments conducted by the yeast's manufacturers, rigorous scientific studies under real winemaking conditions are required. To the best of our knowledge, only one study has addressed the use of Cremanti® encapsulated yeasts, researching the effect of inoculation format (free and encapsulated yeasts) on the amino acid content of sparkling wines [8]. However, other topics of interest such as their volatile composition and sensory characteristics have not been analyzed. In this context, the aim of our present study was to examine the potential application of these immobilized cells for the winemaking of sparkling wines during two consecutive vintages. The chemical and sensory properties of the sparkling wines elaborated with immobilized and free yeasts were compared.

2. Materials and Methods

2.1. Chemicals and Standards

All chemicals and reagents were analytical quality grade and purchased from Panreac (Madrid, Spain), except for the extraction solvent (dichloromethane) and internal standards (4-methyl-2-pentanol and 3-octanol) used for volatile compounds which were provided by Merck (Darmstadt, Germany), including methanol, 1-butanol, isoamyl alcohols, benzyl alcohol, 2-phenyl-ethanol, 1-hexanol, ethyl butyrate, ethyl hexanoate, ethyl octanoate, phenylethyl acetate, and hexanoic, octanoic and decanoic acids. Isobutanol, 1-propanol, *trans*-3-hexenol, *cis*-3-hexenol, furfural, benzaldehyde, ethyl lactate, diethyl succinate, isoamyl acetate, and butyric, isobutyric and isovaleric acids were provided by Fluka (Buchs, Switzerland). R-pantolactone, γ-butyrolactone, and 4-vinyl-guaiacol were acquired from Sigma-Aldrich (Steinheim, Germany).

2.2. Winemaking Process

The study was carried out using the Verdejo white grape variety cultivated in Castilla y León (Spain) from two consecutive vintages. Four different Verdejo base wines from different vineyards were elaborated on each vintage at the experimental wine cellar of the Agricultural Engineering College (University of Valladolid, Palencia, Spain).

Firstly, the white base wines were elaborated following the traditional white winemaking process, by standard technology for the production of white still wines [8]. Nearly 200 kg of Verdejo grapes were destemmed/crushed, pressed and the must was fermented in four stainless steel tanks (100 L). The sulfur dioxide concentrations of the musts were adjusted to 30 mg/L by addition of potassium metabisulphite. The musts were inoculated with *Saccharomyces cerevisiae* (Anchor VIN 13, Erbslöh Cavis, Germany, 20 g/hL). The density and temperature of the musts-wines were monitored during the fermentation letting them proceed at 16–18 °C for about 3 weeks until the sugar concentrations were reduced to less than 3 g/L. The clarifications were performed for 1 week at 4 °C by adding bentonite (70 g/hL, Superbenton DC, Dal Cin Gildo Spa, Milan, Italy) as fining agent, then the wines were racked and stored at 4 °C.

The sparkling wines were elaborated using the above-mentioned Verdejo base wines following the traditional method (second fermentation in bottles). A scheme of the experiments carried out is shown in Figure 1. The second fermentation was conducted by the same *Saccharomyces cerevisiae bayanus* yeast strain as (i) immobilized yeasts (Cremanti®, Erbslöh-Cavis, Germany) or (ii) free yeasts (Erbslöh-Cavis, Germany). Depending on the type of yeast, the protocol for the manufacturing of sparkling wines was different. For the batches with the free yeasts (labeled as FC wines), sucrose (to reach a final concentration of sugar of 24 g/L), yeasts (15 g/hL), and a fining agent (alginate and bentonite, Sekt-Klar plus, Erbslöh, Germany, 60 mL/hL) were added to the base wines. When the immobilized yeasts were employed (labeled as IC wines), the bottles were previously sterilized (dry heat at 100 °C, 24 h) and sucrose (to reach a final concentration of sugar of 24 g/L) was added to the wine. Then, the wine with sucrose was filtered through a 0.45 µm membrane (Opticap® XL10, Millipore, Billerica, MA, USA) under sterile conditions. A dose of 1.75 g of immobilized yeasts per 0.75-L bottle was employed. In all the experiments, the second fermentation and aging were carried out in 0.75-L bottles at a temperature of 14–15 °C for 9 months.

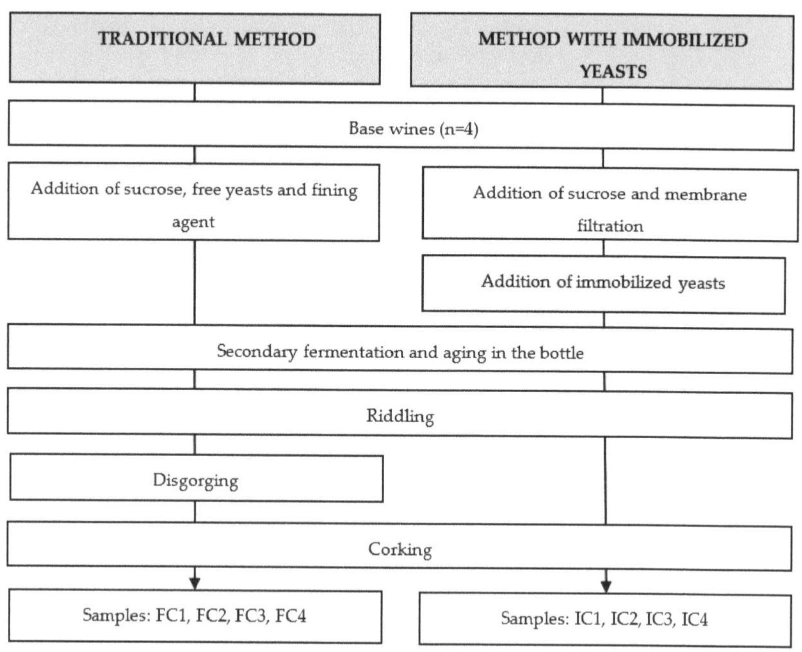

Figure 1. Scheme of the experiments carried out using free and immobilized cells to produce sparkling wines.

2.3. Analytical Procedures

The methods described by the OIV [17] were employed for the determination of pH, total acidity, volatile acidity, reducing sugars, and alcoholic strength.

Total proteins were quantified by the colorimetric method described by Bradford [18]. The formol titration method was employed for the quantification of free amino nitrogen [19]. Neutral polysaccharides were evaluated using the method described by Segarra et al. [20].

Total polyphenols, total hydroxycinnamates, and total flavonols were quantified by measuring the absorbances at 280, 320, and 365 nm, respectively, in quartz cuvettes in a UV–vis spectrophotometer (Lan Optics 2000 UV, Labolan, Spain) [21]. Color intensity was determined at 420 nm. Turbidity of the wines was determined using a turbidimeter (Hanna HI-83749, Woonsocket, RI, USA).

Frequencies of bubble formation using a stroboscope (PCE-OM 15, PCE-Ibérica, Albacete, Spain) were evaluated to determine the effervescence of the sparkling wines [22]. Briefly, 125 mL of sparkling wine was poured into a cylindrical classical crystal flute and three minutes after, the bubble trains were lit with the stroboscope. At a determined frequency of strobe lighting, a bubble train appears motionless, corresponding with its frequency of formation. Experiments were performed at room temperature (20 $\pm$ 2 °C) and using the same glass. This parameter was evaluated in five bubble trains using three bottles (3 $\times$ 5 measurements). The yeast cell concentration of the sparkling wines was determined by counting with a Neubauer camera.

All analyses were carried out in duplicate from three different bottles (3 $\times$ 2 measurements) and from each sample.

2.4. Volatile Compound Analysis

The volatile composition of the wines produced with immobilized and free cells from the first vintage was determined. Each sparkling wine sample was analyzed in triplicate. Methanol and higher alcohols were determined by direct injection (with 4-methyl-2-pentanol as internal standard, 1 g/L) into a Hewlett Packard 5890 gas chromatograph (Agilent Technologies, Palo Alto, CA, USA) with flame ionization detector (FID) following the method described by Bertrand and Ribéreau-Gayon [23]. Some alcohols (alcohols with six carbon atoms, and others), aldehydes, esters, fatty acids, acetates, lactones, sulfur compounds, and volatile phenols were extracted as follows: a sample of 100 mL of wine and 2 mL of 3-octanol (10 mg/L), added as an internal standard, were extracted with 10, 5, and 5 mL of dichloromethane [24]. The organic extract was dried with anhydrous sodium sulfate and concentrated to 0.5 mL with a low stream of nitrogen prior to analysis by gas chromatography (FID or mass spectrometric detection –GC-MS–).

A capillary column HP-Innowax (60 m $\times$ 0.25 mm i.d., 0.25 µm film thickness) was employed. Operating conditions were the following: injector and detector (FID) temperatures, 220 °C; column temperature, programmed from 45 (1 min) to 230 °C at 3 °C/min and final isotherm of 25 min; carrier gas pressure (Helium), 18 psi. An amount of 2 µL of wine sample or 1 µL of sample of the organic extract was injected in splitless mode (30 s).

A Hewlett-Packard model 5890 Series II hyphenated with an HP 5970 mass spectrometer (Agilent Technologies, Palo Alto, CA, USA) was also used. Three microliters of the organic extract were injected in splitless mode (purge time, 30 s; purge rate, 70) on a capillary column and the chromatographic conditions were the same as described above. The spectrometric conditions were electronic impact (ionization energy, 70 eV) and source temperature, 250 °C. The acquisition was made in scanning mode (mass range, 30–300 amu, 1.9 spectrum/s).

Identification of volatile compounds by GC-FID was performed through the retention times of the pure commercial standard compounds. Identification of volatile compounds by GC-MS was confirmed by comparing their mass spectra (MS Chemstation Wiley 7N library) and their retention times with those of the pure compounds. Volatile compounds were quantified using standard reference compounds. Stock standard solutions were prepared for all compounds (1 g/L for isoamyl alcohols and 100 mg/L for the rest of the volatile compounds) in 12% (*v*/*v*) aqueous ethanol and then diluted. Five known amounts of each analyte (in the intervals: 0–300 mg/L for isoamyl alcohols; 0–50 mg/L for methanol, 1-propanol, isobutanol, 2-phenyl-ethanol, and ethyl lactate; 0–20 mg/L for lactones; 0–10 mg/L for fatty acids and diethyl succinate; 0–5 mg/L for 1-butanol, benzyl alcohol, 1-hexanol, and 4-vinyl-guaiacol; 0–0.500 mg/L for acetates; 0–0.250 mg/L for *trans-* and *cis*-3-hexenol, furfural, benzaldehyde, and ethyl esters) were subjected to the same treatment (direct injection or previous liquid–liquid extraction) as that for the sparkling wines to obtain the calibration curves (relative peak areas with respect to the peak area of corresponding internal standard), and the quantification was carried out by interpolation. The correlation coefficients (R2 values) of the calibration curves were above 0.999 for all compounds and the concentrations of volatile compounds in sample wines were expressed as mg/L For the compounds for which pure substances were not available in our laboratory, the content was referred to as a function of the normalized area (%) with respect to the internal standard (3-octanol).

2.5. *Sensory Analysis*

2.5.1. Consumers' Sensory Test

Hedonic sensory evaluation of sparkling wines from the first vintage was conducted with 140 consumer volunteers from 18 to 65 years old of various socioeconomic backgrounds.

Consumer tests were carried out in the Sensory Science Laboratory of the Agricultural Engineering College at the University of Valladolid, Palencia (Spain) in individual booths. The sparkling wines were evaluated on the basis of acceptance of their color, odor, flavor, hitch in mouth, persistence, and overall liking on a 9-point hedonic scale. The scale of values ranged from "like extremely" to "dislike extremely" corresponding to the highest and lowest scores of "9" and "1", respectively.

The consumers tasted 8 samples of the sparkling wines (FC1–4, IC1–4) from the first vintage sequentially and monadically served. Samples were presented on plastic glasses coded with 3-digit random numbers and served in a randomized order. Water was available for rinsing.

2.5.2. Descriptive Sensory Analysis

A panel of ten volunteer panelists participated in the study. All panelists had previously participated in sparkling wine descriptive sensory analysis studies, and their performance had been assessed. The descriptive sensory analysis was carried out in the Sensory Science Laboratory of the Agricultural Engineering College at the University of Valladolid, Palencia (Spain).

The sensory evaluation of the sparkling wines (FC1–4, IC1–4) from the second vintage was performed using a questionnaire composed of 14 descriptors grouped into visual descriptors (color intensity), olfactory descriptors (odor intensity, fruity, floral, yeast, and dough) and flavor descriptors (flavor intensity, bitter, acid, sweet, volume, astringency, carbonic, and amount of carbonic). All attributes were evaluated on a 10 cm unstructured line scale from 0 to 10.

The sparkling wines were evaluated in duplicate by each assessor using a randomized complete block design. The samples were assessed in individual booths in coded wine-tasting glasses.

2.6. *Statistical Analysis*

Analytical data were treated by applying variance analysis (ANOVA). The LSD (least significant difference) test determines statistically significant differences between the means. Confidence intervals of 95% or significant level of $\alpha = 0.05$ were used.

Principal component analysis (PCA) was performed in order to establish a relationship between the different sensory and chemical data and the studied sparkling wines from the two vintages. The PCA was performed only with those variables (chemical and sensory) in which statistically significant differences were found after performing a one-way ANOVA test.

These analyses were performed with the computer program Statgraphics® Plus for Windows 4.0 (Statistical Graphics Corp., Rockville, MD, USA).

3. Results and Discussion

3.1. Chemical Comparison of Sparkling Wines Elaborated with Free and Immobilized Yeasts

As an example, Figure 2 shows the kinetics of sparkling wine fermentations carried out with free and immobilized cells from the first vintage. Both the free and immobilized yeasts displayed similar carbonic acid production kinetics; however, the second fermentation of the immobilized cells started five days earlier, and so did free and immobilized cells in the second vintage. Similar behavior was observed at the second fermentation with immobilized yeast strains in alginate beads [25] and chitosan–calcium–alginate double-layer microcapsules [4]. Some authors reported an increased ethanol tolerance of immobilized yeast cells, facilitating the start of the second fermentation. Moreover, they suggested that this phenomenon can be attributed to cell encapsulation by a protective layer of gel material or modified fatty acid concentration in cell membranes due to oxygen diffusion limitations caused by the matrix of immobilization [12].

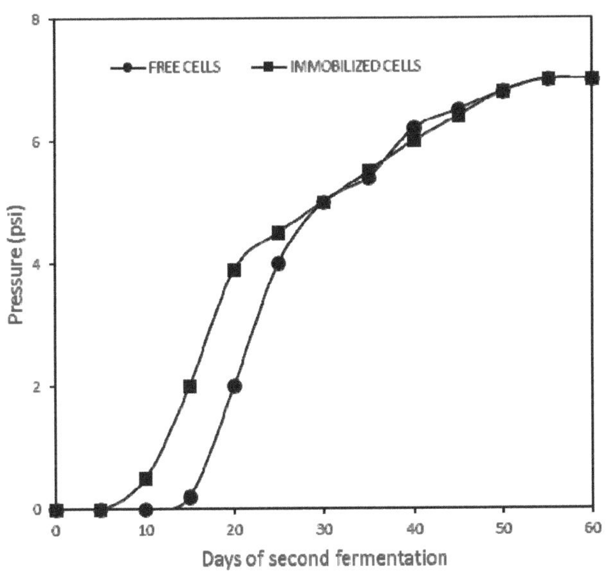

Figure 2. Evolution of the carbon dioxide pressure in the bottle during the second fermentation using free and immobilized yeasts in first vintage.

The main enological parameters of the sparkling wines elaborated with free and immobilized yeasts from the two vintages are shown in Table 1. The results showed that there were no significant differences in pH, total and volatile acidities, reducing sugar, and alcoholic strength between both inoculation formats (free and immobilized yeasts). Some authors have also reported that the use of immobilized yeasts did not affect the enological parameters of the sparkling wines [4,9,13,26]. Moreover, it has been reported that the Ca–alginate entrapment technique did not cause notable changes in the main analytical composition of bottle-fermented sparkling wines [27–29]. Contrarily, all the parameters, except for the volatile acidity, were dependent on the vintage year. The sparkling wines from the second vintage were more acidic, less alcoholic, and a little sweeter than those from the first vintage.

Table 1. Enological parameters of the sparkling wines elaborated with free and immobilized cells from the two vintages.

Variable	Free Cells		Immobilized Cells		Significance	
	First Vintage	Second Vintage	First Vintage	Second Vintage	Method	Vintage
pH	3.51 ± 0.04 [b]	3.03 ± 0.03 [a]	3.34 ± 0.04 [b]	3.09 ± 0.03 [a]	ns	*
Total acidity (g tartaric ac./L)	7.84 ± 0.25 [a]	8.03 ± 0.21 [b]	7.62 ± 0.25 [a]	8.80 ± 0.20 [b]	ns	*
Volatile acidity (g acetic ac./L)	0.35 ± 0.05 [a]	0.28 ± 0.04 [a]	0.29 ± 0.03 [a]	0.27 ± 0.05 [a]	ns	ns
Reducing sugars (g/L)	2.06 ± 0.40 [a]	4.03 ± 0.32 [b]	2.06 ± 0.32 [a]	4.58 ± 0.25 [b]	ns	*
Alcoholic strength (%, v/v)	12.63 ± 0.25 [b]	11.79 ± 0.21 [a]	12.85 ± 0.25 [b]	11.71 ± 0.20 [a]	ns	*
Turbidity (NTU)	88.39 ± 8.54 [c]	54.50 ± 1.29 [b]	6.14 ± 3.03 [a]	4.27 ± 0.44 [a]	*	*
Number of yeasts/mL	975 ± 104 [b]	3262 ± 423 [c]	0 ± 0 [a]	0 ± 0 [a]	*	*
Effervescence (Hz)	20.2 ± 1.2 [a]	20.2 ± 1.4 [a]	19.8 ± 0.8 [a]	19.8 ± 0.9 [a]	ns	ns

ns: not significant effect; * $p < 0.05$. Values with different letters indicate statistically significant difference ($p < 0.05$).

On the other hand, an important point for the industrial application of immobilized yeasts is the study of the cell retention capacity of the matrix employed for yeast immobilization. For that, the turbidity and the number of yeasts were evaluated in the sparkling wines after 9 months of aging (Table 1). The immobilized yeast samples displayed lower turbidity values than the free yeasts ones and no cells were detected in the sparkling wines elaborated with immobilized yeasts. Our results evidence a satisfactory behavior of the matrix employed in the immobilization process. The degree of effervescence of the sparkling wines was also studied by analyzing the values of the frequency of bubble formation (Table 1). This parameter indicates that the wines with more effervescence have higher frequencies of bubble formation. Puig-Pujol et al. [9] indicated that secondary fermentation conducted with immobilized yeast cells resulted in better foam quality than fermentation with traditional free cells. In our study, no significant differences were found in the values of the frequency of bubble formation between the sparkling wines elaborated with free and immobilized cells.

The amino acid concentration, which is influenced by both its consumption by yeasts and its release during the autolysis of yeasts, is considered one of the most important factors affecting the organoleptic quality of sparkling wines [4,30]. For this reason, the concentration of free amino nitrogen in different sparkling wines was analyzed. As was found by López et al. [7], the content of free amino nitrogen in the sparkling wines produced with free yeasts was lower than that found with immobilized yeasts (Figure 3) for both vintages. These data are also in accordance with Bozdogan and Canbas [31] who described slightly higher free amino acid values of the sparkling wines elaborated with *Sacharomyces bayanus* EC 1118 immobilized in alginate gels than that obtained from free yeasts. Prokes et al. [8] also reported higher levels of free amino acids in sparkling wines elaborated with alginate encapsulated cells. This fact can be associated with a physiological response to the process of immobilization, which could cause an increase in the autolysis process and/or a reduction in the consumption of free amino acids by yeasts along the aging period, finally increasing their levels in the sparkling wines [3]. Some authors reported a more rapid decay of immobilized yeasts than their free counterparts. Free cells died completely in 9–12 months, whereas live immobilized cells were not detected after 6 months or earlier. Early decay determined the more rapid autolysis of immobilized yeasts [26]. Moreover, an increase in the content of amino acids in the sparkling wines elaborated with immobilized yeasts was observed. This is an indication of earlier onset and higher intensity of autolysis in immobilized cells, in agreement with their more rapid decay [13].

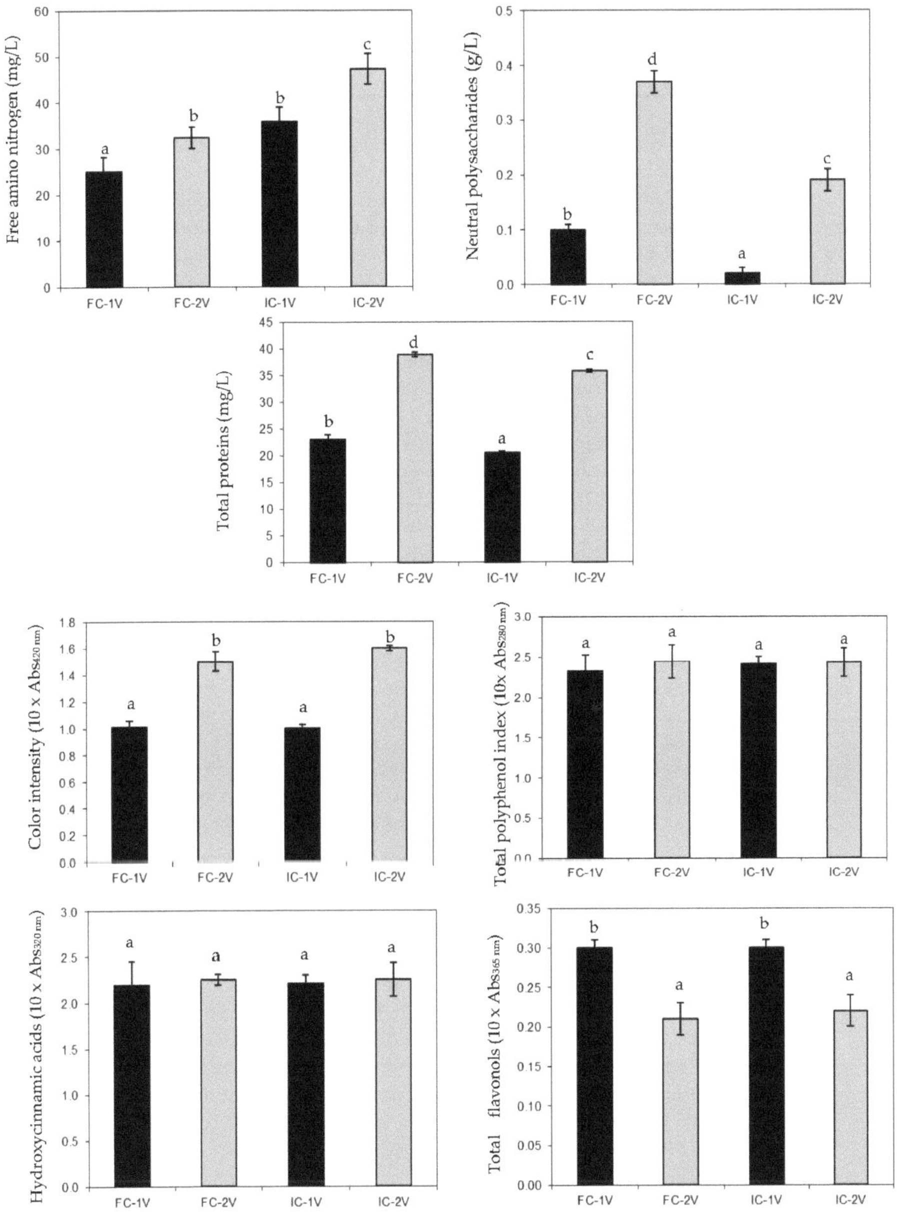

Figure 3. Concentration of free amino nitrogen, neutral polysaccharides, and total proteins, and color intensity, total polyphenol index, and levels of hydroxycinnamic acids and total flavonols of the sparkling wines elaborated with free (FC) and immobilized cells (IC) from the two vintages (1V and 2V). Different letters indicate statistically significant difference ($p < 0.05$).

On the other hand, the study of the level of neutral polysaccharides in sparkling wines was also used to evaluate the degree of autolysis, for it quantifies mainly glucans and mannoproteins from the yeast cell walls [3,32]. The statistical analysis showed that the neutral polysaccharides were lower in the wines elaborated with immobilized yeasts than those with free yeasts (Figure 3) for both vintages. An increase in the yeast autolysis due to the immobilization process, previously postulated, should increase the content of neutral polysaccharides in the sparkling wines elaborated with immobilized yeasts [27]. However, this effect was not observed in this study. This may be due to the limited transfer of the polysaccharides (released during yeast autolysis) from the inside of the matrix of immobilization to the sparkling wine. Moreover, this effect could be more intense in immobilized yeasts prepared with a protective coat which decreases the mass exchange [13].

The protein composition of sparkling wines changes during aging due to the process of yeast autolysis [33]. Figure 3 shows the concentration of total proteins in the sparkling wines after nine months of aging. There were statistically significant differences between treatments (free and immobilized yeasts) for this parameter. For both vintages, the concentration of total proteins of sparkling wines elaborated with free yeasts was higher than that observed for immobilized yeasts. These results are in agreement with those found for neutral polysaccharides. Probably, the porosity of the matrix of immobilization did not allow a free transfer of the yeast proteins to the wine.

The color intensity, the total polyphenol index, and the levels of hydroxycinnamic acids and flavonols analysis were carried out in order to know the possibility of the matrix of alginate to absorb color and/or phenolic compounds in its surface during the aging time. No significant differences were found for these parameters between the sparkling wines elaborated with free or immobilized cells (Figure 3). Puig-Pujol et al. [9] did not detect significant differences in color intensity among the sparkling wines made with free and immobilized yeast cells.

The aromatic composition constitutes an important factor in the production of high-quality sparkling wines. Alcohols, ethyl esters, acetates, lactones, sulfur, or volatile phenol compounds were determined in the Verdejo sparkling wines from the first vintage (Table 2).

In general, the aromatic composition of the wines did not vary significantly according to the yeast-inoculating format. These results are according to other authors [6,9,29]. No statistically significant differences were found between both types of wines, with the exception of only two volatile compounds (isobutyric acid and benzyl alcohol). A higher level of isobutyric acid was observed in the sparkling wines produced by immobilized cells, while a higher concentration of benzyl alcohol was found in samples elaborated with free cells. However, these two substances were present at levels below their respective odor thresholds, 2.3 mg/L [34] and 10 mg/L [35], respectively, and can therefore not be considered as odor actives; although these compounds could add a synergy effect to wine aroma with rancid or buttery and fruity notes, respectively. Unlike our data, Benucci et al. [4] and Costa et al. [11] found significant differences in wine aromatic composition related to yeast-inoculating format. The first study reported lower concentrations of total aldehydes, phenols, and ketones, and higher concentrations of acids in sparkling wines inoculated with chitosan–alginate encapsulated yeasts, while the second study observed lower levels of fatty acids in sparkling wines produced with alginate encapsulated yeasts compared to the free ones. However, López et al. [7] noted that the yeast strain inoculated to base wine had a higher impact on the volatile composition of sparkling wines than the type of inoculum (free or immobilized) used.

Table 2. Volatile compounds (mg/L) of the sparkling wines elaborated with free and immobilized cells from first vintage.

	Free Cells	Immobilized Cells	Significance
Methanol	30 ± 1	30 ± 2	ns
1-propanol	25 ± 7	27 ± 5	ns
Isobutanol	19 ± 3	21 ± 4	ns
1-butanol	0.79 ± 0.39	0.55 ± 0.10	ns
Isoamyl alcohols	192 ± 12	191 ± 28	ns
Benzyl alcohol	2.57 ± 0.41	1.53 ± 0.32	*
2-phenyl-ethanol	40 ± 13	38 ± 16	ns
1-hexanol	0.60 ± 0.10	0.55 ± 0.08	ns
trans-3-hexenol	0.18 ± 0.04	0.15 ± 0.02	ns
cis-3-hexenol	0.09 ± 0.01	0.08 ± 0.00	ns
1-pentanol [a]	7 ± 2	7 ± 1	ns
1-heptanol [a]	26 ± 4	30 ± 3	ns
2,3-butanediol [a]	528 ± 173	421 ± 64	ns
1,2-propanediol [a]	12 ± 3	10 ± 1	ns
Furfural	0.14 ± 0.01	0.14 ± 0.02	ns
Benzaldehyde	0.02 ± 0.01	0.02 ± 0.01	ns
Ethyl buryrate	0.13 ± 0.02	0.13 ± 0.02	ns
Ethyl hexanoate	0.15 ± 0.03	0.15 ± 0.03	ns
Ethyl octanoate	0.18 ± 0.05	0.16 ± 0.04	ns
9-ethyl decanoate [a]	30 ± 8	20 ± 4	ns
Ethyl lactate	31 ± 9	25 ± 3	ns
Ethyl pyruvate [a]	130 ± 58	82 ± 48	ns
Diethyl succinate	7.5 ± 3.7	8.0 ± 3.4	ns
Isoamyl acetate	0.35 ± 0.13	0.35 ± 0.10	ns
Phenylethyl acetate	0.18 ± 0.15	0.20 ± 0.05	ns
Butyric acid	1.53 ± 0.50	1.35 ± 0.10	ns
Isobutyric acid	0.32 ± 0.12	0.54 ± 0.11	*
Isovaleric acid	1.22 ± 0.70	1.24 ± 0.72	ns
Hexanoic acid	3.6 ± 1.7	4.7 ± 0.7	ns
Octanoic acid	7.5 ± 2.2	8.4 ± 1.8	ns
Decanoic acid	2.88 ± 3.54	0.64 ± 0.02	ns
γ-butyrolactone	10.0 ± 2.6	9.2 ± 0.9	ns
R-pantolactone	0.33 ± 0.07	0.33 ± 0.10	ns
4-vinyl-guaiacol	1.70 ± 0.65	1.75 ± 0.21	ns
Methionol [a]	98 ± 36	83 ± 21	ns
Thiazole [a]	364 ± 54	378 ± 85	ns

[a] Expressed as normalized peak area; ns: not significant effect; * $p < 0.05$.

3.2. Sensory and Chemical Characteristics of the Sparkling Wines from the Two Vintages

In the first vintage a consumer study was carried out, but due to the good results obtained, in the second vintage a sensory characterization of the sparkling wines was carried out using a panel of trained tasters. No statistically significant differences were found in the sensory characteristics between the sparkling wines treated with immobilized and free yeasts (Figure 4). Only the sparkling wines elaborated with immobilized yeasts presented higher color acceptability (first vintage) and smell of dough (second vintage) than the sparkling wines elaborated with free yeasts.

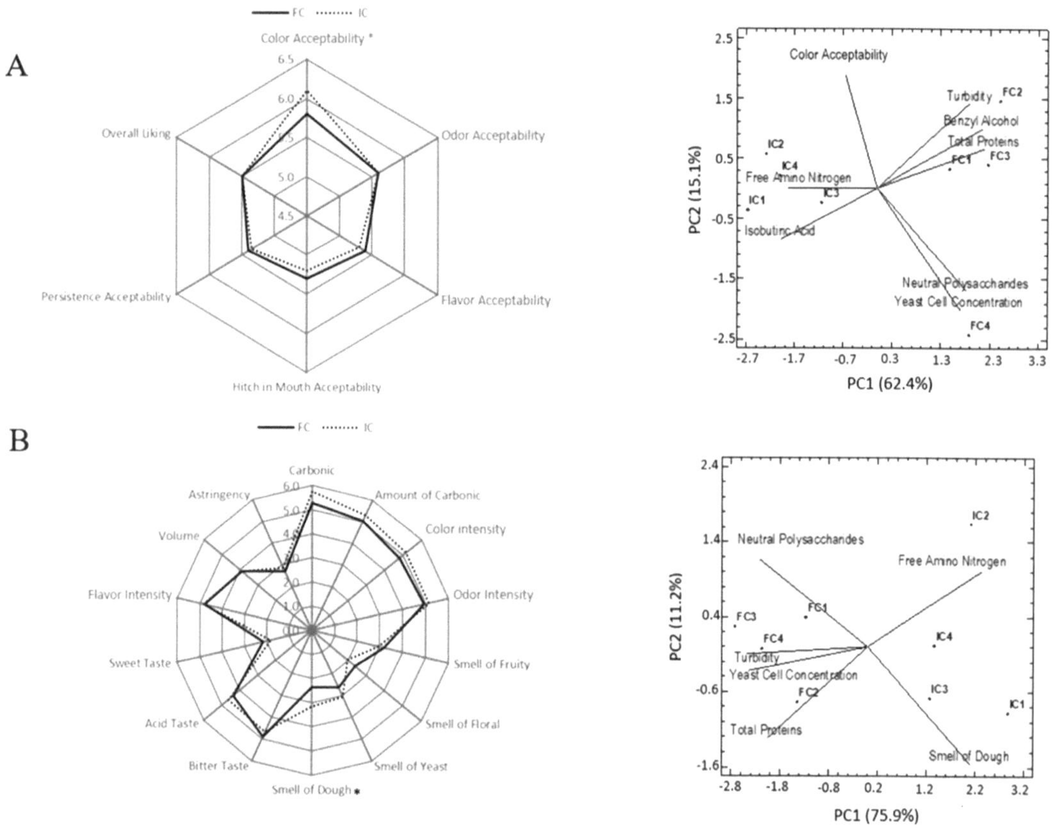

Figure 4. Sensory diagrams and principal component analysis of samples of sparkling wines elaborated with free (FC) and immobilized yeasts (IC) from (**A**) first and (**B**) second vintages by using chemical and sensorial variables. The asterisk indicates statistically significant differences for $p = 0.05$.

Figure 4 also shows the dispersion diagram of the samples of wine from first and second vintage after 9 months of aging, and the chemical and sensory characteristics in the first two main components. For the first vintage (Figure 4A) the two main components can explain 77.5% of the total variance. The first principal component (PC1) explains 62.4% of the variability data and the second component (PC2) explains 15.1% of the variance. In Figure 4B the sparkling wines from the second vintage were located in the vectorial dimension defined by the two principal components, which explained 87.1% of the total variance. In both vintages the defined space of the two principal components clearly differentiates the sparkling wines elaborated using free (FC1, FC2, FC3, FC4) and immobilized yeasts (IC1, IC2, IC3, IC4).

The sparkling wines from both vintages elaborated using immobilized yeasts are characterized by a higher content of free amino nitrogen and smaller values of neutral polysaccharides and total proteins. Regarding the aromatic compounds, the sparkling wines produced with free yeasts (first vintage) showed higher values of benzyl alcohol and lower isobutyric acid. However, from a sensory viewpoint, the consumers did not observe these chemical differences. As pointed out above, there was only a significant difference in the color acceptability of the wines, being higher this attribute for the sparkling wines with immobilized yeasts. The lower turbidity observed in these sparkling wines could be related to the highest acceptability in the color attribute.

In an attempt to test if the chemical differences detected in these wines (according to the type of inoculation) had a real influence on their organoleptic characteristics, a descriptive sensory analysis was carried out with the sparkling wines from the second vintage. As noted previously, the sparkling wines elaborated with immobilized yeasts showed only a higher smell of dough than the wines with free yeasts (Figure 4B). In accordance with our results, Benucci et al. [4] and López et al. [7] observed similar sensory profiles between sparkling wines elaborated with free and encapsulated yeasts. Probably, the higher smell of dough found in wines obtained with immobilized yeast-inoculating format is due to a higher degree of autolysis, as it was discussed above. Finally, the higher levels of neutral polysaccharides and proteins found in the sparkling wines elaborated with free yeasts were probably not enough to detect by the judges' differences in volume between both types of sparkling wines.

4. Conclusions

The use of immobilized yeasts in the champenoise winemaking method makes it possible to reduce and simplify the procedure of riddling and disgorging since rapid sediment of encapsulated cells takes place during the remuage phase. The matrix of immobilization was stable during the nine months of aging and no release of yeasts to wine was observed. Moreover, the kinetic behavior of the immobilized yeasts was excellent. On the other hand, the use of immobilized yeasts did not have either effect on the main enological parameters or on the effervescence of the sparkling wines. The data show that the immobilized cells give differences just in a few analytical results. The wines elaborated with immobilized yeasts presented a higher level of free amino nitrogen, and a lower number of neutral polysaccharides and total proteins. The volatile composition of both types of sparkling wines was also very similar, with the exception of the content of isobutyric acid and benzyl alcohol. Finally, the organoleptic characteristics of sparkling wines do not reveal important disparities. The consumers only found differences in color acceptability and the judges in the smell of dough of both fermentation systems.

Author Contributions: Conceptualization, E.F.-F., J.M.R.-N. and J.V.-C.; methodology, E.F.-F., J.M.R.-N., J.V.-C. and E.F.-L.; validation, E.F.-F., J.M.R.-N. and J.V.-C.; formal analysis, E.F.-F., J.M.R.-N. and J.V.-C.; investigation, E.F.-F., J.M.R.-N. and J.V.-C.; resources, E.F.-F., J.M.R.-N. and J.V.-C.; data curation, E.F.-F., J.M.R.-N., J.V.-C. and E.F.-L.; writing—original draft preparation, E.F.-F., J.M.R.-N. and J.V.-C.; writing—review and editing, E.F.-F., J.M.R.-N., J.V.-C. and E.F.-L.; visualization, E.F.-F., J.M.R.-N. and J.V.-C.; supervision, E.F.-F., J.M.R.-N., J.V.-C. and E.F.-L.; project administration, E.F.-F., J.M.R.-N. and J.V.-C. All authors have read and agreed to the published version of the manuscript.

Funding: This research received no external funding.

Informed Consent Statement: Informed consent was obtained from all subjects involved in the study.

Data Availability Statement: Not applicable.

Acknowledgments: The authors would like to thank Jesús Yuste Bombín and ITACyL (Instituto Tecnológico y Agrario de Castilla y León) for the grapes used in this study, and Mónica Martínez Pérez, María Montserrat Marcos and Alberto de la Iglesia for the analytical work.

Conflicts of Interest: The authors declare no conflict of interest.

References

1. Alexandre, H.; Guilloux-Benatier, M. Yeast autolysis in sparkling wine—A review. *Aust. J. Grape Wine Res.* **2006**, *12*, 119–127. [CrossRef]
2. Torresi, S.; Frangipane, M.T.; Anelli, G. Biotechnologies in sparkling wine production. Interesting approaches for quality improvement: A review. *Food Chem.* **2011**, *129*, 1232–1241. [CrossRef] [PubMed]
3. Pozo-Bayón, M.; Martínez-Rodríguez, A.; Pueyo, E.; Moreno-Arribas, M.V. Chemical and biochemical features involved in sparkling wine production: From a traditional to an improved winemaking technology. *Trends Food Sci. Technol.* **2009**, *20*, 289–299. [CrossRef]

4. Benucci, I.; Cerreti, M.; Maresca, D.; Mauriello, G.; Esti, M. Yeast cells in double layer calcium alginate-chitosan microcapsules for sparkling wine production. *Food Chem.* **2019**, *300*, 125174. [CrossRef]
5. Coloretti, F.; Zambonelli, C.; Tini, V. Characterization of flocculent *Saccharomyces* interspecific hybrids for the production of sparkling wines. *Food Microbiol.* **2006**, *23*, 672–676. [CrossRef]
6. Divies, C.; Cachon, R.; Cavin, J.F.; Prevost, H. Theme 4—Immobilized cell technology in wine production. *Crit. Rev. Biotechnol.* **1994**, *14*, 135–153. [CrossRef]
7. López, N.; Peinado, R.A.; Puig-Pujol, A.; Mauricio, J.C.; Moreno, J.; García-Martínez, T. Influence of two yeast strains in free, bioimmobilized or immobilized with alginate forms on the aromatic profile of long aged sparkling wines. *Food Chem.* **2018**, *250*, 22–29. [CrossRef]
8. Prokes, K.; Baron, M.; Mlcek, J.; Jurikova, T.; Adamkova, A.; Ercisli, S.; Sochor, J. The influence of traditional and immobilized yeast on the amino-acid content of sparkling wine. *Fermentation* **2022**, *8*, 36. [CrossRef]
9. Puig-Pujol, A.; Bertrán, E.; García-Martínez, T.; Capdevila, F.; Mínguez, S.; Mauricio, J.C. Application of a new organic yeast immobilization method for sparkling wine production. *Am. J. Enol. Vitic.* **2013**, *64*, 386–394. [CrossRef]
10. Bidan, P.; Divies, C.; Dupuy, P. Procédé Perfectionné de Préparation de Vins Mousseux. French Patent FR2432045A1, 26 July 1978.
11. Costa, G.P.; Nicolli, K.P.; Welke, J.E.; Manfroi, V.; Zini, C.A. Volatile profile of sparkling wines produced with the addition of mannoproteins or lees before second fermentation performed with free and immobilized yeasts. *J. Braz. Chem. Soc.* **2018**, *29*, 1866–1875. [CrossRef]
12. Kourkoutas, Y.; Bekatorou, A.; Banat, I.M.; Marchant, R.; Koutinas, A.A. Immobilization technologies and support materials suitable in alcohol beverages production: A review. *Food Microbiol.* **2004**, *21*, 377–397. [CrossRef]
13. Martynenko, N.; Gracheva, I. Physiological and biochemical characteristics of immobilized champagne yeasts and their participation in champagnizing processes: A review. *Appl. Biochem. Microbiol.* **2003**, *39*, 439–445. [CrossRef]
14. *International Oenological Codex. Chapter I. Products Used in Oenology*; (Oeno 33/2000, Oeno 410/2010); Organisation Internationale de la Vigne et du Vin (OIV): Paris, France, 2013.
15. Bozdogan, A.; Canbas, A. Influence of yeast strain, immobilisation and ageing time on the changes of free amino acids and amino acids in peptides in bottle-fermented sparkling wines obtained from *Vitis vinifera* cv. Emir. *Int. J. Food Sci. Technol.* **2011**, *46*, 1113–1121. [CrossRef]
16. Godia, F.; Casas, C.; Sola, C. Application of immobilized yeast-cells to sparkling wine fermentation. *Biotechnol. Prog.* **1991**, *7*, 468–470. [CrossRef]
17. *Compendium of International Methods of Analysis of Wines and Musts*; Organisation Internationale de la Vigne et du Vin (OIV): Paris, France, 2013.
18. Bradford, M.M. Rapid and sensitive method for quantitation of microgram quantities of protein utilizing principle of protein-dye binding. *Anal. Biochem.* **1976**, *72*, 248–254. [CrossRef]
19. Shively, C.E.; Henick-Kling, T. Comparison of two procedures for assay of free amino nitrogen. *Am. J. Enol. Vitic.* **2001**, *52*, 400–401.
20. Segarra, I.; Lao, C.; López-Tamames, E.; De La Torre-Boronat, M.C. Spectrophotometric methods for the analysis of polysaccharide levels in winemaking products. *Am. J. Enol. Vitic.* **1995**, *46*, 564–570.
21. Andrés Lacueva, C.; Lamuela Raventós, R.M.; Buxaderas, S.; de la Torre Boronat, M.D.C. Influence of variety and aging on foaming properties of cava (sparkling wine). 2. *J. Agric. Food Chem.* **1997**, *45*, 2520–2525. [CrossRef]
22. Liger-Belair, G.; Marchal, R.; Robillard, B.; Vignes-Adler, M.; Maujean, A.; Jeandet, P. Study of effervescence in a glass of champagne: Frequencies of bubble formation, growth rates, and velocities of rising bubbles. *Am. J. Enol. Vitic.* **1999**, *50*, 317–323.
23. Bertrand, A.; Ribéreau-Gayon, P. Determination of volatile components of wine by gas-phase chromatography. *Ann. Falsif. Expert Chim. Toxicol.* **1970**, *63*, 148–156.
24. Armada, L.; Falqué, E. Repercussion of the clarification treatment agents before the alcoholic fermentation on volatile composition of white wines. *Eur. Food Res. Technol.* **2007**, *225*, 553–558. [CrossRef]
25. Fleet, G.H. Wine yeasts for the future. *FEMS Yeast Res.* **2008**, *8*, 979–995. [CrossRef] [PubMed]
26. Efremenko, E.N.; Stepanov, N.; Martinenko, N.N.; Gracheva, I.M. Cultivation conditions preferable for yeast cells to be immobilized into poly (vinyl alcohol) and used in bottled sparkling wine production. *Chem. Ind. Chem. Eng. Quart.* **2006**, *12*, 18–23. [CrossRef]
27. Busova, K.; Magyar, I.; Janky, F. Effect of immobilized yeasts on the quality of bottle-fermented sparkling wine. *Acta Aliment.* **1994**, *23*, 9–23.
28. Fumi, M.D.; Trioli, G.; Colagrande, O. Preliminary assessment on the use of immobilized yeast-cells in sodium alginate for sparkling wine processes. *Biotechnol. Lett.* **1987**, *9*, 339–342. [CrossRef]
29. Yokotsuka, K.; Yajima, M.; Matsudo, T. Production of bottle-fermented sparkling wine using yeast immobilized in double-layer gel beads or strands. *Am. J. Enol. Vitic.* **1997**, *48*, 471–481.
30. Martínez-Rodríguez, A.J.; Polo, M.C. Effect of the addition of bentonite to the tirage solution on the nitrogen composition and sensory quality of sparkling wines. *Food Chem.* **2003**, *81*, 383–388. [CrossRef]
31. Bozdogan, A.; Canbas, A. The effect of yeast strain, immobilisation, and ageing time on the amount of free amino acids and amino acids in peptides of sparkling wines obtained from cv. Dimrit grapes. *S. Afr. J. Enol. Vitic.* **2012**, *33*, 257–263. [CrossRef]

32. Martínez-Lapuente, L.; Guadalupe, Z.; Ayestarán, B.; Ortega-Heras, M.; Pérez-Magarino, S. Changes in polysaccharide composition during sparkling wine making and aging. *J. Agric. Food Chem.* **2013**, *61*, 12362–12373. [CrossRef]
33. Ferreira, V.; López, R.; Cacho, J.F. Quantitative determination of the odorants of young red wines from different grape varieties. *J. Sci. Food Agric.* **2000**, *80*, 1659–1667. [CrossRef]
34. Ferreira, R.B.; Picarra-Pereira, M.A.; Monteiro, S.; Loureiro, V.B.; Teixeira, A.R. The wine proteins. *Trends Food Sci. Technol.* **2011**, *12*, 230–239. [CrossRef]
35. Maarse, H. *Volatile Compounds in Foods and Beverages*, 1st ed.; Marcel Dekker, Inc.: New York, NY, USA, 1991.

fermentation

MDPI

Review

Advances in *Komagataella phaffii* Engineering for the Production of Renewable Chemicals and Proteins

Clara Vida Galrão Corrêa Carneiro [1,2], Luana Assis Serra [1,2], Thályta Fraga Pacheco [1,2], Letícia Maria Mallmann Ferreira [1,2], Lívia Teixeira Duarte Brandão [2], Mariana Nogueira de Moura Freitas [1,2], Débora Trichez [2] and João Ricardo Moreira de Almeida [1,2,*]

[1] Graduate Program of Microbial Biology, Institute of Biology, University of Brasília, Brasília 70770-901, Brazil
[2] Microbial Genetics and Biotechnology Laboratory, Embrapa Agroenergy, Brasília 70770-901, Brazil
* Correspondence: joao.almeida@embrapa.br

Abstract: The need for a more sustainable society has prompted the development of bio-based processes to produce fuels, chemicals, and materials in substitution for fossil-based ones. In this context, microorganisms have been employed to convert renewable carbon sources into various products. The methylotrophic yeast *Komagataella phaffii* has been extensively used in the production of heterologous proteins. More recently, it has been explored as a host organism to produce various chemicals through new metabolic engineering and synthetic biology tools. This review first summarizes *Komagataella* taxonomy and diversity and then highlights the recent approaches in cell engineering to produce renewable chemicals and proteins. Finally, strategies to optimize and develop new fermentative processes using *K. phaffii* as a cell factory are presented and discussed. The yeast *K. phaffii* shows an outstanding performance for renewable chemicals and protein production due to its ability to metabolize different carbon sources and the availability of engineering tools. Indeed, it has been employed in producing alcohols, carboxylic acids, proteins, and other compounds using different carbon sources, including glycerol, glucose, xylose, methanol, and even CO_2.

Keywords: *Komagataella phaffii*; *Pichia pastoris*; strain engineering; bioeconomy

Citation: Carneiro, C.V.G.C.; Serra, L.A.; Pacheco, T.F.; Ferreira, L.M.M.; Brandão, L.T.D.; Freitas, M.N.d.M.; Trichez, D.; Almeida, J.R.M.d. Advances in *Komagataella phaffii* Engineering for the Production of Renewable Chemicals and Proteins. *Fermentation* **2022**, *8*, 575. https://doi.org/10.3390/fermentation8110575

Academic Editor: Ronnie Willaert

Received: 29 September 2022
Accepted: 20 October 2022
Published: 24 October 2022

Publisher's Note: MDPI stays neutral with regard to jurisdictional claims in published maps and institutional affiliations.

Copyright: © 2022 by the authors. Licensee MDPI, Basel, Switzerland. This article is an open access article distributed under the terms and conditions of the Creative Commons Attribution (CC BY) license (https://creativecommons.org/licenses/by/4.0/).

1. Introduction

The urgency for effective strategies to mitigate climate change is evident. Environmental issues caused by deliberate petroleum exploitation to produce chemicals, energy, and materials are a major concern regarding global warming. In this context, the global economy needs to be transformed faster, and the circular bioeconomy seems to be the right way to achieve it. This new form of economy aims to reuse, recycle, and remanufacture biomass (from agroecological systems, forestry, and urban wastes) to generate valuable products, such as fuels, biomaterials, and fine chemicals [1]. The circular bioeconomy benefits from integrated frameworks to utilize biomass and address its uses to human needs. One of the major areas that compose this framework is biotechnology, which has the potential to lead bioeconomy global advances even further [2]. The extensive range of roles that biotechnology plays alongside bioengineering provides new techniques to modify the DNA of several microorganisms and plants. Such modifications can improve the utilization of lignocellulosic biomass as raw material through the heterologous protein expression and the production of renewable chemicals, biofuels, and materials [3].

Lignocellulosic biomass polymeric carbohydrates, cellulose, and hemicellulose are mainly composed of glucose and xylose. Glucose is a C6 sugar easily consumed by most microorganisms and can be converted into an extensive range of different chemicals. Xylose, a C5 sugar, can be naturally consumed by filamentous fungi, yeasts, and bacteria. However, this pentose is more challenging to metabolize with the same efficiency as glucose. Nonetheless, xylose can also be converted into various chemicals, such as xylitol [4] and xylonic acid [5]. Recently, advances in genetic manipulation techniques have opened up the range of microorganisms that can produce fine chemicals from glucose and xylose with higher yields and productivity. The yeast *Komagataella phaffii* is one example of it. Recent research enabled this yeast to consume these sugars and other substrates and produce various chemicals of industrial interest, as will be discussed in this review.

Komagataella phaffii (previously *Pichia pastoris*) is a well-known methylotrophic yeast. It has many valuable features such as low nutritional requirement, the ability to reach high cell densities (even in acidic culture media), and is widely employed in biotechnological processes to produce heterologous proteins of commercial interest [6]. Therefore, its ability to use methanol as a carbon and energy source, its non-fermentative utilization of glucose under aerobic conditions, and its efficiency to grow on glycerol are some of the main reasons for the preferential choice of this yeast for bioprocess development [7]. These characteristics transformed *K. phaffii* into a suitable host for many industrial applications. More recently, this yeast has been called the "biotech yeast" for its broad utilization as a cell factory for synthetic biology and metabolic engineering aiming for valuable chemical compound productions (Figure 1) [7,8].

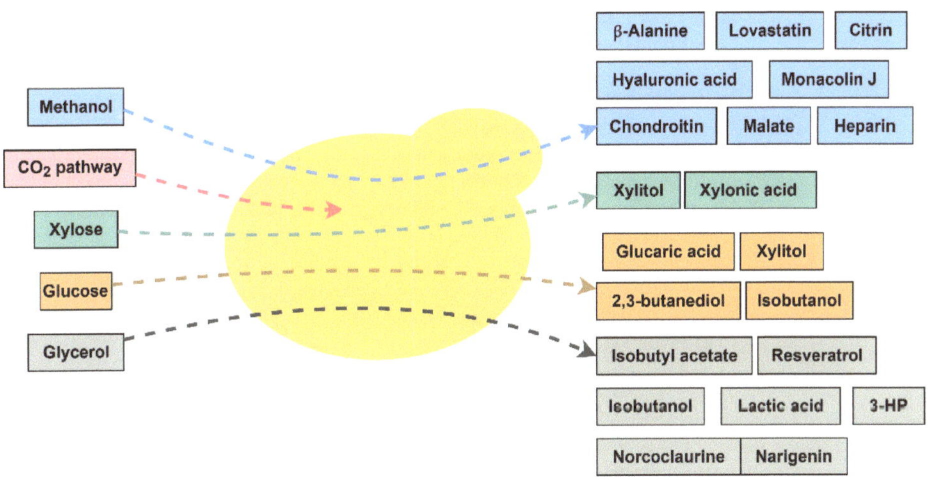

Figure 1. Representative scheme of the different carbon sources and chemicals produced by engineered *K. phaffii* strains. 3-HP = 3-hydroxy propionic acid.

K. phaffii has already been used as a successful host for heterologous protein production. However, optimizing *K. phaffii* to produce chemicals is still necessary, mainly aiming at methanol metabolism, fermentation parameters, and the construction of new genetic tools, such as promoters and plasmids. New methods, platforms, and strategies have been developed to enable more feasible ways to engineer this yeast to address these necessities [9]. Strategies to increase the efficiency rate of target integration genes in the *K. phaffii* genome through the knockout of homologous genes (*ku70* [10] and *Dnl4p* [11]), and also the utilization of CRISPR/Cas9 to facilitate the genome editing of *K. phaffii* [12], have already been used. Several examples of the available tools to enhance *K. phaffii* utilization as a good biotechnology host are summarized in [13].

Previous works have focused on genetic tools and modifications to transform *K. phaffii* into a robust host aiming at biotechnological approaches [7,14,15]. In this review, we outline the recent advances (last five years) in bioengineering of the yeast *K. phaffii* for producing renewable chemicals and proteins. The first section briefly explains *K. phaffii* primordial characteristics and taxonomy. Subsequently, a detailed description of the yeast metabolism of different carbon sources (methanol, glycerol, glucose, and xylose) alongside strategies for the metabolic engineering of the yeast to produce the desired chemicals. The global market size and value (ie., used in this work as the mathematical predictions of the annual volume and value of a product, based on the available data of the past years) of these products is also shown. Additionally, we revise the recent strategies to improve fermentative processes with *K. phaffii*. Finally, the last sections discuss the challenges and prospects of *K. phaffii* industrial utilization in producing renewable chemicals.

2. *Komagataella* Taxonomy and Diversity

The obligate aerobic yeast *Komagataella phaffii* is a non-pathogenic certified and generally recognized as a safe (GRAS) microorganism. It is classified in the Saccharomycetales order and Saccharomycetaceae family. For being capable of using methanol as the only carbon source, *K. phaffii* is methylotrophic and was anteriorly known as *Pichia pastoris*. Phylogenetic studies placed *P. pastoris* in the genus *Komagataella* proposed by Yamada et al. (1995) after the analysis of partial sequences of rRNAs subunits (18S and 26S) of the 12 strains of methanol assimilating yeasts [16]. Supporting this previous study, Kurtzman and Robnett (1998) compared gene sequences (D1/D2 LSU rRNA) of about 500 species of Ascomycetous yeasts. The multigenic sequence analysis sustained the phylogenetic position of *Pichia pastoris* in the *Komagataella* genus [17].

The first species of *P. pastoris* (described initially as *Zygosaccharomyces pastoris*) isolated from a chestnut tree was described in 1919 by Guilliermond [18]. *Komagataella* genus is currently composed of seven species, with most of them (including *K. phaffii*) isolated from tree exudates in North America and Europe (Table 1) [19]. All species have spherical to ovoid shapes white/cream colonies, and during asexual reproduction, haploid cells multiply via multilateral budding and do not have pseudohyphae or true hyphae. In sexual reproduction (diploid cells), the ascospore is hat-shaped, ranging from 1 to 4 spores (that can be conjugated or not) [20]. The *Komagataella* species can grow at high cell densities using methanol, glucose, or glycerol as a carbon source with a doubling time of 2 to 3 h, and ammonium can be used as a nitrogen source during growth [19,20]. Recently, it was demonstrated that some species from *Komagataella*, including *K. phaffii*, possess the capacity to grow on xylose [21].

Table 1. *Komagataella* species diversity and its characteristics.

Type Species	Strain	Genome Size (Mb)	Isolation	Origin	Carbon Sources	Ref
K. pastoris	*CBS 704*	9.6	*Aesculus* species	France	Glucose Glycerol Methanol Ethanol Xylose	ASM170810v1 [A]
K. phaffii	*CBS 7435*	9.4	*Quercus velutina*	California, USA		ASM170808v1 [A]
K. ulmi	*CBS 12361*	9.6	*Ulmus americana*	Illinois, USA		[21]
K. kurtzmanii	CBS 12817	9.6	*Fir flux*	Arizona, USA		
K. mondaviorum	CBS 15017	9.5	*Populus deltoides*	California, USA		
K. pseudopastoris	CBS 9187	10.6	*Salix alba*	Hungary		
K. populi	CBS 12362	9.3	*Populus deltoides*	Illinois, USA		

[A]: NCBI identifier. Table based on [21].

Since all species have similar phenotypic and fermentative features, they cannot be distinguished using the conventional taxonomy, for example, morphological tests, that are usually used for yeasts. Therefore, gene sequence analysis (D1/D2 LSU rRNA, ITS, EF-1α) and other gene markers tracking are necessary for the correct species placement and identification [20,22]. Several genome sequencing and transcriptome studies have been conducted throughout the years [23–25]. A recent study sequenced the genome of all seven type species of *Komagataella* and its strains (a total of 25 isolates), identifying strains capable of growing using xylose as the sole carbon source, and which have a higher tolerance to stress conditions, such as alkaline pH [21].

The genomes of the two most studied species of the genus, *K. pastoris* and *K. phaffii* (previously both species were classified as *P. pastoris*), have close genome structures to *K. ulmi* and *K. kurtzmanii*, respectively. All species are capable of growing in glucose, glycerol, methanol, and ethanol. All 25 strains could grow at acidic pH (4.0) after 7 days, whereas at pH 9.0, all the species were strongly affected. Only the *K. pseudopastoris* (anteriorly known as *P. pseudopastoris*) and *K. kurtzmanii* CBS 12,817 strains were able to adapt and had better growth when compared to the other species [21]. Commonly, non-engineered yeasts in this genus do not present exponential growth on xylose; however, as indicated in the study of Heistinger et al. (2022), all seven species are capable of using this sugar and grow at slow rates. The species *K. pastoris* CBS 704 and *K. populi* CBS 12,362 showed the best growth on xylose [21].

Despite the initial studies on *Komagataella* diversity, the strains of *K. phaffii* X33 (prototrophic strain) and GS115 (HIS^- phenotype) are the ones most used in biotechnological applications, and are mainly used in studies of heterologous expression [22]. In recent years, there have been impressive methodological developments to model and analyze the metabolism of *K. phaffii* and engineer its genome and metabolic pathways, including improvements in genome modification with homologous recombination and target-guided gene cloning using CRISPR/Cas9 [14]. Indeed, the information about the genetic and physiological profile of this specie and the availability of genetic tools to manipulate it has been crucial for works on metabolic engineering and recombinant protein production studies. The most recent data will be presented and discussed in the following sections.

3. Production of Renewable Chemicals from Glycerol by *K. phaffii*

3.1. Metabolism of Glycerol

Glycerol is a carbon source widely used by microorganisms through central carbon metabolism [26]. Due to its higher reduction degree than glucose, the primary carbon source for most microorganisms, glycerol can be used to produce reduced fuel and chemicals at higher yields [27]. In addition, as a significant side stream byproduct in biodiesel production, there has been a growing interest in using glycerol to generate value-added products within the global scenario of renewable and environment-friendly energy, waste reduction, and a more sustainable society [26]. However, its industrial utilization by natural microorganisms is usually limited by the production titers, rates, or yields.

K. phaffii utilizes glycerol as a carbon source very efficiently. Glycerol conversion connects to central carbon metabolism in glycolysis via glyceraldehyde-3-phosphate formation (Figure 2). For such, it is phosphorylated to glycerol 3-phosphate by the cytoplasmic glycerol kinase, sequentially glycerol 3-phosphate dehydrogenase oxidates glycerol-3P to form dihydroxy acetone phosphate (DHAP). Finally, DHAP is converted into glyceraldehyde-3-phosphate by triose phosphate isomerase and enters glycolysis (Figure 2) [28]. Despite the oxidative pathway being preferred in glycerol assimilation, the conversion to glyceraldehyde-3-phosphate can also converge to gluconeogenesis by the activity of a cytosolic aldolase (*FBA1-2*) that catalyzes the reversible reaction of dihydroxyacetone phosphate and glyceraldehyde 3-phosphate into fructose 1,6-bisphosphate (F-1,6BP). The F-1,6BP is converted into Fructose-6P (F-6P) by a phosphofructokinase (*FBP1*), and the F-6P can enter the pentose phosphate pathway (PPP), an important pathway to regenerate the reducing cofactor NADPH [29].

Glycerol is regularly used as the primary initial carbon source for *K. phaffii* cell growth, which allows yeast growth rates similar to those obtained on glucose [30]. In contrast to *S. cerevisiae*, *K. phaffii* can maintain its respiratory metabolism and all its related pathways, such as oxidative phosphorylation, electron transport chain, oxidative balance cofactors, and ATP generation, even in the presence of an excess of glucose, and also demonstrates similar biomass production and substrate uptake kinetics when fed with glucose and glycerol [25]. The efficient use of glycerol as a carbon source by *K. phaffii* is mainly due to the active transport of glycerol via an H^+/glycerol carrier system that allows the utilization of an electrochemical gradient of H^+ protons across the membrane for the cellular influx of glycerol [31].

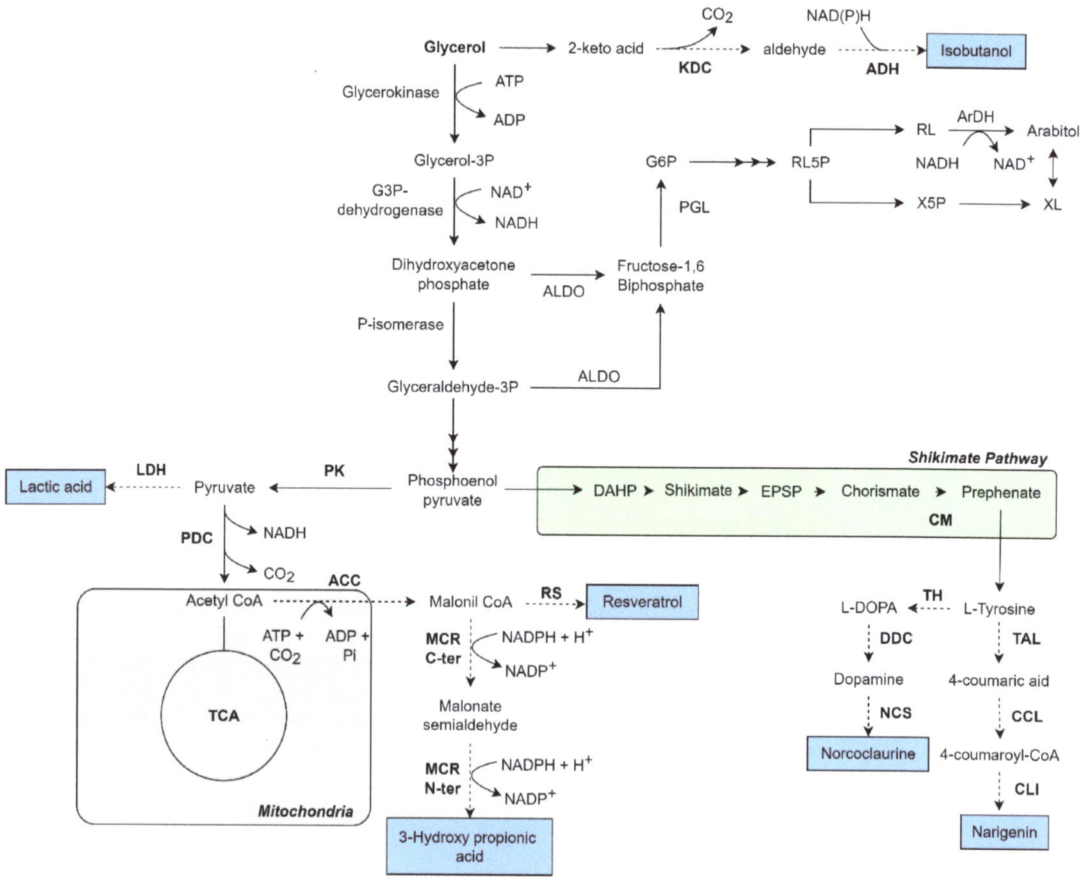

Figure 2. Metabolic pathway to glycerol assimilation and production of a variety of chemicals by *K. phaffii*. Black arrows represent the native glycerol metabolism; dotted arrows represent an overexpressed/heterologous pathway. Bold enzyme names represent an overexpressed/heterologous enzyme, and the blue boxes represent the final product. Some reactions are summarized for better comprehension. Additionally, not all cofactors are represented in the figure. G6P: Glucose-6-phosphate; RL5P: Ribulose-5-phosphate; RL: Ribulose; X5P: Xylulose-5-phosphate; XL: Xylulose; DHAP: 3-deoxy-D-arabinoheptulosonate-7-phosphate; EPSP: 5-O-(1-carboxyvinyl)-3-phosphoshikimate; KDC: Keto-acid decarboxylase; ADH: Alcohol dehydrogenase; ArDH: Arabitol dehydrogenase; ALDO: Adolase; PGL: Phosphoglucose isomerase; LDH: Lactate dehydrogenase; PDC: Pyruvate decarboxylase; PK: Pyruvate kinase; ACC: Acetyl-CoA carboxylase; RS: Resveratrol synthase; MCR (C-ter/N-ter): Malonyl CoA reductase; CM: Chorismate isomerase; TH: Tyrosine hydroxylase; DOC: DOPA decarboxylase; NCS: Norcoclaurine synthase; TAL: Tyrosine ammonia lyase; CCL: 4-coumarate CoA ligase; CLI: Chalcone isomerase. The figure of own authorship is based on KEGG pathways (www.genome.jp—accessed on 13 September 2022) and in the literature cited in the text.

3.2. *K. phaffii* Engineering for Renewable Chemicals from Glycerol

Lignocellulosic valorization through chemical production from glucose and xylose has been widely evaluated and applied. Similarly, using crude glycerol, a byproduct of biodiesel production, to produce platform and fine chemicals is attractive [26]. Its valorization is an exciting option in developing a glycerol-based integrated biorefinery concept due to the high production volumes of crude glycerol in the biodiesel production chain [26,32]. The chemical composition of crude glycerol is variable and can contain 60–80% glycerol, 10–20% of soap or other organic compounds, and 10–20% methanol [33]. Methanol is toxic for most microbes, except for methylotrophic microorganisms. Thus, as *K. phaffii* is able to grow in methanol as the sole carbon source, this yeast is a very well-fit microorganism for utilizing crude glycerol as a primary carbon source for the production of renewables chemicals [14]. Indeed, new metabolic pathway designs and engineering tools for *K. phaffii* increased in the last few years, mainly towards producing complex (secondary) metabolites from glycerol [14]. *K. phaffii* engineered strains for the production of different chemicals from glycerol, including lactic acid, 3-hydroxypropionic acid (3-HP), isobutanol, isobutyl acetate, resveratrol, norcoclaurine, and narigenin, have been constructed. The details are presented and discussed in the following sections.

3.2.1. Lactic Acid Production from Glycerol by *K. phaffii*

Lactic acid (2-hydroxypropanoic acid) is one of the most important industrially used acids. A wide range of organisms can produce lactic acid, including the lactic acid bacteria (LAB) (e.g., strains from *Lactobacillus* genus), filamentous fungi (e.g., genus *Rhizopus*), yeasts (*Kluyveromyces lactis* and *Saccharomyces cerevisiae*), and also by microalgae (e.g., *Scenedesmus obliquus*) [34,35]. More information about lactic acid production by different microorganisms is revised in [35]. Moreover, lactic acid can also be obtained by chemical synthesis [36].

Lactic acid has several applications, such as in the food industry, primarily as an acidulant and preservative chemical agent; and in the pharmaceutical and chemical industries, mainly as a pH regulator, humectant, and solvent. In addition, its monomeric form can be used to produce biodegradable poly-lactic acid (PLA). The PLA has several industrial applications, such as automobile, packaging, and cosmetic industries [34]. The manufacturing of biodegradable PLA materials, ecologically feasible alternatives to the consistent use of petroleum for plastic production, has contributed immensely to the global attention to the biotechnological production of lactic acid. Indeed, the lactic acid market in 2021 was valued at USD 2.9 billion, and the annual growth expectation is of 8.0% rate until 2030 (www.grandviewresearch.com—accessed on 14 September 2022).

Lactic acid production through the biological route is advantageous compared to the chemical one. The main advantage regarding the biotechnological production of lactic acid is that the microorganisms can be engineered to produce only one isomer of lactic acid (L-lactic acid), which facilitates the subsequent production of PLA [37]. *K. phaffii* is one candidate organism for its production using glycerol as substrate [36,37]. As *K. phaffii* does not produce lactic acid naturally, a biotechnological route for glycerol utilization in fermentative production was proposed (Figure 2). The gene *ldh* that codifies for the enzyme lactate dehydrogenase (LDH) was cloned into the yeast through homologous recombination. Additionally, a new homologous lactate transporter (PAS) was identified and overexpressed to enhance lactic acid production in the LDH-expressing strains. The strains were evaluated by their abilities to produce lactic acid in different oxygen concentrations using glycerol as the sole carbon source. The results demonstrated that a higher lactic acid production could be achieved in limited oxygen conditions, though it can guide the yeast through a fermentative metabolism. Under the best conditions, the engineered strain (GLS) was able to produce lactic acid with a yield of 0.7 g/g (Table 2) when fed with pure glycerol (40 g/L). This was the first reported work concerning *K. phaffii* genetic modification to produce lactic acid from glycerol [36].

Table 2. Production of chemicals from different carbon sources by engineered *K. phaffii* strains.

Substrate	Product	Genetic modification	Process	Production ($g{\cdot}L^{-1}$)	Yield ($g{\cdot}g^{-1}$)	Productivity ($g{\cdot}L^{-1}{\cdot}h^{-1}$)	Reference
Glycerol	Lactic Acid	Expression of lactate dehydrogenase (LDH)	Batch and fed-batch fermentation	-	0.70	-	[36]
	Lactic Acid	Deletion of *ArDH* gene	Batch fermentation	24	0.85	-	[38]
	3-hydroxy propionic acid (3-HP)	Expression of *mcr*	Fed-batch	24.75	0.13	0.54	[39]
Glycerol and glucose	Isobutanol	Over-expression of the keto acid degradation pathway and medium supplementation	Batch	2.22	-	-	[40]
	Isobutyl acetate ester	Expression of an alcohol-O-acyltransferase enzyme	Batch	0.051	-	-	[40]
	Isopentyl acetate	Expression of an alcohol-O-acyltransferase enzyme	Batch	0.024	-	-	[40]
Glucose	Isobutanol	Expression of *LlkivD, ScADH7, PpIlv2, PpIlv3, PpIlv5, PpIlv6*	Shake-flask	2.22	0.22	-	[40]
	2,3-butanediol	Expression of *alsS* and *alsD*	2-L bioreactor, fed-batch	74.5	0.30	0.81	[41]
	Glucaric acid	Expression of mMIOX and *Udh*	Shake flask; fed-batch	6.61	-	-	[42]
	Inositol	Overexpression of native inositol pathway	Fed-batch fermentation	30.7	-	-	[43]
Xylose	Xylitol	Expression of PsXYL1 and *gdh*	-	320 (mM)	0.80	2.44	[44]
	Xylonic acid	Expression of XDH	1-L bioreactor; batch	37	0.96	0.41	[45]
Methanol	β-alanine	Overexpression of ADC and aspDH	1-L bioreactor, fed-batch	5.60	-	-	[46]
	Lactic acid	Multicopy integration of *Ldh*	Batch	3.48	0.22	0.036	[47]
	Malic acid	Overexpression of *Mdh, Pyc* and *SpMae*; Δ*gpi*	Shake-flask; batch	2.79	-	-	[48]
	Lovastatin	Expression of *LovB, LovC, LovG, NpgA, LovA, CPR, LovD, LovF*	5-L bioreactor, fed-batch	0.250	-	-	[49]
	Monacolin J	Expression of *LovB, LovC, LovG, NpgA, LovA, CPR*	5-L bioreactor, fed-batch	0.593	-	-	[49]
	6-Methylsalicylic acid	Overexpression of *atX, npgA*	5-L bioreactor with 3-L operating volume, fed-batch	2.2	-	-	[50]

Table 2. *Cont.*

Substrate	Product	Genetic modification	Process	Production ($g \cdot L^{-1}$)	Yield ($g \cdot g^{-1}$)	Productivity ($g \cdot L^{-1} \cdot h^{-1}$)	Reference
	Chondroitin sulfate	Expression of *kfoA*, *kfoC*, *tuaD*, C4ST, PAS_chr1-4_0253, PAS_chr3_0667	3-L bioreactor, fed-batch	2.1	-	-	[51]
	Heparin	Expression of *tuaD*, *kfiC*, *kfiA*, NDST, C5 epi, 2OST,3OST, 6OST	3-L bioreactor, fed-bach	2.08	-	-	[52]
Methanol and Glucose	Hyaluronic acid	Overexpression of *xhasA2*, *xhasB*, *hasC*, *hasD*, *hasE*	2.5-L bioreactor with 1-L operating volume, fed-batch	0.8–1.7	-	-	[53]

A subsequent study approached a different strategy to increase the lactic acid production by *K. phaffii*. As the pyruvate leads to lactic acid production (Figure 2), the authors performed an overexpression of LDH enzymes in combination with the pyruvate decarboxylase (PDC) deletion. This strategy aimed to guide the pyruvate towards lactic acid production and decrease the formation of by-products (e.g., acetic acid). The best strain produced lactic acid with a yield of 0.65 g/g under oxygen-limited condition. Although the acetic acid production was reduced by 20%, the metabolic flux was guided into arabitol formation. The NADH regeneration necessity can explain the flux deviation to arabitol under oxygen-limited conditions. This cofactor unbalance can lead the glycerol metabolism towards the reoxidation of NADH by the reduction of xylulose or ribulose, producing arabitol as a resulting metabolite (Figure 2) [37].

To further investigate glycerol metabolism and the production of lactic acid and arabitol in *K. phaffii* strains, the lactic acid producer strain GLp was extensively investigated under different growth conditions. Once more, lactic acid production was proven better (5 folds higher) in conditions of oxygen limitation, although the arabitol formation is also better in the same conditions. To overcome this issue, a putative arabitol dehydrogenase gene (*ArDH*) was disrupted. As expected, the gene deletion decreased the arabitol production (50% less), leaving only residual concentrations, and increased the lactic acid production yield by 20% (when compared to previous study (Table 2) [37,38].

3.2.2. 3-Hydroxy-Propionic Acid (3-HP) Production from Glycerol in *K. phaffii*

3-Hydroxy propionic acid or 3-hydroxypropanoate (3-HP) is a three-carbon organic compound and is a lactic acid isomer (2-hydroxypropanoic acid). 3-HP is an important building block, mostly used in producing petroleum-based chemicals such as acrylic acid and biopolymers (poly 3-HP) [39,54]. The 3-HP is a versatile molecule for the production of platform chemicals due to its compositional particularity of having a carboxyl and a hydroxyl group. Besides acrylic acid and poly 3-HP, it can be used to produce a wide range of chemicals (e.g., methyl acrylate, ethyl 3-HP, acrylamide, 1,3-propanediol and malonic acid) [54]. Such a wide range of possibilities involving 3-HP utilization increased its global market value, and in 2021 it was valued at USD 111.74 million, with an annual growth expectation of 4.71%, and possibly reaching USD 147.24 million in 6 years (www.marketwatch.com—accessed on 1 September 2022).

Despite the diverse applications, no biological processes are commercially available to produce 3-HP; on the other hand, several patents have been claimed in the past few years, mostly regarding the conversion of 3-HP into acrylic acid in one step reaction. Some microorganisms can produce 3-HP naturally, including *Lactobacillus* sp., and several microorganisms have been modified to enhance the 3-HP production, although the titers, rates and yields (TRY), are not optimum yet [55]. For a more comprehensive view on 3-HP biological production, see [54], and for process engineering for 3-HP production from glycerol through the last few years, see [55].

Nonetheless, the strain engineering of microorganisms aiming to enhance the 3-HP TRY is still necessary to develop a feasible application of the industrial 3-HP biological process [54]. In this context, *K. phaffii* is a suitable candidate to produce 3-HP from glycerol, as this yeast has a naturally well-established glycerol metabolism. To this end, 3-HP was produced in a strain of *P. pastoris* expressing the genes leads to the 3-HP production through the malonyl-CoA pathway (Figure 2). Three heterologous genes were expressed in *K. phafffii*, encoding for malonyl-CoA reductase from *Chloroflexus aurantiacus* (*mcrCa*), an acetyl-CoA carboxylase from *Yarrowia lipolytica* (*ACCYl*), and a cytosolic NADH kinase from *S. cerevisiae* (*cPOSSc*). As a result, the recombinant strain of *K. phaffii* was able to produce 24.75 g/L of 3-HP, with a yield of 0.13 Cmol/Cmol, and 0.54 g/L/h in a fed-batch fermentation. The authors also detected arabitol as a byproduct at the end of the fermentation [39].

3.2.3. Isobutanol and Isobutyl Acetate Production from Glycerol in *K. phaffii*

The four-carbon aliphatic alcohol isobutanol is a valuable platform compound extensively used in various industries such as food, chemicals, and pharmaceuticals [56]. The production of this compound through fermentation by microorganisms is considered a possibility as a second-generation biofuel. When set side by side with ethanol, the prevailing biofuel produced (first generation), isobutanol, presents elevated ignition power, less corrosive, and lower aqueous miscibility [57]. Furthermore, isobutanol (due to its chemical and physical characteristics) is conformable with the procedures and processes commonly applied in the gasoline industry. Consequently, there is no need for profound changes in infrastructure [58].

The Crabtree-negative methylotrophic yeast, *K. phaffii,* was metabolically engineered to produce isobutanol and isobutyl acetate from glucose and glycerol (Table 2) [40]. For the production of isobutanol, genes of the native amino acid biosynthetic pathway were overexpressed, guiding the carbon flux into the 2-keto acids degradation pathway (commonly known as the Ehrlich pathway). In this route, the enzyme 2-keto acid decarboxylase (KDC) decarboxylated the 2-keto acid into an aldehyde that is reduced into alcohol by an alcohol dehydrogenase (ADH) (Figure 2). The first strategy to increase alcohol production was through the Ehrlich pathway, employing new KDC and ADH enzymes selected to be expressed in *K. phaffii*. For the KDC, two enzymes were screened, one from *L. lactis* (*ARO10*) and the other from *S. cerevisiae* (*TH13*). Two enzymes (both from *S. cerevisiae)* were selected for ADH, *ADH6*, and *ADH7*. Additionally, aiming to enhance the isobutanol production, the media was supplemented with 2-ketoisovalerate, a precursor of the Ehrlich pathway. The heterologous expression of KDC and ADH with supplemented media resulted in the production of 284 mg/L of isobutanol [40].

Aiming to enhance isobutanol production, and considering cost-effective approaches, another strategy to improve the level of this compound was applied. Hereupon, the overexpression of part of the L-valine biosynthetic pathway was performed in the engineered strain, eliminating the supplementation of the costly 2-ketoisovalerate [40]. When this approach was applied in *S. cerevisiae,* the levels of isobutanol generation were not elevated, but in contrast, in *K. phaffii,* the production of isobutanol reached 0.89 g/L. Further, the fine-tuning expression of critical enzymes boosted the isobutanol production to 2.22 g/L, which represents a 43-fold enhancement compared to the first results. Additionally, to explore the versatility of this yeast, a gene (*ATF1* from *S. cerevisiae*) that codes for an alcohol O-acyltransferase was heterologously expressed in *K. phaffii* to produce different volatile esters, which resulted in the production of 51 mg/L of isobutyl acetate ester and 24 mg/L isopentyl acetate ester [40].

3.2.4. Aromatic Secondary Metabolite Production from Glycerol in *K. phaffii*

Plant phenolic compounds, such as stilbenoids, flavonoids, and benzylisoquinoline alkaloids (BIAs), are mostly generated from the amino acid L-tyrosine (L-Tyr). These metabolites have a valuable application in the pharmaceutical and food industries. Considering the commercial interest in these aromatic compounds, producing them at high levels by microorganisms is of great interest to [31].

In that way, engineering yeasts have been studied for the overproduction of the amino acid L-Tyr. An engineered *K. phaffii* L-Tyr strain was obtained to enhance the production of phenolic compounds such as resveratrol, naringenin, norcoclaurine, and reticuline (Figure 2) (Table 2). The overexpression of a phosphate synthase ($ARO4^{K229L}$) and chorismate mutase ($ARO7^{G141S}$), both from *S. cerevisiae*, improved the L-Tyr concentration using glycerol as a carbon source in *K. phaffii* [31]. The $ARO4^{K229L}$ is a phosphate mutase, an enzyme involved in the shikimate pathway, which is a route that synthesizes the amino acids phenylalanine, tyrosine, and tryptophan. The shikimate pathway is shown in Figure 2, where the enzyme chorismate mutase (CM) modifies chorismate into prephenate, the precursor for aromatic amino acids, such as L-tyrosine.

With the enhancement of tyrosine production by the engineered *K. phaffii* strain, the resveratrol production reached 451 mg/L, whereas naringenin was 306 mg/L. When a fed-batch approach was performed with glycerol as a carbon source, the amount of these aromatic metabolites was even more significant, reaching 1825 mg/L for the resveratrol and 1067 mg/L for naringenin. These values are the highest reported in the literature for the two compounds [31]. This work mentioned above was the first to evaluate the potential of *K. phaffii* to produce phenolic metabolites using crude glycerol, confirming the importance and versatility of this yeast to generate a range of compounds of interest.

3.3. *Glycerol Co-Utilization by K. phaffii*

Among the major characteristics of *K. phaffii*, the ability to use glucose and glycerol as carbon sources in fast growth rates, resulting in high cell densities, turns attention to the use of this yeast in different applications [7]. One of the main challenges to producing high amounts of a compound of interest is the carbon distribution, which can go for cell growth and synthesis of the product itself. When a mixed carbon source is utilized, it is possible that the cell metabolism could be reasonably split, redirecting the substrates in pathways for the growth and production of aimed compounds. A new study reported an engineered strain of *K. phaffii*, GS115, that was able to co-utilize glucose and glycerol, redirecting the consumption of glycerol for growth, whereas maintaining the glucose for product generation [59].

For this purpose, the identification of *K. phaffii* potential genes involved in the carbon catabolite repression was performed. In this search, the deletion of the glucose sensor (*gss1*) eliminated the glucose-induced suppression of glycerol use in media containing glucose and glycerol. When the transcriptome of the control (non-deleted strain) and the deleted mutant was compared, it was possible to identify that the genes related to glycerol metabolism were under glucose regulation. Then, the overexpression of these genes related to glycerol utilization pathways (gt1, gut1, and gut2) made it feasible for the mutated *K. phaffii* to co-utilize both substrates. Moreover, the knockout of targeted genes lowered the carbon flux from glucose onto major cell pathways, such as glycolysis [60]. Through the metabolic engineering of *K. phaffii*, the new strain was able to utilize glycerol for cell growth, whereas glucose was used for product formation [60]. The co-utilization of different substrates is an interesting and versatile strategy, highlighting *K. phaffii* as an alternative microbial platform for producing value-added products.

4. Renewables from Sugars and Lignocellulosic Hydrolysates

Lignocellulose is one of the most abundant and promising renewable resources, with a global annual yield of around 1.3 billion tons [61]. It can be obtained from agricultural residues after crop processing, forest residues, energy feedstock residues, municipalities, and others [62]. Lignocellulose consists of a complex tridimensional polymer that composes the cell wall matrix of plant biomass, which is usually a remnant material after the extraction of sugars for first-generation bioethanol production or other processes [63]. It is composed of cellulose (38–50% on average), hemicellulose (23–32%), lignin (15–25%), and small amounts of pectin, nitrogen compounds, and inorganic compounds. This ratio may vary depending on plant biomass source, such as herbs, softwood, or hardwood (Table 3) [64].

Table 3. Composition of different lignocellulosic biomasses and hydrolysates.

Biomass	Solid Concentration	Biomass Composition (Dry Basis %)			Pretreatment	Hydrolysate Composition (g/L)		
		Cellulose	Hemicellulose	Lignin		Acetic acid	Furans	Phenols
Sugarcane bagasse	10%	43.1	31.1	11.4	Hot water	1.1–3.4	0.5–5.1	1.4–2.4
Corn stover	10–20%	37	22.7	18.6	Hot water	2.0–2.8	0.74–8.37	181–246 AU
Wheat straw	30%	30.2	21	17	Steam explosion	0.04–1.01	0.16–2.14	nd
Maple	23%	41	15	29.1	Hot water	13.1	4.1	1.3
Olive tree pruning	20%	25	11.1	16.2	Steam explosion	0.4–4.2	0–3.2	nd

AU: Absorbance Unit. Table based on [65].

Cellulose is a linear polymer of glucose monomers linked by β-(1-4) glycosidic bonds that form long and stable glucan microfibrils, providing support to the plant cell wall. Hemicellulose is a mixture of linear and branched polysaccharides connected by acetyl and methyl groups creating an amorphous structure. Xyloglucan, formed by glucose core chain, and heteroxylan, formed by xylose core chain, are the major components of hemicelluloses in plants. These polymers often present heterogeneous side chains of different hexoses (glucose, galactose, mannose, and fucose), pentoses (xylose and arabinose), and uronic acids (glucuronic acid and galacturonic acid). There may also be side branches of methyl, feruloyl, and acetyl groups [66]. Lignin is a heterogeneous aromatic polymer formed by coupling monomers of monolignols such as coniferyl alcohol, sinapyl alcohol, and p-coumaryl alcohol, in a complex structure linked by either carbon-carbon bond connections or ether linkages. A detailed description of lignocellulose is given at [67].

Because of the high degree of polymerization, the production of bioproducts from lignocellulosic biomass requires pretreatment processes to separate or solubilize complex components and release fermentable sugars, e.g., glucose and xylose. Primary, pretreatment methods are applied towards the degradation of the lignin matrix and the removal of hemicellulose fraction. This also reduces cellulose crystallization and increases its surface to later enzymatic and microbial digestion [68].

Pretreatment processes are generally classified as physical, chemical, physicochemical, or biological, and their combinations. Over the last three decades, several pretreatment methods have been developed to provide fast degradation of lignin and high sugar availability with the efficient hydrolysis of cellulose and hemicellulose., including chemical (organic solvents like acid, alkaline, and others), physical (grinding and milling), physio-chemical (hot water, ammonia fiber explosion, and steam explosion), and biological (enzymatic digestion). For detailed data about pretreatments and their applications, refer to [65,68].

During lignocellulosic biomass pretreatment and hydrolysis processes, the cellulose and hemicellulose breakdown, and the sugars are released into the fermentation media, called hydrolysate. In addition, several bioactive compounds are also released or formed during these processes, which can be grouped as organic acids (acetic acid, formic acid, levulinic acid), furaldehydes (furfural and 5-hydroxymethylfurfural), and phenolic compounds (ferulic acid, coumaric acid, vanillin, syringaldehyde, coniferyl alcohol, and others) [69]. In general, lignocellulose inhibitors damage cell membranes and inhibit key metabolic enzymes, affecting the fermentative bioprocess by inhibiting microbial metabolism [69]. Aldehydes can also generate reactive oxygen species (ROS) and lead to their accumulation in the cytoplasm. These toxic molecules inhibit the central carbon metabolism enzymes, such as alcohol dehydrogenase, pyruvate dehydrogenase, and aldehyde dehydrogenase. A more detailed view of lignocellulose-derived inhibitors formation and their effects are described in [69,70]. Apart from biomass constitution, the type and conditions of pretreatments also determine the nature and abundance of the inhibitors released [65]. Some examples of the composition of lignocellulose hydrolysates are presented in the table (Table 3).

To overcome the negative effects of lignocellulose-derived inhibitors, the cells developed strategies to detoxify these compounds and repair damages. Yeast response to lignocellulose inhibitors involves cellular modifications to transcriptional and proteomic levels, resulting in changes in metabolic fluxes and cell reorganization to increase detoxification mechanisms and cell membrane composition [70,71]. For instance, the detoxification of furaldehydes occurs through the compound's NADH- and NADPH-dependent reduction, which affects the yeast's redox balance and fermentation pathways. In *K. phaffii* and *S. cerevisiae,* the presence of furfural and HMF leads to the increased expression of several classes of genes and also of oxidoreductases encoding genes that reduce aldehydes into less toxic corresponding alcohols [71].

K. phaffii shows a dose-dependent response to furaldehydes with a relatively high tolerance to these lignocellulose-derived molecules, maintaining cell viability in the presence of 1.25 g/L furfural and 0.25 g/L HMF. In the presence of these furaldehydes, *K. phaffii* transcriptome analysis revealed the up-regulation of genes involved in oxidoreduction processes and transmembrane transport [71]. Transcriptional profiling of *K. phaffii* in the presence of acetic acid revealed up-regulation of genes involved in nucleic acid processing, methylation, and Rho protein signal transduction regulation, while the presence of furaldehyde revealed up-regulation of genes involved in oxidoreduction and membrane transport [71]. In addition, *K. phaffii* has been shown to sustain growth in the presence of up to 6 g/L of acetic acid in a defined medium. Moreover, the up regulation of genes involved in transmembrane transport, RNA processing, and regulation of metabolic processes was also reported in *K. phaffii* in the presence of 10% diluted acid hydrolysate. In contrast, cellular viability was completely inhibited in the presence of 30% diluted acid hydrolysate [71].

4.1. *Metabolic Engineering of* K. phaffii *to Produce Renewables Chemicals from Glucose*

To produce high value-added products from lignocellulose-derived sugars, it requires *K. phaffii* to incorporate metabolic adaptations to effectively utilize its most abundant C6 and C5 sugars: glucose and xylose [72]. Bio-based chemicals such as ethanol, butanol, and organic acids, showed a compound annual growth rate (i.e., mean annual growth rate of an investment over a period) of 16.16% (USD 6474 million) in 2017. It is expected to reach around USD 23 million by 2025, with biofuels accounting for the highest revenue [64]. In this context, valorizing renewable feedstock into value-added renewable chemicals is an urgent goal.

Several groups have been working in the metabolic engineering of *K. phaffii* to produce renewable chemicals from glucose and xylose in the last couple of years (Table 2). The strategies employed are presented below.

4.1.1. Isobutanol Production from Glucose in *K. phaffii*

As mentioned above, isobutanol ($C_4H_{10}O$) is a branched chain alcohol with the potential as biofuel in substitution of gasoline for high-performance petrol engines [73]. The production of isobutanol creates the opportunity for a range of downstream products. It can be converted into para-xylene via isobutylene and oxidized to terephthalic acid for the production of PET [74]. Bioproduction has been attained in *S. cerevisiae*; however, its natural susceptibility to the Crabtree effect drives to low isobutanol yield against the preference for ethanol production. In *K. phaffii,* isobutanol production can be achieved using glycerol and glucose as substrates (Table 2). Different to with glycerol, the production of isobutanol in *K. phaffii* using glucose was achieved by exploiting the yeast's endogenous L-valine pathway through overexpression of *PpIlv6*, *PpIlv2*, *PpIlv5*, and *PpIlv3*, under GAP promoter in two integrative expression plasmids and diverting the intermediates into the 2-keto acid degradation pathway through the heterologous expression of keto acid decarboxylase (KDC) from *Lactococcus lactis* (*LlkivD*) and ADH7 from *S. cerevisiae* (*ScADH7*). The highest yield of 22.21 mg/g of isobutanol was obtained directly from glucose (Figure 3). Previous work on engineered *S. cerevisiae* has reached up to 8.49 g/L of isobutanol [75]. In a different approach, the production of isobutanol in *S. cerevisiae* was regulated with an optogenetic system called OptoEXP and OptoINVRT. These tools enabled the gene regulation through light pulses (using as a model the blue light-activated EL222 [76]) and the production of three different chemicals of industrial interest, lactate, 2-methyl-1-butanol and isobutanol, the latter with titers of 8.49 ± 0.31 g/L [75]. More information about the microbial production of isobutanol and its future perspectives can be found in [77].

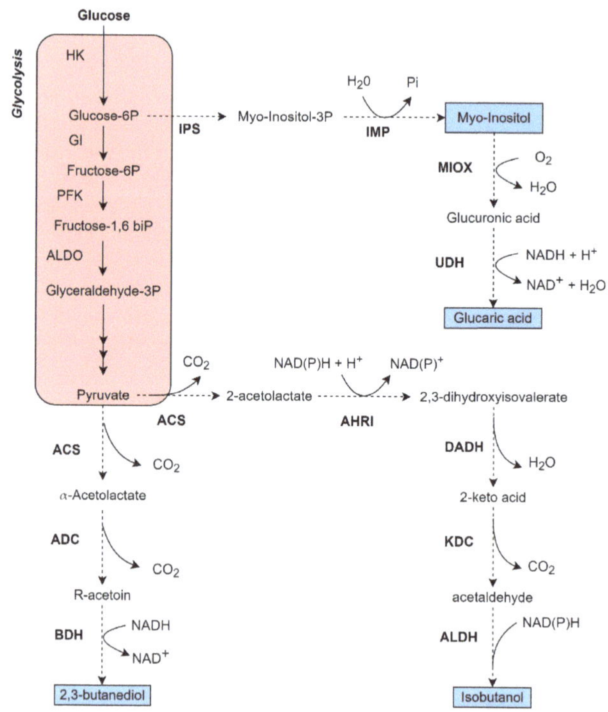

Figure 3. Metabolic pathway to glucose assimilation and production of a variety of chemicals by *K. phaffii*. Black arrows in the red boxes summarize the native glucose metabolism; dotted arrows represent an overexpressed/heterologous pathway. Bold enzyme names represent an overexpressed/heterologous enzyme, and the blue boxes represent the product formation. Some reactions are summarized for better comprehension, also, not all cofactors are represented in the figure. HK: Hexokinase; GI: Glucose isomerase; PFK: Phosphofructokinase; ALDO: Fructose-1,6-biphosphato aldolase; ACS: Acetolactate synthase; ADC: Acetolactate decarboxylase; BDH: Butanediol dehydrogenase; AHRI: Acetohydroxy acid reductor isomerase; DADH: Dihydroxy acid dehydratase; KDC: Keto-acid decarboxylase; ALDH: Aldehyde dehydrogenase; IPS: Inositol phosphate synthase; IMP: Inositol monophosphatase; MIOX: Inositol oxygenase; UDH: Uronate dehydrogenase. The figure of own authorship is based on KEGG pathways (www.genome.jp—accessed on 13 September 2022) and in the literature cited in the text.

4.1.2. 2,3-Butanediol Production from Glucose in *K. phaffii*

2,3-butanediol (2,3-BD) ($C_4H_{10}O_2$) is another bulk platform biochemical that microorganisms can produce from lignocellulose-derived sugars. It can be converted into methyl ethyl ketone, gamma-butyrolactone, and 1,3-butadiene with potential applications in the fuel, polymer, food, and pharmaceutical industries [78]. Heterologous production of 2,3-BD via GRAS microorganisms, such as *E. coli* and *S. cerevisiae*, has been reported, with the highest production of 100 g/L of 2,3-BD in *S. cerevisiae* from glucose and galactose. *K. phaffii* strains were engineered for the production of 2,3-BD isomer (2R, 3R)-2,3-BD with the expression of codon-optimized sequences of acetolactate synthetase (*AlsS*) and acetolactate decarboxylase (*AlsD*) retrieved from *B. subtilis* under pGAP promoter, and also a 2,3-butanediol dehydrogenase (BDH1) from S. cerevisiae was overexpressed in *K. phaffii*, thus, the heterologous BDH1 did not increased the 2,3-BD production. [41]. The highest 2,3-BD production from glucose was achieved in fed-batch optimized cultivation, reaching 74 g/L with yield and productivity of 0.3 and 0.81 g/L/h, respectively, with 99% enantiopurity (Table 2). This result was close to the 100 g/L production of 2,3-BD previously achieved in *S. cerevisiae* by [79] with 0.35 yield and productivity of 0.33 g/L/h with 98% enantiopurity.

4.1.3. Inositol Production from Glucose in *K. phaffii*

Myo-inositol (inositol) ($C_6H_{12}O_6$) is a carbocyclic sugar present in microorganisms, plants, and animals. Dietary inositol has been found to benefit treatments of neurological and endocrine diseases and is also regarded as an essential nutrient for aquatic animals [43,80]. Recently, *K. phaffii* was exploited for inositol production through enhancing its native inositol biosynthesis pathway [43]. The production of 0.71 g/L was obtained by the overexpression of endogenous inositol-3-phosphate synthase (Pp*IPS*) and the knock out of inositol transporters Pp*ITR1* and Pp*ITR2*, and B-subunit of 6-phosphofructokinase gene pfk2. The introduction of heterologous inositol-3-phosphate synthase (*IPS*), inositol monophosphatase (*IMP*) (Figure 3), and the further substitution of promoters *zwf* (encoding glucose-6-phosphate dehydrogenase), *pgi* (encoding glucose-6-phosphate isomerase), and *pfk1* (encoding α-subunit of 6-phosphofructokinase) for glycerol-induced pGUT, could improve inositol production up to 30.71 g/L in high-cell-density fermentation, where glycerol was first used as sole carbon source then glucose was added into the medium for inositol production when wet cell weight reached about 0.2 g/mL (Table 2).

4.1.4. Glucaric Acid Production from Glucose in *K. phaffii*

Glucaric acid ($C_6H_{10}O_8$) is a natural valuable organic acid that can be found in fruits, vegetables, and mammals. Glucaric acid have a range of utilizations in the chemical, pharmaceutical, and food industry and has also been identified as a "top value-added chemical from biomass" for its potential use in the formation of biodegradable detergents and polymers, and metal complexation agents [81]. The co-expression of heterologous mouse MIOX and urinate dehydrogenase (*Udh*) from *P. putida* in *K. phaffii* led to the accumulation of 107 mg/L of glucaric acid from glucose. In contrast, the co-expression of native MIOX and *Udh* produced no glucaric acid from the same carbon source (Figure 3). The optimization of cultivation with the addition of myo-inositol increased glucaric acid concentration from 785.4 mg/L to 1697.6 mg/L in flasks, and production reached 6.61 g/L in fed-batch culture, the highest among other hosts such as *E. coli* and *Granulobacter* sp. (Table 2) [42].

4.2. Metabolic Engineering of K. phaffii to Produce Renewables Chemicals from Xylose

Xylose is another abundant substrate for biotechnological processes. In yeast species that naturally utilize xylose, the metabolism of the pentose starts with its conversion to xylitol, mediated by a NAD(P)H-dependent xylose reductase, followed by the conversion to xylulose, mediated by an NAD^+-dependent xylitol dehydrogenase. Xylulose is therefore phosphorylated by a xylulokinase and enters the pentose phosphate pathway (PPP) as xylulose-5P (Figure 4). An isomerase pathway found mainly in bacteria allows direct conversion of xylose to xylulose via xylose isomerase. In addition, Archaea can assimilate xylose by a non-phosphorylate pathway (NP) converting it into xylonolactone by a xylose dehydrogenase reaction. Xylonolactone will be further converted into xylonate rather spontaneously or with the aid of a xylonolactonase. Xylonate is then dehydrated into 2-keto-3-D-deoxypentanoate to enter the central carbon metabolism [82].

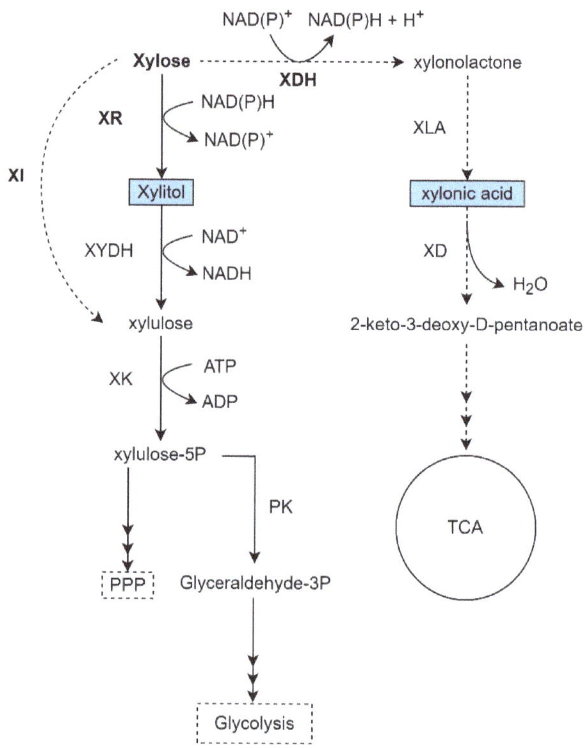

Figure 4. Metabolic pathways to xylose assimilation and production of a variety of chemicals by *K. phaffii.* Black arrows represent the native xylose metabolism, dotted arrows represent an overexpressed/heterologous pathway. Bold enzyme names represent an overexpressed/heterologous enzyme, and the blue boxes represents the product formation. Some reactions are summarized for better comprehension. XR: Xylose reductase; XI: Xylose isomerase; XYDH: Xylitol dehydrogenase; XK: Xylulo kinase; PK: Phosphoketolase; XDH: Xylose dehydrogenase; XLA: Xylono lactonase; XD: Xylonate dehydratase. The figure of own authorship is based on KEGG pathways (www.genome.jp—accessed on 13 September 2022) and in the literature cited in the text.

Despite *Komagataella,* species are generally described as non-xylose utilizing organisms; an examination of *K. phaffii* genome revealed genes encoding for putative XR and a putative XDH homologous to other yeasts, such as *S. cerevisiae* (56% identity) and *Kluyveromyces marxianus* (57% identity) [83]. Recent findings also show that several *Komagataella* species, including *K. phaffii,* can grow on xylose as the sole carbon source, suggesting a hidden pathway of xylose utilization [21]. The introduction of a XI from *Orpinomyces* sp. in *K. phaffii,* in association with adaptive evolution, effectively enhanced the flux of xylose towards PPP (Figure 4), enhancing the consumption rate of xylose from 0.193 g/g to 0.378 g/g after 50 generations [83]. Recent advances in using xylose as a substrate for chemical production in *K. phaffii* refer to xylitol [84] and xylonic acid [45] production (Table 2).

4.2.1. Xylitol Production from Xylose in *K. phaffii*

Xylitol ($C_5H_{12}O_5$) is a natural occurring pentahydroxy sugar alcohol found in a variety of fruits, vegetables, and mushrooms, with high application potential in personal care, pharmaceuticals, and the food industry, as a sweetener that is similar to sucralose but does not require insulin to be metabolized [4,84]; current production is around 200,000 tons [74]. The global market of xylitol exceeded USD 880 million and is expected to expand to over USD 1 billion by 2026 [84]. The biological production of xylitol in yeasts occurs in a single step where xylose is reduced to xylitol by a XR (Figure 4). Recombinant *K. phaffii* strains for xylitol production from xylose and hemicellulose hydrolysate were obtained through (over)expression of XR (*XYL1*) from *P. stipitis* solely or in combination with glucose dehydrogenase (*gdh*) from *Bacillus subtilis*, so that the oxidation of glucose could be used to re-generate NAD(P)H required by the XR reaction [44]. The highest conversion rate by the recombinant strain was 320 mM of xylitol from 400 mM of xylose, with a productivity value of 2.44 g/L/h in a defined medium with both modifications. 300 mM of xylitol from 420 mM of xylose, with a productivity value of 0.46 g/L/h, was obtained in hemicellulose hydrolysate (Table 2).

4.2.2. Xylonic Acid Production from Xylose in *K. phaffii*

Xylonic acid ($C_5H_{10}O_6$) is an organic compound obtained from the microbial oxidation of xylose by naturally or engineered species. It has been reported for serval industrial applications such as a cement additive, plasticizer, cleaner agent, polyamide and polyester co-polymerization, and as a precursor of other value-added chemicals such as ethylene glycol, glycolic acid, and 1,2,4-butanetriol (reviewed in [5]). Several archaea and some bacteria produce xylonic acid by the conversion of xylose to xylonolactone by a xylose dehydrogenase (XDH), which is then converted to xylonic acid in a spontaneous hydration reaction catalyzed by a xylonolactonase (XLA) (Figure 4) [5,82]. Yeasts, on the contrary, are not naturally capable of producing xylonic acid. Therefore, various yeast strains have conquered xylonic acid by the introduction of a heterologous xylose dehydrogenase pathway [5].

The recent construction of *K. phaffii* strains with putative heterologous XDHs was described in [45]. To this end, 11 putative protein sequences of XDH from bacteria and filamentous fungi were identified and overexpressed in *K. phaffii*. Six strains were able to produce xylonic acid. Strains with bacterial XDHs were able to produce tenfold more xylonic acid than the strains with fungal XDHs. The best strain produced up to 37 g/L of xylonic acid in defined medium and up to 11 g/L in sugarcane bagasse hydrolysate, with the highest yield of 0.96 g/g and 0.43 g/g, respectively (Table 2).

5. Renewables from Methanol and CO_2

Methanol is produced in large volumes worldwide, with a production capacity achieving more than 100 million metric tons per year. This C1 nonfood feedstock can be derived from fossil raw materials or renewable resources such as biomass, glycerol, methane, and carbon dioxide, rendering it a flexible and attractive alternative to sugar-based substrates [85]. In addition, methanol is more reduced than most carbohydrates, which can enhance the yields of metabolites produced by microorganisms when using this carbon source as a substrate or co-substrate [86].

Several studies have focused on engineering relevant industrial microorganisms, such as *E. coli* and *S. cerevisiae*, for methanol metabolism [87]. However, implementing the methanol assimilation pathway in non-native methylotrophs is quite challenging. It involves a complete change in the microorganism's lifestyle, requiring, at least, the capacity to build essential cellular metabolites from a single carbon molecule and tolerance to the highly toxic intermediate compounds produced. Despite the achievements, until now, growth performances of the engineered cells on methanol are far from those observed for natural methylotrophs [87].

Several native methylotrophic microorganisms, including *K. phaffii*, *Ogataea (Hansenula) polymorpha*, and *Bacillus methanolicus*, are naturally capable of utilizing C1 compounds as a source of energy and carbon through different metabolic routes [85,88,89]. Among them, *K. phaffii* presents some prominent properties that make this yeast an interesting host for biotechnological processes. For example, we can cite resistance to high concentrations of methanol and other stressful industrial conditions, growth to high cell densities, and ability to express recombinant protein efficiently [7,85]. In addition, several tools are available for synthetic and metabolic engineering for this yeast and well-established fermentation processes [90,91]. Therefore, *K. phaffii* has been widely used for the recombinant protein production of several biopharmaceuticals and industrial enzymes, as demonstrated in this review (as discussed in the next section), and, more recently, has also been employed for the production of chemicals and other valued added compounds, although mainly using sugars as carbon sources as previously presented here [14,85].

In *K. phaffii*, methanol metabolism initiates with its oxidation to formaldehyde. Then, the formaldehyde can follow either an assimilatory or dissimilatory pathway. In the assimilatory pathway, localized in peroxisomes, formaldehyde is condensed to phosphosugars and enters the central metabolism, yielding biomass. In the dissimilatory pathway, localized both in peroxisome and in the cytosol, the formaldehyde is oxidized to CO_2, yielding NADH (Figure 5) [92].

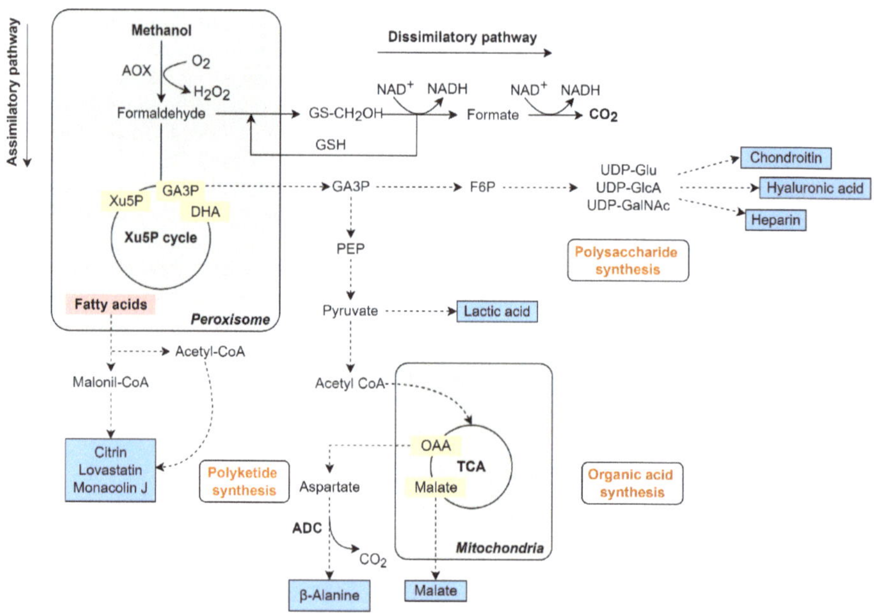

Figure 5. Metabolic pathways that led to methanol assimilation and production of a variety of chemicals by *K. phaffii*. Black arrows represent the native methanol metabolism, dotted arrows represent an overexpressed/heterologous pathway. Bold enzyme names represent an overexpressed/heterologous enzyme, and the blue boxes represents the product formation. Some reactions are summarized for better comprehension, also, not all cofactors are represented in the figure. AOX: Alcohol oxidase; Xu5P: Xylulose-5-phosphate; DHA: Dihydroxyacetone; GAP: Glyceraldehyde-3-phosphate; XuMP: Xylulose-monophosphate cycle; PEP: Phosphoenol pyruvate; F6P: Fructose-6-phosphate; OAA: Oxalacetate; ADC: L-aspartate decarboxylase. The figure of own authorship is based on KEGG pathways (www.genome.jp—accessed on 13 September 2022) and in the literature cited in the text.

The first enzymatic reaction occurs in the peroxisomes, where methanol is oxidized to formaldehyde with the release of hydrogen peroxide (H_2O_2) by the action of the key-enzyme alcohol oxidase (AOX). As H_2O_2 is toxic to the cell, the compartmentalization of the process in the peroxisomes aid in protecting the cell until the catalase converts it into H_2O and O_2 as well as could alleviate the toxicity caused by formaldehyde accumulation (Figure 5) [85]. *K. phaffii* has two isoforms of AOX, encoded by the genes AOX1 and AOX2, which share 97% of identity at the protein level. Among them, AOX1 represents the major AOX protein expressed during growth on methanol, whose content could reach 20% to 30% of the total protein [85].

In the assimilatory pathway, formaldehyde is then combined with xylulose-5-phosphate (X5P) by the enzyme dihydroxyacetone synthase (*Das1* and *Das2* in *K. phaffii*), generating dihydroxyacetone (DHA) and glyceraldehyde-3-phosphate (GAP). A total of 1 mol of GAP are produced from 3 mol of methanol, which is used to generate biomass and energy. X5P is recycled by sugar phosphates interconversions catalyzed by PPP isoenzymes in the peroxisomes [89,92]. This compartmentalized and cyclic C1 assimilation process is called Xylulose-Monophosphate cycle (XuMP) (Figure 5) [89].

Part of the formaldehyde generated in the methanol oxidation reaction follows the dissimilatory pathway. In this case, the formaldehyde flows into the cytosol and reacts with glutathione through the formaldehyde dehydrogenase, S-formylglutathione hydrolase, and formate dehydrogenase enzymes, yielding CO_2 and NADH (Figure 5) [89]. Although the C1 carbon is lost in the process, it is an essential route for energy generation (in the form of the reducing power NADH) and formaldehyde detoxification, thereby affecting the ability of cell growth on methanol [85].

5.1. *K. phaffii Engineering for CO_2 Assimilation*

Exploring the methanol-assimilation pathway, a *K. phaffii* strain able to grow using CO_2 as a sole carbon source, was recently reported [92]. In the synthetic autotroph strain, the peroxisomal pathway for methanol assimilation was rewired to a CO_2-fixation pathway that resembles the Calvin-Benson-Bassham cycle (CBB). For this, eight heterologous genes were expressed and combined with the XuMP: six CBB enzymes (RuBisCO form II encoded by *cbbM, PGK1, TDH3, TPI1, TKL1, PRK*) and two chaperones (*groEL, groES*) to assist RuBisCO folding. In addition, the native genes *DAS1, DAS2,* and AOX1, were deleted to separate the CO_2-fixation machinery from energy generation. Thus, methanol was used to provide energy via the dissimilatory pathway [92]. Moreover, adaptive laboratory evolution (ALE) was employed to increase the growth rates of the strain on CO_2, resulting in a gain of more than 2-fold (from 0.008 h^{-1} to 0.018 h^{-1}) [92].

The Improved phenotypes of these evolved CO_2-assimilating *K. phaffii* yeasts were further investigated [93]. Reverse genetic engineering was used to evaluate the mutations found in the genes encoding for the heterologous phosphoribulokinase (*PRK*) and the native nicotinic acid mononucleotide adenylyltransferase (*NMA1*). The results showed that lower activities of the mutated enzymes, which affect the availability of ATP and other consequences, are beneficial to the autotrophic phenotype. Furthermore, through a second evolution round, it was identified that facilitating peroxisomal import can be another engineering strategy to improve the autotrophic growth of *K. phaffii* [93].

These studies represent a remarkable advance in microbial utilization of one carbon (C1) feedstocks, since synthetic autotrophs can be used as an industrial platform for the production of several value-added chemicals and enzymes. Besides, this approach may have a significant role in the circular bioeconomy, promoting more sustainable biotechnological bioprocesses and also contributing to mitigating the atmospheric greenhouse gas CO_2 [92,93]. However, despite the huge potential of the implementation of the CO_2-assimilating pathway, many challenges should be addressed before it can be successfully explored in industrial production bioprocesses.

5.2. Metabolic Engineering of K. phaffii for Production of Renewables from Methanol

C1 substrates, such as CO_2, methane, and methanol, have been considered next-generation feedstocks since they are relatively inexpensive due to their natural abundance, low production cost, or availability as industrial by-products [94]. In particular, methanol has gained attention as it can be produced directly from methane and CO_2 using renewable resources [87]. Considering that, the native capacity of *K. phaffii* to metabolize methanol has been the focus of several studies to develop biotechnological fermentation processes. Although methanol has been widely employed in several bioprocesses related to recombinant protein expression by *K. phaffii*, only more recently, methanol has been explored as a substrate or co-substrate for the synthesis of various chemical compounds. We can cite, for example, the production of β -alanine, organic acids, alkenes, fatty acids, and pharmaceutical compounds, such as methyl salicylic acid and lovastatin (Table 2) [14,46–50,85].

5.2.1. β-Alanine Production from Methanol in *K. phaffii*

β-alanine is a natural β-amino acid that can be used to synthesize several nitrogen-containing compounds for the food and pharmaceutical industry. Biological production of β -alanine is obtained via L-aspartate decarboxylation. Considering that aspartate is one of the main amino acid pools in *K phaffii*, different bacterial L-aspartate decarboxylase (ADC) genes were expressed in the yeast to evaluate β-alanine production (Figure 5) [46]. Further improvement on β-alanine synthesis was achieved by increasing the ADC copy number and the supply of the C4 precursor aspartate. The engineered strain produced 1.2 g/L of β -alanine in shake-flask culture and reached a titer of 5.6 g/L in a fermenter using a two-stage strategy with high initial biomass (Table 2). This study was the first attempt to produce amino acids in *K. phaffii* using methanol as substrate, indicating the great potential of the yeast for fermentative production of amino acids [46].

5.2.2. Organic Acids Production from Methanol in *K. phaffii*

Malic acid, like other C4 dicarboxylic acids, and lactic acid, are on the top of the most relevant bio-based chemicals that can be produced through microbial processes from renewable resources [74,81]. Both compounds have high commercial value and a broad range of applications in the industrial sector, with a global market size estimated at USD 182.6 million in 2018 for malic acid (https://www.grandviewresearch.com/industry-analysis/malic-acid-market—accessed on 17 August 2022) and USD 2.9 billion in 2021 for lactic acid (https://www.grandviewresearch.com/industry-analysis/lactic-acid-and-poly-lactic-acid-market—accessed on 17 August 2022).

Considering that, *K. phaffii* has been engineered to produce organic acids using methanol [47,48]. The expression of multiple copies of the D-lactate dehydrogenase gene from *Leuconostoc mesenteroides* via gene integration into the rDNA locus of *K. phaffii* and post-transformational vector amplification, led to a D-lactic acid production of 3.48 g/L by the recombinant cells after 96 h of cultivation [47]. In another study, the conversion of methanol into malic acid was reported [48]. Guo et al. [48] evaluated three different malic acid accumulation modules in *K. phaffii*. Among them, the reductive TCA pathway derived from *Rhizopus oryzae* showed the best results for malic acid production. Additional metabolic engineering was also tested to optimize the strain, which included: (i) overexpression of the SpMAE transporter, (ii) deletion of genes related to by-product formation, and (iii) modifications on the methanol metabolism pathway. The resulting strain could produce 2.79 g/L of malic acid from methanol [48]. Although a malic acid titer of 42.28 g/L has been already obtained by a recombinant *K. phaffii* strain during the cultivation on glucose and using methanol as inducer [95], it was the first report of malic acid production using methanol as the only carbon source (Table 2) (Figure 5).

5.2.3. Biopolymers Production from Methanol in *K. phaffii*

Biopolymers have also been produced in *K. phaffii*. Hyaluronic acid is a glycosaminoglycan polysaccharide with remarkable biological activity. Because of that, it is widely used in the medical and pharmaceutical sectors, with a global market valued at USD 9.1 billion in 2019 (reviewed in [96]). The commercial production of hyaluronic acid is mainly from animal tissue extraction and, more recently, by fermentation using the pathogenic bacteria *Streptococcus* sp. [96]. However, attempts have been developed to produce this polysaccharide using other non-pathogenic host strains [53]. Considering that, *K. phaffii* was engineered to express the hyaluronan synthase (*xhasA2*) and UDP-glucose dehydrogenase (*xhasB*) from *Xenopus laevis*, in combination with the overexpression of native genes involved with the hyaluronic acid biosynthesis (*hasC*, *hasD*, *hasE*). The engineered yeasts produced an amount of 0.8–1.7 g/L of hyaluronic acid with a molecular weight from 1.2 to 2.5 MDa (Table 2), using glucose as the main substrate and methanol as co-substrate and inducer (Figure 5) [53].

Exploring the natural occurrence of the biosynthesis pathway of sulfation in the yeast, the production of chondroitin sulfate, another clinically relevant biopolymer, has been proposed and implemented in *K. phaffii* (Figure 5). Chondroitin is a glycosaminoglycan composed of sulfated disaccharides repetitions of glucuronic acid (GlcA) and N-acetylgalactosamine (GalNAc) linked by β (1 → 3) and β (1 → 4) bonds [51]. Initially, an engineered strain was constructed by introducing the chondroitin biosynthesis pathway harboring the genes *kfoC* and *kfoA* from *Escherichia coli*, which led to the synthesis of the intermediate metabolites UDP-GlcA and UDP-GalNAc, and *tuaD* from *Bacillus subtilis*, which encoded a chondroitin polymerase responsible for the polysaccharide formation. The resulting strain produced 5.5 mg/L of chondroitin [51]. For further improvement, combined strategies were employed, such as codon optimization, expression of the chondroitin-4-O-sulfotransferase (C4ST), and increased supply of the 3′-phosphoadenosine-5′-phosphosulfate (PAPS, a universal donor for sulfonated molecules). As a result, the yeast was capable of synthesizing 2.1 g/L of chondroitin sulfate from methanol [51].

Recently, this previously engineered strain [51] was used as a host to express sulfotransferases and an epimerase for in vitro biosynthesis of heparin [52]. Taking into consideration the successful expression of all required enzymes, the authors proposed the construction of a yeast platform for complete biosynthesis of heparin from methanol. The recombinant *K. phaffii*, co-expressing heparosan synthesis enzymes (tuaD, KfiC, KfiA), bifunctional enzyme NDST, sulfotransferases (2-OST, 6-OST and 3-OST), and epimerase C5 epi, showed production of 2.08 g/L of the bioengineered heparin, with a molecular weight of 349 kDa, in fed-batch cultures [52]. Heparin is a linear and unbranched glycosaminoglycan widely used as an anticoagulant drug. The global market was valued at USD 7.3 billion in 2021 and is growing with the increasing worldwide demand (https://www.grandviewresearch.com/industry-analysis/heparin-market—accessed on 17 August 2022).

5.2.4. Polyketides Production from Methanol in *K. phaffii*

The potential of *K. phaffii* as a microbial chassis for polyketide biosynthesis has also been evaluated (Figure 5) [49,50,97]. Polyketides are a class of secondary metabolites produced by several organisms, showing structural diversity and bioactivity, making some interesting pharmaceutical drugs [14,50]. The synthesis of the polyketide 6-methyl salicylic acid (6MSA) was described by Gao et al. [50]. In *K. phaffii*, through overexpression of the 6-methyl salicylic acid synthase gene (*atX*) from *Aspergillus terreus* and phosphopantetheinyl transferase gene (*npgA*) from *Aspergillus nidulans*, a production of 2.2 g/L 6MSA was achieved during cultivation in bioreactor [50]. Later, the same research group reported the synthesis of the polyketide citrinin by *K. phaffii* [97]. For this, the citrinin biosynthetic pathway, which genes are derived from *Monascus purpureus* (*pksCT*, *mpl1*, *mpl2*, and *mpl4*) and *Monascus ruber* (*mpl6*, *mpl7*), was introduced into the yeast together with the expression of the transferase *npgA*-encoding gene. The engineered strain was able to accumulate 0.6 mg/L of the polyketide using methanol as a carbon source and inducer [97].

Lovastatin, a common pharmaceutical drug used in hypercholesterolemia treatment, and its precursor monacolin J, are other examples of polyketides produced in *K. phaffii* [49].

The commercial product is obtained mainly from fermentation using an *Aspergillus terreus* strain. The metabolic engineering of other microbial hosts, such as *K. phaffii*, can be an alternative to reduce the long fermentation period required by filamentous fungi and limit multiple byproduct formation [49]. For production in *K. phaffii*, the complex monacolin J and lovastatin biosynthetic pathways were assembled and implemented in the yeast, resulting in 3.3 and 21 mg/L of lovastatin and monacolin J, respectively. Using a buffered medium, an improvement in product concentration of approximately three times was observed during shake flask cultivation.

Furthermore, the pathway was split into two modules and implemented in two separated strains to reduce the metabolic burden and avoid the accumulation of intermediate metabolites. During bioreactor cultivation, the co-culture system enabled the production of 250 mg/L of lovastatin and 593 mg/L monacolin J using methanol as substrate. These results demonstrated that microbial consortium could be an efficient strategy for metabolite production. However, it depends on cell internalization of the intermediate compounds [14]. This study was the first report of biosynthesis of lovastatin in a heterologous host; additionally, the compound production was comparable to that obtained with the native *A. terreus* strain [49].

As can be seen, several studies demonstrate the potential of *K. phaffii* as a microbial chassis to produce value-added chemicals using methanol as a carbon source. However, methanol fermentation efficiency is still lower than sugar-based processes. To overcome the current proof-of-concept stage, some problems should be addressed, including the cellular toxicity of methanol and the intermediate formaldehyde, as well as the carbon loss (via the dissimilatory pathway) that reduce product yields [85,88]. Therefore, further improvements are required for the industrial application of *K. phaffii* in methanol bioconversions.

6. *Komagataella phaffii* Application in Protein Production

Komagataella phaffii has been one of the most used species for the cost-effective production of recombinant protein, capable of expressing a range of enzymes with industrial and pharmaceutical relevance (Table 4). This fact is due to the ability of this microorganism to grow to very high cell densities (biomass over 100 g/L using methanol) and produce amounts of recombinant protein ranging from milligrams to grams for both research (laboratory) and industry purposes [19,98]. Anteriorly, *Pichia pastoris* (include *K. phaffii* and *K. pastoris* strains) first caught attention by being used as a single-cell protein (SCP) by Phillips Petroleum and commercialized for animal supplementation [72,99]. Later, with more advanced genetic manipulation of *K. phaffii* and the isolation of different promoters, especially the alcohol oxidase genes (AOX), glyceraldehyde-3-phosphate dehydrogenase (GAP), and the 3-phosphoglyceratekinase gene (PGK1), the development of expression vectors, new techniques, and protocols for genetic transformation, culminated in the creation of the current successful heterologous expression system [99–101].

Among the main advantages of recombinant expression in *K. phaffii* are rapid growth in cheaper culture media, scaling up of protein production, easy maintenance, and genetic manipulation. In addition, these organisms are capable of performing most post-translational modifications such as glycosylation (N or O), formation of disulfide bridges, and adequate folding of the protein of interest, allowing the production of a cost-competitive enzyme with minimal processing in the following steps [6,72,99]. Moreover, the choice of a promoter (constitutive or induced), selection marks (dominant or auxotrophic), and signal peptides for extracellular secretion (when desired), are essential for the successful production of the protein of interest. Additionally, integrated vectors into the genome potentially increase the genetic stability of the engineered strains even in continuous and large-scale fermentation processes [99].

Table 4. Examples of heterologous proteins expressed in *K. phaffii* and production yields.

Protein	Expression Vector	Promoter	Production	Yield/Activity	Reference
Malaria vaccine candidate protein	pPICZαA	AOX	3-L bioreactor, fed-batch	62.2 g/L	[102]
Dengue vaccine candidate protein	pPICZ-A	AOX	Shake-flask; fed-batch	15 mg/L	[103]
Chikungunya vaccine candidate protein	pPIC9K	AOX	Shake-flask; fed-batch	60 mg/L	[104]
Tuberculosis vaccine candidate protein	pPICZαA	AOX	Shake-flask; fed-batch	5 μg/mL	[105]
SARS-CoV-2 Spike RBD	pPICZαA	AOX	7-L bioreactor, fed-batch	45 mg/L	[106]
Human proinsulin	pPICZα	AOX	Shake-flask; fed-batch	5 mg/L	[107]
Human epidermal growth	pPIC9K	AOX	Shake-flask; fed-batch	2.27 μm/mL	[108]
Interleukin-1beta	pPICZαA	AOX	Shake-flask; fed-batch	250 mg/L	[109]
Antimicrobial Hispidalin	pPICZαA	AOX	Shake-flask; fed-batch	98.6 μg/mL	[110]
Antimicrobial CecropinA-thanatin	pPICZαA	AOX	Shake-flask; fed-batch	1.061 μmol/L	[111]
Antimicrobial PAF102 peptide	pGAPZA	GAP	Shake-flask; batch	180 mg/L	[112]
Antimicrobial fowlicidins	pPICZαA	AOX	Shake-flask; fed-batch	85.6 mg/L	[113]
Human serum albumin	pPIC9K	AOX	Shake-flask; fed-batch	8.86 g/L	[114]
Endoglucanase	pPink-GAP	GAP	15-L bioreactor, fed-batch	3 to 5 g/L	[115]
Cellobiohydrolase	pPpB1	AOX	1-L bioreactor, fed-batch	6.55 g/L	[116]
β-glucosidases	pPIC3.5K	AOX	5-L bioreactor, fed-batch	403 mg/L	[117]
LPMO	pPICZαA	AOX	Shake-flask; fed-batch	-	[118]
Expansin	pPICZαA	AOX	5-L bioreactor, fed-batch	4.3 mg/L	[119]
Xylanase	pPICZαA	AOX	7.5-L bioreactor, fed-batch	2503 U/mL	[120]
β-xilosidase	pPICZαA	AOX	Shake-flask; fed-batch	0.22 mg/L	[121]
Feruloyl esterase	pGAPZαA	GAP	Shake-flask; fed-batch	-	[122]
Acetyl xylan esterase	pPICZαA	AOX	Shake-flask; fed-batch	1.5 mg/L	[121]
α-L-arabinofuranosidase	pPICZαA	AOX	7.5-L bioreactor, fed-batch	164 U/mL	[123]
Mannase	pPIC9K	AOX	10-L bioreactor, fed-batch	10.47 g/L	[117]
Lipase	pPICZαA	AOX	Shake-flask; fed-batch	145.4 U/mg	[124]

μ: specific growth rate (h^{-1}); qP: specific product formation rate (h^{-1}); qV: volumetric productivity ($g/L^{-1}{\cdot}h^{-1}$); YP/X: product to biomass yield (-); D: dilution rate (h^{-1}); FBA: flux balance analysis; HCDF: High cell density fermentation.

Several studies have reported the production of recombinant proteins in *K. phaffii*, registering more than 500 biopharmaceutical proteins, such as insulin, anti-microbial peptides, human serum albumin (HSA), and epidermal growth factor [98]. These proteins represent a growing market in the therapeutics biotechnology business. Since *K. phaffii* has the status GRAS, a range of approved biopharmaceuticals such as vaccines is being extensively studied for production in this yeast. For instance, the production of protein candidates for subunit vaccines has been expressed in *K. phaffii* to be used against malaria [102], dengue [103], chikungunya [104], tuberculosis [105,125], and SARS [126,127]. Further, the expression of SARS-CoV-2 Spike RBD (receptor binding domain) in *K. phaffii* was scaled-up reaching yields above 45 mg/L of 90% pure protein. This work provided an alternative for the production of neutralizing antibodies and the possible development of SARS-CoV-2 vaccines to combat the COVID-19 pandemic [106].

Additionally, other important therapeutic proteins were produced at high levels in *K. phaffii,* such as biologically active human proinsulin [107] and human epidermal growth [108]. The interleukin-1beta involved in different physiological responses was successfully expressed in *K. phaffii*, and its activity was detected to inhibit the growth of B16 melanoma cells [109]. Currently, different active antimicrobials against pathogens (bacteria and fungi) like Hispidalin [110], the Hybrid cecropinA-thanatin [111], PAF102 peptide (under GAP regulation) [112], and fowlicidins [113] have been produced in *K. phaffii.* Moreover, *K. phaffii* achieved one of the highest concentrations (8.86 g/L) of a human serum albumin, an important protein with different therapeutic functions [114]. All these cited proteins were expressed under the regulation of the AOX promoter unless otherwise specified.

The global market for recombinant proteins is expected to reach USD 1.7 billion by 2026 [128]. In the food, feed, detergent, and clothes industries, many relevant enzymes for biotechnology applications have been produced in *K. phaffii*, such as lignocellulolytic enzymes (responsible for the degradation of main components of the plant cell wall, such as cellulose, hemicellulose, pectin, lignin, and others) [116,129]. A wide range of these enzymes is naturally produced by fungal strains like *Trichoderma* spp., *Penicillium* spp., *Aspergillus* spp., *Humicola* spp., *Thermomyces*, and *Myceliophthora* spp. [130]. To produce purified and high yields of the targeted proteins from these fungi, using heterologous expression in *K. phaffii* is highly favorable. Additionally, an important fact is that *K. phaffii* does not secrete endogenous lignocellulolytic enzymes in significant amounts. Thus, recombinant strains of this yeast can provide heterologous enzyme preparations without the need for many purification steps, reducing the time and cost of the process [99,116].

Cellulases such as endoglucanases [115], cellobiohydrolases [116], and β-glucosidases [131], have been expressed in *K. phaffii* allowing purification, biochemical characterization, and biotechnological applications. These proteins can be used in paper and pulp, textile, and detergent industries, and applied in biorefineries for the bioconversion of lignocellulosic biomass [132]. In this context, to aid in the degradation of cellulose chains, the oxidative enzymes called lytic polysaccharide monooxygenases (LPMOs) and classified as auxiliary activity (AA) proteins have been produced in *K. phaffii* [118,133], as well as expansins (AA proteins), which act in synergy with cellulases to enhance sugar release from the substrate [119].

Likewise, hemicellulases like xylanases and β-xilosidases have been extensively expressed in *K. phaffii*. These enzymes can be applied in the textile, paper, and pulp industries, and hemicellulose deconstruction [120,134,135]. Other enzymes that can be utilized with xylanases and β-xilosidases to increase hemicellulose degradation have also been expressed in this yeast, including feruloyl esterase, acetyl xylan esterase, and α-L-arabinofuranosidases [121–123]. The recombinant mannanases and pectinases produced in *K. phaffii* are very relevant for the food industry and have been used in preparation for prebiotics and juice clarification, respectively [117,136]. Furthermore, successfully recombinant lipases produced in *K. phaffii* expression system are being studied as biocatalysts for the conversion of biomass fatty acids in biodiesel [124,137].

Several proteins of industrial and medical importance have been produced in *K. phaffii* and approved by regulatory institutions so far, noting the efficient expression system established in this yeast. Meaningful progress has been made with improved protein expression levels with different approaches. However, protein production still needs to be optimized and maximized for large-scale processes utilized in industry.

7. Strategies to Develop and Optimize Fermentative Processes

The production of proteins and chemicals in *K. phaffii* can be a powerful tool to address global needs. However, despite the increasing development of this microbial platform, their fermentation processes have still not achieved the same maturity as conventional biochemical processes [138–140]. There is a disparity between a promising laboratory scale and its industrial application; although many studies have demonstrated proof-of-concept pathways for new molecules, a much smaller amount of these works have focused on bioprocess design, optimization strategies, mathematical modeling, or scaling-up processes [139,141–143]. Integrating strain engineering and fermentation strategies is crucial for improving bioprocess efficiency. The technology transfer of bioprocesses in host systems is usually a result of iterative process optimizations and biological re-design cycles [144]. In this sense, a deep understanding of cell factories, process development, optimization, and scaling-up must be part of the same improvement. As strain engineering has been discussed in the previous sections, this one will focus on fermentation characteristics and strategies to produce a maximum amount of product in the minimum process time with quality specifications.

Despite the existence of conditions widely used in the fermentative processes in *K. phaffii*, optimal conditions differ according to the target product, the combination of genetic construction, or whether the production occurs under the control of inducible or constitutive promoters [98,140]. The most successful process development in *K. phaffii* suggests that the optimized process moves away from standard protocols and adapts process parameters to specific strain/product/bioreactor [140,145]. The main factors are discussed in detail in the following and summarized on Table 5.

Table 5. Strategies for process development, simulation, and optimization with *K. phaffii*.

Heterologous Product	Optimization Strategy	Observed Improvement	Ref
Erythropoietin (biopharmaceutical product)	Dynamic flux balance analysis (elementary process function integrated with FBA)	The maximum productivity obtained in optimization is 66% higher than the benchmark experimental study	[146]
Recombinant Human Growth Hormone (rhGH)	Study of sorbitol-methanol co-feeding strategy and results compared with the basic feeding protocol	Under optimal conditions, cell biomass, total protein, and rhGH concentration increased 15%, 99.5%, and 99.4%	[147]
Sea raven antifreeze protein	A model-based approach to optimize qV in a glycerol/methanol mixed-feed continuous stirred-tank reactor	In the optimized conditions, qV was 2.2 mg/L.h, representing a tenfold increase compared with an initial strategy	[148]
Growth and AOX-promoter based recombinant protein expression	Adaptive laboratory evolution to improve growth and recombinant protein production in methanol-based growth media	Evolved populations showed increased μ. A selected clone showed increased product titers ranging from a 2.5-fold increase in shake flask batch culture to a 1.8-fold increase in fed-batch cultivation	[149]
Human 2F5 antigen-binding fragment (Fab)	Development of a novel operational strategy, with carbon-starving periods and elucidation of the μ effects on the protein secretion	Increments up to 50% of both yields and total production were observed. High μ presented an increment up to 8-fold on the production rates	[150]
Human 2F5 antigen-binding fragment (Fab)	Evaluation of a wide range of oxygen-limiting conditions in chemostat cultivations	Specific conditions that lead to the maximum productivity of the process were determined and resulted in an increase of up to 3-fold in qV and $Y_{P/X}$	[151]
Human recombinant alpha 1-antitrypsin (A1AT)	A new control system designed for maintaining the μ_{opt} during the induction phase. The neural network was applied to adjust and optimize the performance of the robust control system	The newly designed μ-stat control technique enhanced production by up to 1.5 and 2.1 folds in comparison with oxygen-limited fed-batch feeding and mixed feed methods, respectively	[152]

Table 5. *Cont.*

Heterologous Product	Optimization Strategy	Observed Improvement	Ref
Human Serum Albumin (HSA)	Dynamic genome-scale metabolic model for glucose-limited, aerobic cultivations for batch and fed-batch cultures	The model suggested that implementation of a decreasing μ during the feed phase of fed-batch culture results in a 25% increase in qV	[153]
Fab fragment (anti-HIV antibody 2F5)	Model for product accumulation in fed-batch, based on iterative calculation, to optimize the time course of the media feed to maximize qV	Good correlation to the optimized model data, and a 2.2-fold improvement of the volumetric productivity in fed-batch optimal profile	[138]
Cytochrome P450 enzymes (CYPs)	Rational optimization criteria to optimize production kinetics in bioreactors	Carbon-limiting strategy at the highest μ maximized qP. In the optimum condition, up to threefold increases in terms of qV and yield were achieved in comparison with initial tests	[154]
Recombinant hepatitis B small surface antigen (rHBsAg)	Optimization of continuous process efficiency by evaluation of D, and comparison with well-established fed-batch mode	Continuous process reaches similar levels of product titer with qV and qP, respectively, about 1.5 and 1.3 times higher than in fed-batch mode	[155]
Candida rugosa lipase	A combination of strain and bioprocess engineering: different gene dosages compared in chemostat cultures with different oxygen-limiting conditions and hypoxic conditions in carbon-limited fed-batch cultures	Increases of up to 9-fold in the production rates were reached when both strain and bioprocess engineering were improved	[156]
Candida antartica lipase	Trade-off between fed-batch and continuous. The influence of the μ was examined on various key bioprocess parameters	In continuous mode, the overall production was 5.8 times greater than the fed-batch process	[155]
Streptomyces ghanaenis L-glutamate oxidase	Study of the effect of feeding strategy on cell growth and enzyme production	The cell density and total enzyme activity were 210 g/L and 118 U/mL, respectively, which represent a 3-fold and 36-fold increase relative to shake flask experiments	[142]
Trichoderma reesei xylanase	HCDF under optimal parameters	HCDF strategy boosts the amount of enzyme by 40.1-fold in comparison to the shake flask fermentation	[142]
Rhizopus oryzae lipase	A numerical optimization of mathematical model which includes cell substrate and product kinetics	The optimal profiles were defined, and bioprocess efficiency improvement was confirmed in terms of a 2.2-fold higher final titer and 3.4-fold higher productivity	[157]
Applicable to different products	Combination of three metabolic models optimized with constraint-based FBA	Simulated values were highly comparable with existing experimental results that outperformed each model. This similarity can be useful to reduce experimental work and costs in optimization process	[158]

7.1. Fermentation Parameters

The efficiency of fermentation processes is strongly influenced by operational parameters such as temperature, pH, dissolved oxygen (DO), medium composition, and osmolality. For *K. phaffii* cultivation, several authors use a similar range of optimal set-point and control strategies for these parameters, mainly based on standard protocols (Table 5) [98,138,142,154].

The required growth temperature for *K. phaffii* is 28–30 °C [98,155]. A temperature greater than 32 °C could cause cell death and reduce protein expression. For heterologous protein production, in some cases, reducing the cultivation temperature may improve protein production by increasing yeast viability (despite decreasing growth rate), decreasing folding stress, and reducing proteolytic activity against the target protein [98,155]. Besides that, decreasing the cultivation temperature increases the oxygen solubility and consequently improves the oxygen transfer rate [155]. Studies focused on the temperature effect indicated that the improvement in efficiency depends on the target product and should be evaluated in each case [155,159].

The pH value for *K. phaffii* cultivation range from 5 to 6.5. pH, values above 8 reduce cell viability [155,160]. The optimum value is strongly dependent on the recombinant protein or metabolite produced. pH value of 5.5 contributes to minimizing the harmful effect of proteases [140,159].

The set point for DO concentration in *K. phaffii* fermentation is usually kept at 20–30% saturated [155]. The higher the cell density in fermentation, the more onerous it is to maintain this concentration [151,155]. The limited oxygen transfer capacity is an essential factor in the *K. phaffii* process development. Due to the low solubility of oxygen in cultivation media, in *K. phaffii* cultivation at high cell densities, the maximum oxygen transfer capacity of the equipment can determine the maximum amount of biomass that can be reached [151]. Moreover, studies related to recombinant protein production observed that oxygen-limiting conditions results in higher protein production [138,151].

The large-scale production of *K. phaffii* also depends on low-cost culture media. New medium formulations have been evaluated for specific studies; no generic composition works in all cases [145,155]. Media optimization can be an effective alternative to increase growth rate and product yield; however, this option is not widely explored for *K. phaffii* [145,160]. The most cited medium for *K. phaffii* fermentation is the basal salt (BSM) medium. Still, some studies report a potential increase of product titers at similar cell densities or improved per-cell productivity by adding nutrients in the basal medium [160].

7.2. Operation Mode

Regarding operation mode, the fed-batch culture, with gradual feeding of the substrate, is the most popular method used for the recombinant production process in *K. phaffii*. This strategy achieves high cell densities and high product titers and prevents substrate inhibition or catabolite repression [9,161]. In fed-batch culture, exponential feeding profiles have been considered the most effective procedure to achieve a pseudo-stationary state and, consequently, constant and controlled specific growth, consumption, and production rates [140,154]. Common feeding strategies such as pulse, constant rates, ramp, or step-based additions, have been considered obsolete since they do not provide the physiological needs of cells and do not allow optimal performance [138,155,157]. Without changing growth parameters, the combination of feeding profiles and carbon-starving is recommended for recombinant protein overexpression in fed-batch cultures [138,150]. Although advantageous in some aspects, this strategy reduces productivity values and is not always better than exponential feeding [154].

The use of high-cell-density fermentation (HCDF) in fed-batch mode has been the most adopted strategy in recent studies to achieve the high-yield of enzymes or chemicals production [140,142]. However, there are still difficulties in establishing a robust HCDF process, mainly when methanol is used as a carbon source and inducer [161]. Since high methanol concentrations are toxic, it must be continuously added and maintained at a proper level in fermentation. Then, cell growth, substrate consumption, and product formation will depend on the methanol feeding strategy [161]. Although HCDF contributes to increased productivity, it can also imply accumulation of protease, especially harmful in heterologous protein production. Some strategies can be adopted to overcome the problem, such as the pH of the fermentation medium controlled in a range of 3.0–7.0, the addition of amino acid-rich supplements, or using a protease-deficient *K. phaffii* host strain [142].

Although the fed-batch operation is the most used method, continuous mode production has become a trend. In continuous culture, fresh medium is continuously added to the bioreactor, and culture broth and cell are removed at the same flow rate. Then, the specific growth rate (μ) is kept close to μ_{max}. This fermentation mode presents advantages compared to fed-batch, such as higher productivity, lower spending on utilities, and the simplification of the process by operation under steady-state, which provides highly robust and reliable data [138,155,162]. Continuous culture is considered the best alternative for obtaining accurate physiological data for reliable strain characterization and process [155]. However, using the continuous operation mode in the industry also presents some challenges, including a higher risk of genetic instability and contamination, demand for a higher level of automation, and other product-specific problems [138,155]. As far as we know, for *K. phaffii*, successful continuous fermentations have been reported on the laboratory scale [15,154,155,159], but there are no continuous biomanufacturing processes.

In the process development of continuous fermentation, the dilution rate (D), the same value of μ, defined by mass balance, is the most critical parameter to be considered [163]. The most reported type of cell growth kinetics is described by models based on the Monod equation [140,155]. However, equation parameters or a pattern of $Y_{P/X}$ (product to biomass yield) behavior as a function of μ is challenging to be compared since different strains using different substrates and metabolic routes are evaluated.

Regardless of the strategy used for process optimization, some factors must be considered as design criteria. Depending on strain engineering and culture conditions, an optimum specific growth rate (μ_{opt}) can be determined for better performance. The maximization of conventional parameters must be considered, for example, final titer and specific product formation rate (qP), usually for high added-value products, and $Y_{P/S}$ (product to substrate yield), typically for low-cost products [138,157,162].

Usually, a high initial cell concentration and a low specific growth rate are recommended for higher product titers, since growth-associated biomass is a by-product and has to be controlled not to exceed the limit concentration [138,140,162]. Based on this, the relationship between qP and μ becomes crucial information to compare efficiency in different conditions. This relationship is considered by some authors the key factor in bioprocess development [138,140,155]. An empirical understanding of the relationship between growth and product formation is usually necessary, since lacking a theoretical base [140]. This observation of product formation kinetics would define conditions to control the fed-batch process.

The recombinant protein production is usually growth-associated, and the optimum qP is reached at high μ [138,154]. When applicable, positive growth associated with production (specific product formation rate increases with the specific growth rate) is the better approach to evolving strains while maintaining high titers and yield [138,140,162]. This strategy was used to couple succinate overproduction with growth through the deletion of genes in *E. coli* [141] and to improve an extracellular human granulocyte-macrophage colony-stimulating factor expression in *K. phaffii* [140]. Genome-scale models have allowed to couple growth and production for most metabolites [141]. When it is impossible to biochemically couple target metabolism with growth dependence, or the growth coupling is limited to some cultivation conditions, engineered strain, with increased tolerance to stresses and inhibitors, is mandatory [141]. In this case, to preserve strain performance, the selection of strains with improved growth and loss of production must be avoided over the generations [138]. One possible alternative to ensure stable strain performance, and decoupling growth and production, is to activate product pathways only after cell density reaches the stationary phase [141].

The maximum specific growth rate (μ_{max}) is another crucial parameter for bioprocess development since it defines the maximum substrate consumption and, consequently, the upper limit of substrate assimilation in the case of the fed-batch mode [140,162]. Usually, strains engineered to produce heterologous proteins present μ_{max} lower than values observed for a non-engineered strain [140]. One more critical factor is the relationship between yield and productivity. One of these variables cannot be individually optimized to the detriment of the other [138,157]. Mathematical tools to define a global optimum solution based on goal criteria can be helpful, especially when production kinetics are nonlinear.

One frequent issue in process development that must be considered is the chemical inhibition by media, metabolites, or products, mainly when employing complex and crude carbon sources. In these cases, a selected strain is unlikely to provide all the desired properties for an optimum process naturally. Adaptive laboratory evolution (ALE) can be an alternative to overcome these physiological limitations and offer new targets for the next design step [164,165]. ALE has been shown to be useful in many studies to improve the strains of *K. phaffii* for different purposes [155,165]. Although ALE is a powerful method to accelerate the development of efficient recombinant hosts, this evolution can drive the population towards a low-production high-fitness phenotype [164,165]. The combination of growth-coupled production with ALE has been responsible for high-production strains that preserve phenotype [141]. Combining multi-omic studies may also help in this selection stage [165].

As may be seen, some parameters of development are common to bioprocess in *K. phaffii*. However, the challenges, in general, are specific to the product and must be considered particularly in each case [9,138].

7.3. Mathematical Models

Despite the advances, the process development and optimization for biological systems are usually empirical. Mathematical models can reduce experimental work and costs, proposing favorable and non-intuitive operating conditions, and designing optimizing control strategies. These models may be considered and integrated to direct the process to the best performance at macroscopic or microscopic levels.

At a microscopic level, two types of mathematical models are available for comprehending cellular metabolism: the kinetic model and the genome-scale metabolic model [158]. The kinetic model can be a limited approach for understanding cellular metabolism because enzyme kinetics parameters of all reactions are unavailable [158]. Then, the use of genome-scale modeling (GEM), which depends on genome sequence data, is very helpful in identifying engineering targets and developing efficient bioprocess strategies. Several GEMs are available for *K. phaffii*, with constant refinement and evolution [90,153,158].

For a macroscopic approach, the established techniques of design of experiments (DoE), artificial neural networks (ANN), and genetic algorithms (GA), have proven to be robust for fermentation process optimization [166]. Usually, these techniques are associated with fermentation media development or early process optimization [166]. The combination of GEM and DoE could determine the effect of genetic and culture conditions in *K. phaffii* to produce thaumatin, a sweetener [158].

The dynamic macroscopic models of bioprocesses, based on mass and energy balances, are essential to understanding how cells behave to changes in environmental conditions and to optimizing bioreactor fermentation; however, a limited number of these models are available for *K. phaffii*, usually with simple formulations [153,167]. Recently, an extensive macroscopic bioreactor model was constructed for *K. phaffii* [167], which describes substrates and other medium components, biomass, total protein, and off-gas components in different operating conditions.

Model predictive control (MPC) has gained prominence for bioprocess optimization [152]. This practice considers dynamic and static interactions between input, output, and disturbance variables, and uses a predictive process model to synchronize the control estimate with the optimum set points calculations [152]. Simulation in *K. phaffii* fermentation process, a highly nonlinear system, employing the MPC method, has been shown to be accurate in describing experimental data [167]. Still, it remains underused for this kind of bioprocess.

Despite the advances, observing all limitations in integrating cells, metabolites, and processes, with precision in predicting all responses, remain a challenge [164,167]. Gradually, multipurpose models are beginning to be used as an efficient tool in systematic metabolic engineering and bioprocess optimization.

Table 5 summarizes strategies discussed above applied to the development, simulation, and optimization of processes with *K. phaffii*. Although some techniques are still incipient for this host system, there is a clear need to apply efforts in this area to establish industrial fermentation processes with *K. phaffii*.

8. Challenges for Industrial Processes with *K. phaffii*

Currently, *K. phaffii* is responsible for more than 300 licensed industrial processes and 70 different types of commercial products, such as enzymes and biopharmaceuticals [138,145]. The www.pichia.com accessed on 13 September 2022 website provides a list of products on the market or in the industry development stage using this yeast. Compared to its use as a protein expression system, the *K. phaffii* as a cell factory for the production of chemicals has been less exploited, despite increasing use in recent years.

Until then, the underestimated use of *K. phaffii* as a cell factory or expression system in industrial-scale processes could be mainly related to challenges such as economic competitiveness, low titers and yields, and scaling-up adversities. Furthermore, despite the numerous advantages of producing chemicals using bio-based methods, there are few cases in which bioprocessing competes with the fossil-based chemical industry in terms of time and cost [164].

As highlighted earlier, an efficient bioprocess in *K. phaffii* and other microorganisms requires an optimized and stable host strain with improved metabolic rates; and an effective and reproducible fermentation process, operating with a high titer, yield, and productivity. However, to produce marketed chemicals or proteins, it is also mandatory a successful scaling-up of a bioprocess, designed to be scalable, without loss of efficiency or significant increase in cost; a technically and cost-effective downstream on a large-scale process; a rigorously and constantly monitored and controlled process. All these factors are intrinsically interconnected and must be considered together to establish the industrial fermentation process [15,141].

Despite the recent advances in *K. phaffii* industrial utilization, the scaling-up process also has some drawbacks due to strain limitations. Unlike *S. cerevisiae*, *K. phaffii* strains are mainly prototype, which means that they have few genetic markers, they change their mating type by aggregating, and there are just a few strains used in industry and research, bringing limitations to its genetic pool [15]. Another critical metabolic limitation is glucose uptake. The glucose uptake is reduced and does not exceed the yeast's respiratory capacities. Compared to *S. cerevisiae*, the maximum glucose uptake is almost tenfold lower in *K. phaffii*, which may reduce process efficiency [14]. Presumably, the strict glucose uptake and high PPP flux can lead to the production of complex undesired secondary metabolites as fermentative byproducts, such as terpenoids and carotenoids [15]. Thus, the recent advances in enhancing cellular function, development of tools, promoters, and markers, and designing new expression platforms, allowed the construction of various strains capable of producing renewable chemicals (Table 2). However, the limitations in carbon assimilation (consumption capacity and rates) and byproduct formation are still the main drawbacks in the industrial use of recombinant *K. phaffii* due to low production titers, yields, and/or productivity.

Scale-up is considered the most critical bottleneck for contemporary biotechnology to establish in a large-scale industry [143,144]. Usually, process engineering does not receive the same interest or develop at the same speed as biological systems' genetic and metabolic engineering. The financial investment to scale up a bioprocess is typically more significant than the cost to develop a lab-scale production. This can cost USD 100 million to USD 1 billion and take 3–10 years to scale the transition [144]. Thus, several lab-scale bioprocesses cannot overcome the Death Valley of process development.

Industrial-scale bioreactors are substantially different from laboratory cultures in several aspects. Then, if a process optimization on a small scale is not well designed, it cannot address the real problem of the industrial scale. Previous scale-up criteria for fermentation processes were based on keeping constant process engineering parameters, such as power input/volume ratio (P/V), impeller tip speed, or volumetric mass transfer coefficient (kLa), and thus have not resulted in universally valid standards for successful scale-up [138,144]. In this context, the most successful technology transfers to the full-scale fermentation process resulting from scale-down models, where strains are evaluated under critical aspects of the large-scale fermentation environment [139,141,144,168]. This systematic method of bioprocess and strain engineering for scale-up requires a large-scale view of production conditions on a small scale to define combinations of a host, path, and process, that will meet the requirements of industrial production. The successful use of this approach considers successive tiers to experimental evaluation, but it starts using the same raw materials and feeding algorithms as used on an industrial scale. Even on a laboratory scale, industrial-scale and process control aspects must be considered to minimize scale-up risk. The selection of parameters for a well-established evaluation is particular to the process, strain, and product type.

Even when systems biology and process development were planned with increasing bioreactor volume, other factors not previously observed can become an obstacle. Examples of these differences in scale are: the hydrostatic pressure gradient of the reactor, which can influence enzyme activities, metabolic flux, or the concentration of dissolved gases in the medium; or poor mixing and dead zones created with increasing bioreactor volume, which results in the gradients of parameters such as pH, temperature, dissolved oxygen and the concentration of nutrients; mass transfer limitations; yeast metabolism generates a large amount of heat and therefore need efficient cooling. These stress conditions can trigger genetic and physiological responses in the host strain and lead to difficulties in monitoring and controlling large-scale cultivation [138,141,151,164]. In this sense, a rigorous study for transitioning the bioprocess to the industrial scale is a mandatory stage. Using multi-omic characterization and the integration of fluid dynamics and cell metabolism, or building mathematic multiscale models, can be helpful, as discussed previously [164].

Regarding the downstream, the fermentation process strongly influences this stage by requiring the separation of high cell density and the purification of desired product out of a complex matrix. The downstream must be considered part of the bioprocess, and its development and optimization must use the same logic applied to the process development [139,141,144,166]. In the same way, the technical and economic feasibility of downstream processing will also depend on the product and should be evaluated in each case [138,143,154].

A better understanding of the interaction between strain and bioprocess and the ability to use new and sophisticated tools has brought us closer to the market availability of several new renewable molecules produced by *K. phaffii*.

9. Conclusions

The *Komagataella* species are versatile yeasts capable of producing a range of bio compounds of interest and heterologous proteins. With the increasing number of strains and new molecular tools and methodologies, these yeasts are becoming even more competitive hosts when compared to others. For instance, *K. phaffii* has the advantage of being eukaryotic (when compared to bacteria) and presents undeveloped potential regarding fermentative cultivation (compared with *S. cerevisiae*). Besides this fact, *K. phaffii* is one of the major microbial platforms for the production of recombinant proteins, that are constantly employed in different industries, such as pharmaceuticals, food, feed, detergents, and others. More recent results demonstrate *K. phaffii* feasibility to produce complex chemicals via metabolic engineering. As can be seen, several studies highlight the potential of *K. phaffii* for being a microbial chassis for the production of value-added compounds using glycerol, glucose, xylose, and methanol as carbon source. Despite the great advancements in the field, further improvements are required for the industrial application of *K. phaffii* in methanol bioconversions.

Author Contributions: Conceptualization, C.V.G.C.C. and J.R.M.d.A.; writing—original draft preparation, C.V.G.C.C., D.T., L.M.M.F., L.T.D.B., L.A.S., M.N.d.M.F. and T.F.P. writing—review and editing, C.V.G.C.C. and J.R.M.d.A.; visualization, C.V.G.C.C.; funding acquisition, J.R.M.d.A. All authors have read and agreed to the published version of the manuscript.

Funding: This research was funded by Foundation for Research Support of the Federal District (FAP_DF) grant number 00193-00000762/2021-92.

Institutional Review Board Statement: Not applicable.

Informed Consent Statement: Not applicable.

Data Availability Statement: Not applicable.

Conflicts of Interest: The authors declare no conflict of interest.

References

1. Leong, H.Y.; Chang, C.K.; Khoo, K.S.; Chew, K.W.; Chia, S.R.; Lim, J.W.; Chang, J.S.; Show, P.L. Waste Biorefinery towards a Sustainable Circular Bioeconomy: A Solution to Global Issues. *Biotechnol. Biofuels* **2021**, *14*, 1–15. [CrossRef]
2. Muscat, A.; de Olde, E.M.; Ripoll-Bosch, R.; Van Zanten, H.H.E.; Metze, T.A.P.; Termeer, C.J.A.M.; van Ittersum, M.K.; de Boer, I.J.M. Principles, Drivers and Opportunities of a Circular Bioeconomy. *Nat. Food* **2021**, *2*, 561–566. [CrossRef]
3. Kardung, M.; Cingiz, K.; Costenoble, O.; Delahaye, R.; Heijman, W.; Lovrić, M.; van Leeuwen, M.; M'barek, R.; van Meijl, H.; Piotrowski, S.; et al. Development of the Circular Bioeconomy: Drivers and Indicators. *Sustainability* **2021**, *13*, 413. [CrossRef]
4. Carneiro, C.V.G.; Silva, F.C.D.P.E.; Almeida, J.R. Xylitol Production: Identification and Comparison of New Producing Yeasts. *Microorganisms* **2019**, *7*, 484. [CrossRef]
5. Trichez, D.; Carneiro, C.V.G.C.; Braga, M.; Almeida, J.R.M. Recent Progress in the Microbial Production of Xylonic Acid. *World J. Microbiol. Biotechnol.* **2022**, *38*, 127. [CrossRef] [PubMed]
6. Ata, Ö.; Ergün, B.G.; Fickers, P.; Heistinger, L.; Mattanovich, D.; Rebnegger, C.; Gasser, B. What Makes *Komagataella Phaffii* Non-Conventional? *FEMS Yeast Res.* **2021**, *21*, foab059. [CrossRef]
7. Bernauer, L.; Radkohl, A.; Lehmayer, L.G.K.; Emmerstorfer-Augustin, A. Komagataella Phaffii as Emerging Model Organism in Fundamental Research. *Front. Microbiol.* **2021**, *11*, 1–16. [CrossRef]
8. Chiang, C.-J.; Ho, Y.-J.; Hu, M.-C.; Chao, Y.-P. Rewiring of Glycerol Metabolism in Escherichia Coli for Effective Production of Recombinant Proteins. *Biotechnol. Biofuels* **2020**, *13*, 205. [CrossRef]
9. Yang, Z.; Zhang, Z. Engineering Strategies for Enhanced Production of Protein and Bio-Products in Pichia Pastoris. *Biotechnol. Adv.* **2018**, *36*, 182–195. [CrossRef]
10. Näätsaari, L.; Mistlberger, B.; Ruth, C.; Hajek, T.; Hartner, F.S.; Glieder, A. Deletion of the Pichia Pastoris KU70 Homologue Facilitates Platform Strain Generation for Gene Expression and Synthetic Biology. *PLoS ONE* **2012**, *7*, e39720. [CrossRef]
11. Ito, Y.; Watanabe, T.; Aikawa, S.; Nishi, T.; Nishiyama, T.; Nakamura, Y.; Hasunuma, T.; Okubo, Y.; Ishii, J.; Kondo, A. Deletion of DNA Ligase IV Homolog Confers Higher Gene Targeting Efficiency on Homologous Recombination in Komagataella Phaffii. *FEMS Yeast Res.* **2018**, *18*, foy074. [CrossRef]
12. Weninger, A.; Hatzl, A.-M.; Schmid, C.; Vogl, T.; Glieder, A. Combinatorial Optimization of CRISPR/Cas9 Expression Enables Precision Genome Engineering in the Methylotrophic Yeast Pichia Pastoris. *J. Biotechnol.* **2016**, *235*, 139–149. [CrossRef]
13. Fischer, J.E.; Glieder, A. Current Advances in Engineering Tools for Pichia Pastoris. *Curr. Opin. Biotechnol.* **2019**, *59*, 175–181. [CrossRef] [PubMed]
14. Peña, D.A.; Gasser, B.; Zanghellini, J.; Steiger, M.G.; Mattanovich, D. Metabolic Engineering of Pichia Pastoris. *Metab. Eng.* **2018**, *50*, 2–15. [CrossRef] [PubMed]
15. Gasser, B.; Mattanovich, D. A Yeast for All Seasons—Is Pichia Pastoris a Suitable Chassis Organism for Future Bioproduction? *FEMS Microbiol. Lett.* **2018**, *365*, fny181. [CrossRef] [PubMed]
16. Yamada, Y.; Matsuda, M.; Maeda, K.; Mikata, K. The Phylogenetic Relationships of Methanol-Assimilating Yeasts Based on the Partial Sequences of 18S and 26S Ribosomal RNAs: The Proposal of Komagataella Gen. Nov. (Saccharomycetaceae). *Biosci. Biotechnol. Biochem.* **1995**, *59*, 439–444. [CrossRef]
17. Kurtzman, C.P.; Robnett, C.J. Identification and Phylogeny of Ascomycetous Yeasts from Analysis of Nuclear Large Subunit (26S) Ribosomal DNA Partial Sequences. *Antonie Van Leeuwenhoek* **1998**, *73*, 331–371. [CrossRef]
18. Guilliermond, A. Zygosaccharomyces Pastori, Nouvelle Espèce de Levures à Copulation Hétérogamique. *Bull. Société Mycol. Fr.* **1920**, *36*, 203–211.
19. Heistinger, L.; Gasser, B.; Mattanovich, D. Microbe Profile: Komagataella Phaffii: A Methanol Devouring Biotech Yeast Formerly Known as Pichia Pastoris. *Microbiology* **2020**, *166*, 614–616. [CrossRef] [PubMed]
20. Kurtzman, C.P.; Fell, J.W.; Boekhout, T. *The Yeasts—A Taxonomy Study*; Elsevier Science: Amsterdam, The Netherlands, 2011; Volume 5, ISBN 978-0-08-093127-2.

21. Heistinger, L.; Dohm, J.C.; Paes, B.G.; Koizar, D.; Troyer, C.; Ata, Ö.; Steininger-Mairinger, T.; Mattanovich, D. Genotypic and Phenotypic Diversity among Komagataella Species Reveals a Hidden Pathway for Xylose Utilization. *Microb. Cell Factories* **2022**, *21*, 70. [CrossRef] [PubMed]
22. Kurtzman, C.P. Description of Komagataella Phaffii Sp. Nov. and the Transfer of Pichia Pseudopastoris to the Methylotrophic Yeast Genus Komagataella. *Int. J. Syst. Evol. Microbiol.* **2005**, *55*, 973–976. [CrossRef] [PubMed]
23. De Schutter, K.; Lin, Y.-C.; Tiels, P.; Van Hecke, A.; Glinka, S.; Weber-Lehmann, J.; Rouzé, P.; Van de Peer, Y.; Callewaert, N. Genome Sequence of the Recombinant Protein Production Host Pichia Pastoris. *Nat. Biotechnol.* **2009**, *27*, 561–566. [CrossRef] [PubMed]
24. Love, K.R.; Shah, K.A.; Whittaker, C.A.; Wu, J.; Bartlett, M.C.; Ma, D.; Leeson, R.L.; Priest, M.; Borowsky, J.; Young, S.K.; et al. Comparative Genomics and Transcriptomics of Pichia Pastoris. *BMC Genom.* **2016**, *17*, 550. [CrossRef] [PubMed]
25. Mattanovich, D.; Graf, A.; Stadlmann, J.; Dragosits, M.; Redl, A.; Maurer, M.; Kleinheinz, M.; Sauer, M.; Altmann, F.; Gasser, B. Genome, Secretome and Glucose Transport Highlight Unique Features of the Protein Production Host Pichia Pastoris. *Microb. Cell Factories* **2009**, *8*, 29. [CrossRef] [PubMed]
26. Almeida, J.R.M.; Fávaro, L.C.L.; Quirino, B.F. Biodiesel Biorefinery: Opportunities and Challenges for Microbial Production of Fuels and Chemicals from Glycerol Waste. *Biotechnol. Biofuels* **2012**, *5*, 48. [CrossRef] [PubMed]
27. da Silva, G.P.; Mack, M.; Contiero, J. Glycerol: A Promising and Abundant Carbon Source for Industrial Microbiology. *Biotechnol. Adv.* **2009**, *27*, 30–39. [CrossRef]
28. Tomàs-Gamisans, M.; Ødum, A.S.R.; Workman, M.; Ferrer, P.; Albiol, J. Glycerol Metabolism of Pichia Pastoris (Komagataella Spp.) Characterised by 13C-Based Metabolic Flux Analysis. *New Biotechnol.* **2019**, *50*, 52–59. [CrossRef] [PubMed]
29. Prielhofer, R.; Cartwright, S.P.; Graf, A.B.; Valli, M.; Bill, R.M.; Mattanovich, D.; Gasser, B. Pichia Pastoris Regulates Its Gene-Specific Response to Different Carbon Sources at the Transcriptional, Rather than the Translational, Level. *BMC Genom.* **2015**, *16*, 167. [CrossRef] [PubMed]
30. Çelik, E.; Ozbay, N.; Oktar, N.; Çalık, P. Use of Biodiesel Byproduct Crude Glycerol as the Carbon Source for Fermentation Processes by Recombinant *Pichia Pastoris*. *Ind. Eng. Chem. Res.* **2008**, *47*, 2985–2990. [CrossRef]
31. Kumokita, R.; Bamba, T.; Inokuma, K.; Yoshida, T.; Ito, Y.; Kondo, A.; Hasunuma, T. Construction of an L-Tyrosine Chassis in *Pichia Pastoris* Enhances Aromatic Secondary Metabolite Production from Glycerol. *ACS Synth. Biol.* **2022**, *11*, 2098–2107. [CrossRef]
32. Chen, Z.; Liu, D. Toward Glycerol Biorefinery: Metabolic Engineering for the Production of Biofuels and Chemicals from Glycerol. *Biotechnol. Biofuels* **2016**, *9*, 205. [CrossRef] [PubMed]
33. Luo, Z.; Miao, J.; Luo, W.; Li, G.; Du, Y.; Yu, X. Crude Glycerol from Biodiesel as a Carbon Source for Production of a Recombinant Highly Thermostable β-Mannanase by Pichia Pastoris. *Biotechnol. Lett.* **2018**, *40*, 135–141. [CrossRef]
34. Abdel-Rahman, M.A.; Tashiro, Y.; Sonomoto, K. Recent Advances in Lactic Acid Production by Microbial Fermentation Processes. *Biotechnol. Adv.* **2013**, *31*, 877–902. [CrossRef] [PubMed]
35. Abedi, E.; Hashemi, S.M.B. Lactic Acid Production—Producing Microorganisms and Substrates Sources-State of Art. *Heliyon* **2020**, *6*, e04974. [CrossRef] [PubMed]
36. de Lima, P.B.A.; Mulder, K.C.L.; Melo, N.T.M.; Carvalho, L.S.; Menino, G.S.; Mulinari, E.; de Castro, V.H.; dos Reis, T.F.; Goldman, G.H.; Magalhães, B.S.; et al. Novel Homologous Lactate Transporter Improves L-Lactic Acid Production from Glycerol in Recombinant Strains of Pichia Pastoris. *Microb. Cell Factories* **2016**, *15*, 158. [CrossRef]
37. Melo, N.; Mulder, K.; Nicola, A.; Carvalho, L.; Menino, G.; Mulinari, E.; Parachin, N. Effect of Pyruvate Decarboxylase Knockout on Product Distribution Using Pichia Pastoris (Komagataella Phaffii) Engineered for Lactic Acid Production. *Bioengineering* **2018**, *5*, 17. [CrossRef]
38. Tamires Moreira Melo, N.; Pontes, G.C.; Procópio, D.P.; de Gois e Cunha, G.C.; Eliodório, K.P.; Costa Paes, H.; Basso, T.O.; Parachin, N.S. Evaluation of Product Distribution in Chemostat and Batch Fermentation in Lactic Acid-Producing Komagataella Phaffii Strains Utilizing Glycerol as Substrate. *Microorganisms* **2020**, *8*, 781. [CrossRef]
39. Fina, A.; Brêda, G.C.; Pérez-Trujillo, M.; Freire, D.M.G.; Almeida, R.V.; Albiol, J.; Ferrer, P. Benchmarking Recombinant *Pichia Pastoris* for 3-hydroxypropionic Acid Production from Glycerol. *Microb. Biotechnol.* **2021**, *14*, 1671–1682. [CrossRef]
40. Siripong, W.; Wolf, P.; Kusumoputri, T.P.; Downes, J.J.; Kocharin, K.; Tanapongpipat, S.; Runguphan, W. Metabolic Engineering of Pichia Pastoris for Production of Isobutanol and Isobutyl Acetate. *Biotechnol. Biofuels* **2018**, *11*, 1. [CrossRef]
41. Yang, Z.; Zhang, Z. Production of (2R, 3R)-2,3-Butanediol Using Engineered Pichia Pastoris: Strain Construction, Characterization and Fermentation. *Biotechnol. Biofuels* **2018**, *11*, 35. [CrossRef]
42. Liu, Y.; Gong, X.; Wang, C.; Du, G.; Chen, J.; Kang, Z. Production of Glucaric Acid from Myo-Inositol in Engineered Pichia Pastoris. *Enzym. Microb. Technol.* **2016**, *91*, 8–16. [CrossRef] [PubMed]
43. Zhang, Q.; Wang, X.; Luo, H.; Wang, Y.; Wang, Y.; Tu, T.; Qin, X.; Su, X.; Huang, H.; Yao, B.; et al. Metabolic Engineering of Pichia Pastoris for Myo-Inositol Production by Dynamic Regulation of Central Metabolism. *Microb. Cell Factories* **2022**, *21*, 112. [CrossRef] [PubMed]
44. Louie, T.M.; Louie, K.; DenHartog, S.; Gopishetty, S.; Subramanian, M.; Arnold, M.; Das, S. Production of Bio-xylitol from Xylose by an Engineered P.Pastoris. *Microb. Cell Factories* **2021**, *20*, 50. [CrossRef] [PubMed]

45. Ramos, T.G.S.; Justen, F.; Carneiro, C.V.G.C.; Honorato, V.M.; Franco, P.F.; Vieira, F.S.; Trichez, D.; Rodrigues, C.M.; Almeida, J.R.M. Xylonic Acid Production by Recombinant Komagataella Phaffii Strains Engineered with Newly Identified Xylose Dehydrogenases. *Bioresour. Technol. Rep.* **2021**, *16*, 100825. [CrossRef]
46. Miao, L.; Li, Y.; Zhu, T. Metabolic Engineering of Methylotrophic Pichia Pastoris for the Production of β-Alanine. *Bioresour. Bioprocess.* **2021**, *8*, 89. [CrossRef]
47. Yamada, R.; Ogura, K.; Kimoto, Y.; Ogino, H. Toward the Construction of a Technology Platform for Chemicals Production from Methanol: D-Lactic Acid Production from Methanol by an Engineered Yeast Pichia Pastoris. *World J. Microbiol. Biotechnol.* **2019**, *35*, 37. [CrossRef]
48. Guo, F.; Dai, Z.; Peng, W.; Zhang, S.; Zhou, J.; Ma, J.; Dong, W.; Xin, F.; Zhang, W.; Jiang, M. Metabolic Engineering of *Pichia Pastoris* for Malic Acid Production from Methanol. *Biotechnol. Bioeng.* **2021**, *118*, 357–371. [CrossRef]
49. Liu, Y.; Tu, X.; Xu, Q.; Bai, C.; Kong, C.; Liu, Q.; Yu, J.; Peng, Q.; Zhou, X.; Zhang, Y.; et al. Engineered Monoculture and Co-Culture of Methylotrophic Yeast for de Novo Production of Monacolin J and Lovastatin from Methanol. *Metab. Eng.* **2018**, *45*, 189–199. [CrossRef]
50. Gao, L.; Cai, M.; Shen, W.; Xiao, S.; Zhou, X.; Zhang, Y. Engineered Fungal Polyketide Biosynthesis in Pichia Pastoris: A Potential Excellent Host for Polyketide Production. *Microb. Cell Factories* **2013**, *12*, 77. [CrossRef]
51. Jin, X.; Zhang, W.; Wang, Y.; Sheng, J.; Xu, R.; Li, J.; Du, G.; Kang, Z. Biosynthesis of Non-Animal Chondroitin Sulfate from Methanol Using Genetically Engineered *Pichia Pastoris*. *Green Chem.* **2021**, *23*, 4365–4374. [CrossRef]
52. Zhang, Y.; Wang, Y.; Zhou, Z.; Wang, P.; Xi, X.; Hu, S.; Xu, R.; Du, G.; Li, J.; Chen, J.; et al. Synthesis of Bioengineered Heparin by Recombinant Yeast *Pichia Pastoris*. *Green Chem.* **2022**, *24*, 3180–3192. [CrossRef]
53. Jeong, E.; Shim, W.Y.; Kim, J.H. Metabolic Engineering of Pichia Pastoris for Production of Hyaluronic Acid with High Molecular Weight. *J. Biotechnol.* **2014**, *185*, 28–36. [CrossRef] [PubMed]
54. Kumar, V.; Ashok, S.; Park, S. Recent Advances in Biological Production of 3-Hydroxypropionic Acid. *Biotechnol. Adv.* **2013**, *31*, 945–961. [CrossRef] [PubMed]
55. de Fouchécour, F.; Sánchez-Castañeda, A.-K.; Saulou-Bérion, C.; Spinnler, H.É. Process Engineering for Microbial Production of 3-Hydroxypropionic Acid. *Biotechnol. Adv.* **2018**, *36*, 1207–1222. [CrossRef]
56. Tashiro, Y.; Rodriguez, G.M.; Atsumi, S. 2-Keto Acids Based Biosynthesis Pathways for Renewable Fuels and Chemicals. *J. Ind. Microbiol. Biotechnol.* **2015**, *42*, 361–373. [CrossRef]
57. Weber, C.; Farwick, A.; Benisch, F.; Brat, D.; Dietz, H.; Subtil, T.; Boles, E. Trends and Challenges in the Microbial Production of Lignocellulosic Bioalcohol Fuels. *Appl. Microbiol. Biotechnol.* **2010**, *87*, 1303–1315. [CrossRef]
58. Wess, J.; Brinek, M.; Boles, E. Improving Isobutanol Production with the Yeast Saccharomyces Cerevisiae by Successively Blocking Competing Metabolic Pathways as Well as Ethanol and Glycerol Formation. *Biotechnol. Biofuels* **2019**, *12*, 173. [CrossRef]
59. Wang, X.; Zhao, X.; Luo, H.; Wang, Y.; Wang, Y.; Tu, T.; Qin, X.; Huang, H.; Bai, Y.; Yao, B.; et al. Metabolic Engineering of *Komagataella Phaffii* for Synergetic Utilization of Glucose and Glycerol. *Yeast* **2022**, *39*, 412–421. [CrossRef]
60. Wang, B.; He, M.; Wang, X.; Tang, H.; Zhu, X. A Multi-Model Predictive Control Method for the Pichia Pastoris Fermentation Process Based on Relative Error Weighting Algorithm. *Alex. Eng. J.* **2022**, *61*, 9649–9660. [CrossRef]
61. Baruah, J.; Nath, B.K.; Sharma, R.; Kumar, S.; Deka, R.C.; Baruah, D.C.; Kalita, E. Recent Trends in the Pretreatment of Lignocellulosic Biomass for Value-Added Products. *Front. Energy Res.* **2018**, *6*, 141. [CrossRef]
62. Zabed, H.; Sahu, J.N.; Suely, A.; Boyce, A.N.; Faruq, G. Bioethanol Production from Renewable Sources: Current Perspectives and Technological Progress. *Renew. Sustain. Energy Rev.* **2017**, *71*, 475–501. [CrossRef]
63. Chen, H. Chemical Composition and Structure of Natural Lignocellulose. In *Biotechnology of Lignocellulose*; Springer: Dordrecht, The Netherlands, 2014; pp. 25–71. ISBN 978-94-007-6897-0.
64. Chandel, A.K.; Garlapati, V.K.; Jeevan Kumar, S.P.; Hans, M.; Singh, A.K.; Kumar, S. The Role of Renewable Chemicals and Biofuels in Building a Bioeconomy. *Biofuels Bioprod. Bioref.* **2020**, *14*, 830–844. [CrossRef]
65. Kim, D. Physico-Chemical Conversion of Lignocellulose: Inhibitor Effects and Detoxification Strategies: A Mini Review. *Molecules* **2018**, *23*, 309. [CrossRef]
66. Liu, C.-G.; Xiao, Y.; Xia, X.-X.; Zhao, X.-Q.; Peng, L.; Srinophakun, P.; Bai, F.-W. Cellulosic Ethanol Production: Progress, Challenges and Strategies for Solutions. *Biotechnol. Adv.* **2019**, *37*, 491–504. [CrossRef] [PubMed]
67. Zhou, Z.; Liu, D.; Zhao, X. Conversion of Lignocellulose to Biofuels and Chemicals via Sugar Platform: An Updated Review on Chemistry and Mechanisms of Acid Hydrolysis of Lignocellulose. *Renew. Sustain. Energy Rev.* **2021**, *146*, 111169. [CrossRef]
68. Chen, H.; Liu, J.; Chang, X.; Chen, D.; Xue, Y.; Liu, P.; Lin, H.; Han, S. A Review on the Pretreatment of Lignocellulose for High-Value Chemicals. *Fuel Process. Technol.* **2017**, *160*, 196–206. [CrossRef]
69. Almeida, J.R.; Modig, T.; Petersson, A.; Hähn-Hägerdal, B.; Lidén, G.; Gorwa-Grauslund, M.F. Increased Tolerance and Conversion of Inhibitors in Lignocellulosic Hydrolysates BySaccharomyces Cerevisiae. *J. Chem. Technol. Biotechnol.* **2007**, *82*, 340–349. [CrossRef]
70. Sjulander, N.; Kikas, T. Origin, Impact and Control of Lignocellulosic Inhibitors in Bioethanol Production—A Review. *Energies* **2020**, *13*, 4751. [CrossRef]
71. Paes, B.G.; Steindorff, A.S.; Formighieri, E.F.; Pereira, I.S.; Almeida, J.R.M. Physiological Characterization and Transcriptome Analysis of Pichia Pastoris Reveals Its Response to Lignocellulose-Derived Inhibitors. *AMB Expr.* **2021**, *11*, 2. [CrossRef]

72. Bustos, C.; Quezada, J.; Veas, R.; Altamirano, C.; Braun-Galleani, S.; Fickers, P.; Berrios, J. Advances in Cell Engineering of the Komagataella Phaffii Platform for Recombinant Protein Production. *Metabolites* **2022**, *12*, 346. [CrossRef]
73. Lamsen, E.N.; Atsumi, S. Recent Progress in Synthetic Biology for Microbial Production of C3–C10 Alcohols. *Front. Microbiol.* **2012**, *3*, 196. [CrossRef] [PubMed]
74. Jong, E.d.; Stichnothe, H.; Bell, G.; Jørgensen, H. *Task 42-Bio-Based Chemicals: A 2020 Update*; IEA Bioenergy: Vienna, Austria, 2020.
75. Zhao, E.M.; Zhang, Y.; Mehl, J.; Park, H.; Lalwani, M.A.; Toettcher, J.E.; Avalos, J.L. Optogenetic Regulation of Engineered Cellular Metabolism for Microbial Chemical Production. *Nature* **2018**, *555*, 683–687. [CrossRef] [PubMed]
76. Motta-Mena, L.B.; Reade, A.; Mallory, M.J.; Glantz, S.; Weiner, O.D.; Lynch, K.W.; Gardner, K.H. An Optogenetic Gene Expression System with Rapid Activation and Deactivation Kinetics. *Nat. Chem. Biol.* **2014**, *10*, 196–202. [CrossRef] [PubMed]
77. Lakshmi, N.M.; Binod, P.; Sindhu, R.; Awasthi, M.K.; Pandey, A. Microbial Engineering for the Production of Isobutanol: Current Status and Future Directions. *Bioengineered* **2021**, *12*, 12308–12321. [CrossRef] [PubMed]
78. Hazeena, S.H.; Sindhu, R.; Pandey, A.; Binod, P. Lignocellulosic Bio-Refinery Approach for Microbial 2,3-Butanediol Production. *Bioresour. Technol.* **2020**, *302*, 122873. [CrossRef]
79. Lian, J.; Chao, R.; Zhao, H. Metabolic Engineering of a Saccharomyces Cerevisiae Strain Capable of Simultaneously Utilizing Glucose and Galactose to Produce Enantiopure (2R,3R)-Butanediol. *Metab. Eng.* **2014**, *23*, 92–99. [CrossRef]
80. Li, Y.; Han, P.; Wang, J.; Shi, T.; You, C. Production of *Myo*-inositol: Recent Advance and Prospective. *Biotechnol. Appl. Biochem.* **2022**, *69*, 1101–1111. [CrossRef]
81. Braga, M.; Ferreira, P.M.; Almeida, J.R.M. Screening Method to Prioritize Relevant Bio-based Acids and Their Biochemical Processes Using Recent Patent Information. *Biofuels Bioprod. Bioref.* **2021**, *15*, 231–249. [CrossRef]
82. Francois, J.M.; Alkim, C.; Morin, N. Engineering microbial pathways for production of bio-based chemicals from lignocellulosic sugars: Current status and perspectives. *Biotechnol. Biofuels* **2020**, *13*, 118. [CrossRef]
83. Li, P.; Sun, H.; Chen, Z.; Li, Y.; Zhu, T. Construction of Efficient Xylose Utilizing Pichia Pastoris for Industrial Enzyme Production. *Microb. Cell Factories* **2015**, *14*, 22. [CrossRef]
84. Araújo, D.; Costa, T.; Freitas, F. Biovalorization of Lignocellulosic Materials for Xylitol Production by the Yeast Komagataella Pastoris. *Appl. Sci.* **2021**, *11*, 5516. [CrossRef]
85. Guo, Y.; Liao, Y.; Wang, J.; Ma, C.; Qin, J.; Feng, J.; Li, Y.; Wang, X.; Chen, K. Methylotrophy of *Pichia Pastoris*: Current Advances, Applications, and Future Perspectives for Methanol-Based Biomanufacturing. *ACS Sustain. Chem. Eng.* **2022**, *10*, 1741–1752. [CrossRef]
86. Whitaker, W.B.; Sandoval, N.R.; Bennett, R.K.; Fast, A.G.; Papoutsakis, E.T. Synthetic Methylotrophy: Engineering the Production of Biofuels and Chemicals Based on the Biology of Aerobic Methanol Utilization. *Curr. Opin. Biotechnol.* **2015**, *33*, 165–175. [CrossRef] [PubMed]
87. Peiro, C.; Vicente, C.M.; Jallet, D.; Heux, S. From a Hetero- to a Methylotrophic Lifestyle: Flash Back on the Engineering Strategies to Create Synthetic Methanol-User Strains. *Front. Bioeng. Biotechnol.* **2022**, *10*, 907861. [CrossRef]
88. Cotton, C.A.; Claassens, N.J.; Benito-Vaquerizo, S.; Bar-Even, A. Renewable Methanol and Formate as Microbial Feedstocks. *Curr. Opin. Biotechnol.* **2020**, *62*, 168–180. [CrossRef] [PubMed]
89. Rußmayer, H.; Buchetics, M.; Gruber, C.; Valli, M.; Grillitsch, K.; Modarres, G.; Guerrasio, R.; Klavins, K.; Neubauer, S.; Drexler, H.; et al. Systems-Level Organization of Yeast Methylotrophic Lifestyle. *BMC Biol.* **2015**, *13*, 80. [CrossRef]
90. De, S.; Mattanovich, D.; Ferrer, P.; Gasser, B. Established Tools and Emerging Trends for the Production of Recombinant Proteins and Metabolites in *Pichia Pastoris*. *Essays Biochem.* **2021**, *65*, 293–307. [CrossRef]
91. Zahrl, R.J.; Peña, D.A.; Mattanovich, D.; Gasser, B. Systems Biotechnology for Protein Production in Pichia Pastoris. *FEMS Yeast Res.* **2017**, *17*, fox068. [CrossRef]
92. Gassler, T.; Sauer, M.; Gasser, B.; Egermeier, M.; Troyer, C.; Causon, T.; Hann, S.; Mattanovich, D.; Steiger, M.G. The Industrial Yeast Pichia Pastoris Is Converted from a Heterotroph into an Autotroph Capable of Growth on CO_2. *Nat. Biotechnol.* **2020**, *38*, 210–216. [CrossRef]
93. Gassler, T.; Baumschabl, M.; Sallaberger, J.; Egermeier, M.; Mattanovich, D. Adaptive Laboratory Evolution and Reverse Engineering Enhances Autotrophic Growth in Pichia Pastoris. *Metab. Eng.* **2022**, *69*, 112–121. [CrossRef]
94. Zhang, C.; Ottenheim, C.; Weingarten, M.; Ji, L. Microbial Utilization of Next-Generation Feedstocks for the Biomanufacturing of Value-Added Chemicals and Food Ingredients. *Front. Bioeng. Biotechnol.* **2022**, *10*, 874612. [CrossRef] [PubMed]
95. Zhang, T.; Ge, C.; Deng, L.; Tan, T.; Wang, F. C4-Dicarboxylic Acid Production by Overexpressing the Reductive TCA Pathway. *FEMS Microbiol. Lett.* **2015**, *362*, fnv052. [CrossRef] [PubMed]
96. Ucm, R.; Aem, M.; Lhb, Z.; Kumar, V.; Taherzadeh, M.J.; Garlapati, V.K.; Chandel, A.K. Comprehensive Review on Biotechnological Production of Hyaluronic Acid: Status, Innovation, Market and Applications. *Bioengineered* **2022**, *13*, 9645–9661. [CrossRef]
97. Xue, Y.; Kong, C.; Shen, W.; Bai, C.; Ren, Y.; Zhou, X.; Zhang, Y.; Cai, M. Methylotrophic Yeast Pichia Pastoris as a Chassis Organism for Polyketide Synthesis via the Full Citrinin Biosynthetic Pathway. *J. Biotechnol.* **2017**, *242*, 64–72. [CrossRef] [PubMed]
98. Karbalaei, M.; Rezaee, S.; Farsiani, H. Pichia Pastoris: A Highly Successful Expression System for Optimal Synthesis of Heterologous Proteins. *J. Cell. Physiol.* **2020**, *235*, 5867–5881. [CrossRef]
99. Ahmad, M.; Hirz, M.; Pichler, H.; Schwab, H. Protein Expression in Pichia Pastoris: Recent Achievements and Perspectives for Heterologous Protein Production. *Appl. Microbiol. Biotechnol.* **2014**, *98*, 5301–5317. [CrossRef]

100. Cereghino, J.L.; Cregg, J.M. Heterologous Protein Expression in the Methylotrophic Yeast *Pichia Pastoris*. *FEMS Microbiol. Rev.* **2000**, *24*, 45–66. [CrossRef]
101. de Almeida, J.R.M.; de Moraes, L.M.P.; Torres, F.A.G. Molecular Characterization of the 3-Phosphoglycerate Kinase Gene (PGK1) from the Methylotrophic Yeast Pichia Pastoris. *Yeast* **2005**, *22*, 725–737. [CrossRef]
102. Spiegel, H.; Schinkel, H.; Kastilan, R.; Dahm, P.; Boes, A.; Scheuermayer, M.; Chudobová, I.; Maskus, D.; Fendel, R.; Schillberg, S.; et al. Optimization of a Multi-Stage, Multi-Subunit Malaria Vaccine Candidate for the Production in *Pichia Pastoris* by the Identification and Removal of Protease Cleavage Sites: Optimization of a Malaria Vaccine Candidate. *Biotechnol. Bioeng.* **2015**, *112*, 659–667. [CrossRef]
103. Tripathi, L.; Mani, S.; Raut, R.; Poddar, A.; Tyagi, P.; Arora, U.; de Silva, A.; Swaminathan, S.; Khanna, N. Pichia Pastoris-Expressed Dengue 3 Envelope-Based Virus-like Particles Elicit Predominantly Domain III-Focused High Titer Neutralizing Antibodies. *Front. Microbiol.* **2015**, *6*, 1005. [CrossRef]
104. Saraswat, S.; Athmaram, T.N.; Parida, M.; Agarwal, A.; Saha, A.; Dash, P.K. Expression and Characterization of Yeast Derived Chikungunya Virus Like Particles (CHIK-VLPs) and Its Evaluation as a Potential Vaccine Candidate. *PLoS Negl. Trop. Dis.* **2016**, *10*, e0004782. [CrossRef] [PubMed]
105. Farsiani, H.; Mosavat, A.; Soleimanpour, S.; Sadeghian, H.; Akbari Eydgahi, M.R.; Ghazvini, K.; Sankian, M.; Aryan, E.; Jamehdar, S.A.; Rezaee, S.A. Fc-Based Delivery System Enhances Immunogenicity of a Tuberculosis Subunit Vaccine Candidate Consisting of the ESAT-6:CFP-10 Complex. *Mol. BioSyst.* **2016**, *12*, 2189–2201. [CrossRef] [PubMed]
106. Argentinian AntiCovid Consortium. Structural and Functional Comparison of SARS-CoV-2-Spike Receptor Binding Domain Produced in Pichia Pastoris and Mammalian Cells. *Sci. Rep.* **2020**, *10*, 21779. [CrossRef] [PubMed]
107. Baeshen, M.N.; Bouback, T.A.F.; Alzubaidi, M.A.; Bora, R.S.; Alotaibi, M.A.T.; Alabbas, O.T.O.; Alshahrani, S.M.; Aljohani, A.A.M.; Munshi, R.A.A.; Al-Hejin, A.; et al. Expression and Purification of C-Peptide Containing Insulin Using *Pichia Pastoris* Expression System. *BioMed Res. Int.* **2016**, *2016*, 3423685. [CrossRef] [PubMed]
108. Eissazadeh, S.; Moeini, H.; Dezfouli, M.G.; Heidary, S.; Nelofer, R.; Abdullah, M.P. Production of Recombinant Human Epidermal Growth Factor in Pichia Pastoris. *Braz. J. Microbiol.* **2017**, *48*, 286–293. [CrossRef]
109. Li, P.; Yang, G.; Geng, X.; Shi, J.; Li, B.; Wang, Z.; Zhang, Q.; Yang, Y.; Xu, C. High-Level Secretory Expression and Purification of Recombinant Human Interleukin 1 Beta in Pichia Pastoris. *Protein Pept. Lett.* **2016**, *23*, 763–769. [CrossRef] [PubMed]
110. Meng, D.-M.; Li, W.-J.; Shi, L.-Y.; Lv, Y.-J.; Sun, X.-Q.; Hu, J.-C.; Fan, Z.-C. Expression, Purification and Characterization of a Recombinant Antimicrobial Peptide Hispidalin in Pichia Pastoris. *Protein Expr. Purif.* **2019**, *160*, 19–27. [CrossRef]
111. Liu, Z.; Zhu, M.; Chen, X.; Yang, G.; Yang, T.; Yu, L.; Hui, L.; Wang, X. Expression and Antibacterial Activity of Hybrid Antimicrobial Peptide CecropinA-Thanatin in Pichia Pastoris. *Front. Lab. Med.* **2018**, *2*, 23–29. [CrossRef]
112. Popa, C.; Shi, X.; Ruiz, T.; Ferrer, P.; Coca, M. Biotechnological Production of the Cell Penetrating Antifungal PAF102 Peptide in Pichia Pastoris. *Front. Microbiol.* **2019**, *10*, 1472. [CrossRef]
113. Xing, L.-W.; Tian, S.-X.; Gao, W.; Yang, N.; Qu, P.; Liu, D.; Jiao, J.; Wang, J.; Feng, X.-J. Recombinant Expression and Biological Characterization of the Antimicrobial Peptide Fowlicidin-2 in Pichia Pastoris. *Exp. Ther. Med.* **2016**, *12*, 2324–2330. [CrossRef]
114. Zhu, W.; Gong, G.; Pan, J.; Han, S.; Zhang, W.; Hu, Y.; Xie, L. High Level Expression and Purification of Recombinant Human Serum Albumin in Pichia Pastoris. *Protein Expr. Purif.* **2018**, *147*, 61–68. [CrossRef] [PubMed]
115. Várnai, A.; Tang, C.; Bengtsson, O.; Atterton, A.; Mathiesen, G.; Eijsink, V.G. Expression of Endoglucanases in Pichia Pastoris under Control of the GAP Promoter. *Microb. Cell Factories* **2014**, *13*, 57. [CrossRef] [PubMed]
116. Mellitzer, A.; Weis, R.; Glieder, A.; Flicker, K. Expression of Lignocellulolytic Enzymes in Pichia Pastoris. *Microb. Cell Factories* **2012**, *11*, 61. [CrossRef] [PubMed]
117. Liu, Z.; Ning, C.; Yuan, M.; Fu, X.; Yang, S.; Wei, X.; Xiao, M.; Mou, H.; Zhu, C. High-Efficiency Expression of a Superior β-Mannanase Engineered by Cooperative Substitution Method in Pichia Pastoris and Its Application in Preparation of Prebiotic Mannooligosaccharides. *Bioresour. Technol.* **2020**, *311*, 123482. [CrossRef]
118. Rodrigues, K.B.; Macêdo, J.K.A.; Teixeira, T.; Barros, J.S.; Araújo, A.C.B.; Santos, F.P.; Quirino, B.F.; Brasil, B.S.A.F.; Salum, T.F.C.; Abdelnur, P.V.; et al. Recombinant Expression of Thermobifida Fusca E7 LPMO in Pichia Pastoris and Escherichia Coli and Their Functional Characterization. *Carbohydr. Res.* **2017**, *448*, 175–181. [CrossRef]
119. Matsuyama, K.; Sunagawa, N.; Igarashi, K. Mutation of Cysteine Residues Increases Heterologous Expression of Peach Expansin in the Methylotrophic Yeast Pichia Pastoris. *Plant Biotechnol.* **2020**, *37*, 397–403. [CrossRef] [PubMed]
120. Miao, T.; Basit, A.; Liu, J.; Zheng, F.; Rahim, K.; Lou, H.; Jiang, W. Improved Production of Xylanase in Pichia Pastoris and Its Application in Xylose Production From Xylan. *Front. Bioeng. Biotechnol.* **2021**, *9*, 690702. [CrossRef]
121. Juturu, V.; Aust, C.; Wu, J.C. Heterologous Expression and Biochemical Characterization of Acetyl Xylan Esterase from Coprinopsis Cinerea. *World J. Microbiol. Biotechnol.* **2013**, *29*, 597–605. [CrossRef]
122. Zhang, S.-B.; Wang, L.; Liu, Y.; Zhai, H.-C.; Cai, J.-P.; Hu, Y.-S. Expression of Feruloyl Esterase A from Aspergillus Terreus and Its Application in Biomass Degradation. *Protein Expr. Purif.* **2015**, *115*, 153–157. [CrossRef]
123. Zheng, F.; Liu, J.; Basit, A.; Miao, T.; Jiang, W. Insight to Improve α-L-Arabinofuranosidase Productivity in Pichia Pastoris and Its Application on Corn Stover Degradation. *Front. Microbiol.* **2018**, *9*, 3016. [CrossRef]
124. Adina, S.R.; Suwanto, A.; Meryandini, A.; Puspitasari, E. Expression of Novel Acidic Lipase from Micrococcus Luteus in Pichia Pastoris and Its Application in Transesterification. *J. Genet. Eng. Biotechnol.* **2021**, *19*, 55. [CrossRef] [PubMed]

125. Teng, X.; Chen, X.; Zhu, K.; Xu, H. Immunogenicity of Heparin-Binding Hemagglutinin Expressed by Pichia Pastoris GS115 Strain. *Iran. J. Basic Med. Sci.* **2018**, *21*, 219–224. [CrossRef] [PubMed]
126. Chen, W.-H.; Du, L.; Chag, S.M.; Ma, C.; Tricoche, N.; Tao, X.; Seid, C.A.; Hudspeth, E.M.; Lustigman, S.; Tseng, C.-T.K.; et al. Yeast-Expressed Recombinant Protein of the Receptor-Binding Domain in SARS-CoV Spike Protein with Deglycosylated Forms as a SARS Vaccine Candidate. *Hum. Vaccines Immunother.* **2014**, *10*, 648–658. [CrossRef]
127. Chen, W.-H.; Chag, S.M.; Poongavanam, M.V.; Biter, A.B.; Ewere, E.A.; Rezende, W.; Seid, C.A.; Hudspeth, E.M.; Pollet, J.; McAtee, C.P.; et al. Optimization of the Production Process and Characterization of the Yeast-Expressed SARS-CoV Recombinant Receptor-Binding Domain (RBD219-N1), a SARS Vaccine Candidate. *J. Pharm. Sci.* **2017**, *106*, 1961–1970. [CrossRef]
128. Markets Recombinant Proteins Market by Product, Application, End User-Global Forecasts (2022–2026). Available online: https://www.marketsandmarkets.com/Market-Reports/recombinant-proteins-market-70095015.html (accessed on 8 August 2022).
129. Haon, M.; Grisel, S.; Navarro, D.; Gruet, A.; Berrin, J.-G.; Bignon, C. Recombinant Protein Production Facility for Fungal Biomass-Degrading Enzymes Using the Yeast Pichia Pastoris. *Front. Microbiol.* **2015**, *6*, 1002. [CrossRef]
130. Steindorff, A.S.; Serra, L.A.; Formighieri, E.F.; de Faria, F.P.; Poças-Fonseca, M.J.; de Almeida, J.R.M. Insights into the Lignocellulose-Degrading Enzyme System of *Humicola Grisea* Var. *Thermoidea* Based on Genome and Transcriptome Analysis. *Microbiol. Spectr.* **2021**, *9*, e01088–e01121. [CrossRef]
131. Liu, W.; Xiang, H.; Zhang, T.; Pang, X.; Su, J.; Liu, H.; Ma, B.; Yu, L. Development of a New High-Cell Density Fermentation Strategy for Enhanced Production of a Fungus β-Glucosidase in Pichia Pastoris. *Front. Microbiol* **2020**, *11*, 1988. [CrossRef] [PubMed]
132. Ejaz, U.; Sohail, M.; Ghanemi, A. Cellulases: From Bioactivity to a Variety of Industrial Applications. *Biomimetics* **2021**, *6*, 44. [CrossRef]
133. Gaber, Y.; Rashad, B.; Hussein, R.; Abdelgawad, M.; Ali, N.S.; Dishisha, T.; Várnai, A. Heterologous Expression of Lytic Polysaccharide Monooxygenases (LPMOs). *Biotechnol. Adv.* **2020**, *43*, 107583. [CrossRef]
134. He, J.; Yu, B.; Zhang, K.; Ding, X.; Chen, D. Expression of Endo-1, 4-Beta-Xylanase from Trichoderma Reesei in Pichia Pastorisand Functional Characterization of the Produced Enzyme. *BMC Biotechnol.* **2009**, *9*, 56. [CrossRef]
135. Michelin, M.; Polizeli, M.d.L.T.M.; Ruzene, D.S.; Silva, D.P.; Vicente, A.A.; Jorge, J.A.; Terenzi, H.F.; Teixeira, J.A. Xylanase and β-Xylosidase Production by Aspergillus Ochraceus: New Perspectives for the Application of Wheat Straw Autohydrolysis Liquor. *Appl. Biochem. Biotechnol.* **2012**, *166*, 336–347. [CrossRef] [PubMed]
136. Pagnonceli, J.; Rasbold, L.M.; Rocha, G.B.; Silva, J.L.C.; Kadowaki, M.K.; Simão, R.C.G.; Maller, A. Biotechnological Potential of an Exo-polygalacturonase of the New Strain *Penicillium Janthinellum* VI2R3M: Biochemical Characterization and Clarification of Fruit Juices. *J. Appl. Microbiol.* **2019**, *127*, 1706–1715. [CrossRef] [PubMed]
137. Sena, R.O.; Carneiro, C.; Moura, M.V.H.; Brêda, G.C.; Pinto, M.C.C.; Fé, L.X.S.G.M.; Fernandez-Lafuente, R.; Manoel, E.A.; Almeida, R.V.; Freire, D.M.G.; et al. Application of Rhizomucor Miehei Lipase-Displaying Pichia Pastoris Whole Cell for Biodiesel Production Using Agro-Industrial Residuals as Substrate. *Int. J. Biol. Macromol.* **2021**, *189*, 734–743. [CrossRef] [PubMed]
138. García-Ortega, X.; Cámara, E.; Ferrer, P.; Albiol, J.; Montesinos-Seguí, J.L.; Valero, F. Rational Development of Bioprocess Engineering Strategies for Recombinant Protein Production in Pichia Pastoris (Komagataella Phaffii) Using the Methanol-Free GAP Promoter. Where Do We Stand? *New Biotechnol.* **2019**, *53*, 24–34. [CrossRef]
139. Hill, P.; Benjamin, K.; Bhattacharjee, B.; Garcia, F.; Leng, J.; Liu, C.-L.; Murarka, A.; Pitera, D.; Rodriguez Porcel, E.M.; da Silva, I.; et al. Clean Manufacturing Powered by Biology: How Amyris Has Deployed Technology and Aims to Do It Better. *J. Ind. Microbiol. Biotechnol.* **2020**, *47*, 965–975. [CrossRef]
140. Looser, V.; Bruhlmann, B.; Bumbak, F.; Stenger, C.; Costa, M.; Camattari, A.; Fotiadis, D.; Kovar, K. Cultivation Strategies to Enhance Productivity of Pichia Pastoris: A Review. *Biotechnol. Adv.* **2015**, *33*, 1177–1193. [CrossRef]
141. Wehrs, M.; Tanjore, D.; Eng, T.; Lievense, J.; Pray, T.R.; Mukhopadhyay, A. Engineering Robust Production Microbes for Large-Scale Cultivation. *Trends Microbiol.* **2019**, *27*, 524–537. [CrossRef]
142. Duman-Özdamar, Z.E.; Binay, B. Production of Industrial Enzymes via Pichia Pastoris as a Cell Factory in Bioreactor: Current Status and Future Aspects. *Protein J.* **2021**, *40*, 367–376. [CrossRef]
143. Brooks, S.M.; Alper, H.S. Applications, Challenges, and Needs for Employing Synthetic Biology beyond the Lab. *Nat. Commun.* **2021**, *12*, 1390. [CrossRef]
144. Crater, J.S.; Lievense, J.C. Scale-up of Industrial Microbial Processes. *FEMS Microbiol. Lett.* **2018**, *365*, fny138. [CrossRef]
145. Safder, I.; Khan, S.; Islam, I.-u.; Ali, M.K.; Bibi, Z.; Waqas, M. *Pichia Pastoris* Expression System: A Potential Candidate to Express Protein in Industrial and Biopharmaceutical Domains. *Biomed. Lett.* **2018**, *4*, 1–14.
146. Emenike, V.N.; Schenkendorf, R.; Krewer, U. Model-Based Optimization of Biopharmaceutical Manufacturing in Pichia Pastoris Based on Dynamic Flux Balance Analysis. *Comput. Chem. Eng.* **2018**, *118*, 1–13. [CrossRef]
147. Azadi, S.; Mahboubi, A.; Naghdi, N.; Solaimanian, R.; Mortazavi, S.A. Evaluation of Sorbitol-Methanol Co-Feeding Strategy on Production of Recombinant Human Growth Hormone in Pichia Pastoris. *Iran J. Pharm. Res.* **2017**, *16*, 1555–1564. [PubMed]
148. D'anjou, M.C.; Daugulis, A.J. A Rational Approach to Improving Productivity in RecombinantPichia Pastoris Fermentation. *Biotechnol. Bioeng.* **2001**, *72*, 1–11. [CrossRef]
149. Moser, J.W.; Prielhofer, R.; Gerner, S.M.; Graf, A.B.; Wilson, I.B.H.; Mattanovich, D.; Dragosits, M. Implications of Evolutionary Engineering for Growth and Recombinant Protein Production in Methanol-Based Growth Media in the Yeast Pichia Pastoris. *Microb. Cell Factories* **2017**, *16*, 49. [CrossRef]

150. Garcia-Ortega, X.; Adelantado, N.; Ferrer, P.; Montesinos, J.L.; Valero, F. A Step Forward to Improve Recombinant Protein Production in Pichia Pastoris : From Specific Growth Rate Effect on Protein Secretion to Carbon-Starving Conditions as Advanced Strategy. *Process Biochem.* **2016**, *51*, 681–691. [CrossRef]
151. Garcia-Ortega, X.; Valero, F.; Montesinos-Seguí, J.L. Physiological State as Transferable Operating Criterion to Improve Recombinant Protein Production in *Pichia Pastoris* through Oxygen Limitation: Physiological State Control to Improve Protein Specific Productivity in *Pichia Pastoris* with Oxygen Limitation. *J. Chem. Technol. Biotechnol* **2017**, *92*, 2573–2582. [CrossRef]
152. Rathore, A.S.; Mishra, S.; Nikita, S.; Priyanka, P. Bioprocess Control: Current Progress and Future Perspectives. *Life* **2021**, *11*, 557. [CrossRef]
153. Saitua, F.; Torres, P.; Pérez-Correa, J.R.; Agosin, E. Dynamic Genome-Scale Metabolic Modeling of the Yeast Pichia Pastoris. *BMC Syst. Biol.* **2017**, *11*, 27. [CrossRef]
154. Garrigós-Martínez, J.; Weninger, A.; Montesinos-Seguí, J.L.; Schmid, C.; Valero, F.; Rinnofner, C.; Glieder, A.; Garcia-Ortega, X. Scalable Production and Application of Pichia Pastoris Whole Cell Catalysts Expressing Human Cytochrome P450 2C9. *Microb. Cell Factories* **2021**, *20*, 90. [CrossRef]
155. Nieto-Taype, M.A.; Garcia-Ortega, X.; Albiol, J.; Montesinos-Seguí, J.L.; Valero, F. Continuous Cultivation as a Tool Toward the Rational Bioprocess Development With Pichia Pastoris Cell Factory. *Front. Bioeng. Biotechnol.* **2020**, *8*, 632. [CrossRef] [PubMed]
156. Gasset, A.; Garcia-Ortega, X.; Garrigós-Martínez, J.; Valero, F.; Montesinos-Seguí, J.L. Innovative Bioprocess Strategies Combining Physiological Control and Strain Engineering of Pichia Pastoris to Improve Recombinant Protein Production. *Front. Bioeng. Biotechnol.* **2022**, *10*, 818434. [CrossRef] [PubMed]
157. Ponte, X.; Barrigón, J.M.; Maurer, M.; Mattanovich, D.; Valero, F.; Montesinos-Seguí, J.L. Towards Optimal Substrate Feeding for Heterologous Protein Production in *Pichia Pastoris (Komagataella Spp)* Fed-Batch Processes under P_{AOX1} Control: A Modeling Aided Approach: Optimal Feeding for *Pichia Pastoris*. *J. Chem. Technol. Biotechnol.* **2018**, *93*, 3208–3218. [CrossRef]
158. Patra, P.; Das, M.; Kundu, P.; Ghosh, A. Recent Advances in Systems and Synthetic Biology Approaches for Developing Novel Cell-Factories in Non-Conventional Yeasts. *Biotechnol. Adv.* **2021**, *47*, 107695. [CrossRef]
159. Berrios, J.; Flores, M.-O.; Díaz-Barrera, A.; Altamirano, C.; Martínez, I.; Cabrera, Z. A Comparative Study of Glycerol and Sorbitol as Co-Substrates in Methanol-Induced Cultures of *Pichia Pastoris*: Temperature Effect and Scale-up Simulation. *J. Ind. Microbiol. Biotechnol.* **2017**, *44*, 407–411. [CrossRef]
160. Matthews, C.B.; Kuo, A.; Love, K.R.; Love, J.C. Development of a General Defined Medium for *Pichia Pastoris*. *Biotechnol. Bioeng.* **2018**, *115*, 103–113. [CrossRef]
161. Liu, W.-C.; Inwood, S.; Gong, T.; Sharma, A.; Yu, L.-Y.; Zhu, P. Fed-Batch High-Cell-Density Fermentation Strategies for *Pichia Pastoris* Growth and Production. *Crit. Rev. Biotechnol.* **2019**, *39*, 258–271. [CrossRef]
162. Nieto-Taype, M.A.; Garrigós-Martínez, J.; Sánchez-Farrando, M.; Valero, F.; Garcia-Ortega, X.; Montesinos-Seguí, J.L. Rationale-based Selection of Optimal Operating Strategies and Gene Dosage Impact on Recombinant Protein Production in *Komagataella Phaffii* (*Pichia Pastoris*). *Microb. Biotechnol.* **2020**, *13*, 315–327. [CrossRef]
163. de Macedo Robert, J.; Garcia-Ortega, X.; Montesinos-Seguí, J.L.; Guimaraes Freire, D.M.; Valero, F. Continuous Operation, a Realistic Alternative to Fed-Batch Fermentation for the Production of Recombinant Lipase B from Candida Antarctica under the Constitutive Promoter PGK in Pichia Pastoris. *Biochem. Eng. J.* **2019**, *147*, 39–47. [CrossRef]
164. Campbell, K.; Xia, J.; Nielsen, J. The Impact of Systems Biology on Bioprocessing. *Trends Biotechnol.* **2017**, *35*, 1156–1168. [CrossRef]
165. Guirimand, G.; Kulagina, N.; Papon, N.; Hasunuma, T.; Courdavault, V. Innovative Tools and Strategies for Optimizing Yeast Cell Factories. *Trends Biotechnol.* **2021**, *39*, 488–504. [CrossRef] [PubMed]
166. Abt, V.; Barz, T.; Cruz-Bournazou, M.N.; Herwig, C.; Kroll, P.; Möller, J.; Pörtner, R.; Schenkendorf, R. Model-Based Tools for Optimal Experiments in Bioprocess Engineering. *Curr. Opin. Chem. Eng.* **2018**, *22*, 244–252. [CrossRef]
167. Hong, M.S.; Velez-Suberbie, M.L.; Maloney, A.J.; Biedermann, A.; Love, K.R.; Love, J.C.; Mukhopadhyay, T.K.; Braatz, R.D. Macroscopic Modeling of Bioreactors for Recombinant Protein Producing *Pichia Pastoris* in Defined Medium. *Biotechnol. Bioeng.* **2021**, *118*, 1199–1212. [CrossRef] [PubMed]
168. Narayanan, H.; Luna, M.F.; Stosch, M.; Cruz Bournazou, M.N.; Polotti, G.; Morbidelli, M.; Butté, A.; Sokolov, M. Bioprocessing in the Digital Age: The Role of Process Models. *Biotechnol. J.* **2020**, *15*, 1900172. [CrossRef]

fermentation

Review

Nonconventional Yeasts Engineered Using the CRISPR-Cas System as Emerging Microbial Cell Factories

Jongbeom Park [1], In Jung Kim [2,*] and Soo Rin Kim [1,3,*]

1 School of Food Science and Biotechnology, Kyungpook National University, Daegu 41566, Republic of Korea
2 Department of Applied Biosciences, Graduate School, Kyungpook National University, Daegu 41566, Republic of Korea
3 Research Institute of Tailored Food Technology, Kyungpook National University, Daegu 41566, Republic of Korea
* Correspondence: ij0308@knu.ac.kr (I.J.K.); soorinkim@knu.ac.kr (S.R.K.); Tel.: +82-53-950-6654 (I.J.K.); +82-53-950-7769 (S.R.K.)

Abstract: Because the petroleum-based chemical synthesis of industrial products causes serious environmental and societal issues, biotechnological production using microorganisms is an alternative approach to achieve a more sustainable economy. In particular, the yeast *Saccharomyces cerevisiae* is widely used as a microbial cell factory to produce biofuels and valuable biomaterials. However, product profiles are often restricted due to the Crabtree-positive nature of *S. cerevisiae*, and ethanol production from lignocellulose is possibly enhanced by developing alternative stress-resistant microbial platforms. With desirable metabolic pathways and regulation in addition to strong resistance to diverse stress factors, nonconventional yeasts (NCY) may be considered an alternative microbial platform for industrial uses. Irrespective of their high industrial value, the lack of genetic information and useful gene editing tools makes it challenging to develop metabolic engineering-guided scaled-up applications using yeasts. The recently developed clustered regularly interspaced short palindromic repeats (CRISPR)-associated protein (Cas) system is a powerful gene editing tool for NCYs. This review describes the current status of and recent advances in promising NCYs in terms of industrial and biotechnological applications, highlighting CRISPR-Cas9 system-based metabolic engineering strategies. This will serve as a basis for the development of novel yeast applications.

Keywords: nonconventional yeast; genome editing; metabolic engineering; CRISPR-Cas9 system

Citation: Park, J.; Kim, I.J.; Kim, S.R. Nonconventional Yeasts Engineered Using the CRISPR-Cas System as Emerging Microbial Cell Factories. *Fermentation* **2022**, *8*, 656. https://doi.org/10.3390/fermentation8110656

Academic Editor: Ronnie Willaert

Received: 24 October 2022
Accepted: 17 November 2022
Published: 19 November 2022

Publisher's Note: MDPI stays neutral with regard to jurisdictional claims in published maps and institutional affiliations.

Copyright: © 2022 by the authors. Licensee MDPI, Basel, Switzerland. This article is an open access article distributed under the terms and conditions of the Creative Commons Attribution (CC BY) license (https://creativecommons.org/licenses/by/4.0/).

1. Introduction

Petroleum-derived chemical production has detrimental effects on the environment and exhibits industrial noncompatibility associated with cost-effectiveness due to multiple labor-intensive processes [1]. To overcome this problem, biotransformation using microorganisms can be considered an alternative approach [1,2], which has the advantage of rapid growth rate and easy cultivation of microorganisms under laboratory conditions [3]. In particular, eukaryotic yeasts represent robust microbial cell factories owing to their simple structure and ability to grow on various substrates, as well as the relatively simple gene editing techniques used to manipulate their genomes [4].

Saccharomyces cerevisiae is the most widely used eukaryotic platform in bioprocesses [5,6]. *S. cerevisiae* has long been a model organism for fundamental biological research and industrial applications because of its ease of handling and safety as a generally recognized as safe (GRAS) strain [7]. Moreover, its genetics are well understood, and tools for manipulating it are well established; thus, numerous specialized strains and plasmids are available. However, as a Crabtree-positive organism, the carbon flux in *S. cerevisiae* is mainly directed toward the ethanol fermentation pathway. This preference for ethanol production often limits its utilization as a host when nonethanol products are to be synthesized. Therefore, other suitable yeasts, known as nonconventional yeasts (NCY), should be considered as alternative hosts.

Approximately 1500 NCY species have been identified to date [8], each of which exhibits unique genetics, physiology, and characteristics. They often have excellent potential for industrial uses that are not feasible with *S. cerevisiae*. Thus, NCY can be considered as promising eukaryotic hosts alternative to *S. cerevisiae* to overcome or improve its limitations. Specifically, NCYs that are Crabtree-negative can diversify the profile of industrially useful products. In addition, a much higher capacity of NCYs for pentose phosphate pathway relative to *S. cerevisiae* is also advantageous feature when synthesizing products using NCYs as the cell factory by increasing the available pool of cofactors and precursors [9]. Moreover, a high tolerance against multiple stress factors, such as heat, low pH, and salt, can extend yeasts' utility. Advancements in genetic and metabolic engineering technologies will facilitate NCY-based scaled-up bioprocesses [8,10].

Despite their beneficial traits, genetic information and manipulation tools for many NCYs are lacking compared with those for *S. cerevisiae* [11]. As most organisms have not been thoroughly analyzed for safety and lack genome sequencing data with the limited information on the exact gene loci and protein functions, it is challenging to develop and apply suitable gene editing tools, together with establishing transformation protocols and selectable marker genes [11].

The clustered regularly interspaced short palindromic repeats (CRISPR)-associated protein (Cas) system is an adaptive immune system of bacteria and archaea that protects them from invasion by foreign genetic elements [12]. As a gene editing tool, the CRISPR-Cas system is based on a simple single guide RNA (sgRNA)/DNA hybrid that recognizes specific target DNA, providing a simple-to-design methodology, which has more sophisticated, accurate, and cheaper gene editing capabilities compared with traditional methods like zinc finger nucleases (ZFNs) and transcription activator-like effector nucleases (TALENs) [13]. Thus, this revolutionary CRISPR-Cas9 system has been applied to various yeast strains, such as *S. cerevisiae*, *Pichia pastoris*, *Kluyveromyces marxianus*, and *Yarrowia lipolytica*; among these, most studies have been conducted on *S. cerevisiae* [14,15]. Although numerous studies have reported the CRISPR-guided metabolic engineering of *S. cerevisiae*, only few studies have explored the applicability of this system in NCYs [16]. To achieve commercial scale bioproduction using NCYs as cell factories, it is necessary to develop highly efficient and convenient engineered strains [10,17].

In this review, we provide an overview of NCYs with excellent potential for industrial applications. In particular, the biotechnological applications of engineered NCYs using the CRISPR-Cas9 system are highlighted. The advances and challenges of CRISPR–Cas9-mediated biotechnology for NCYs are also discussed.

2. Industrial Value of NCYs

NCYs have several advantages over *S. cerevisiae* from an industrial viewpoint [18]. *S. cerevisiae* is often directed toward ethanol synthesis (due to its Crabtree-positive effect), restricting product diversification. In contrast, NCYs may have desired metabolic pathways, enabling product profile expansion. The ability to resist various stresses is a key benefit in industrial bioprocesses. For example, ethanol production from lignocellulose can be enhanced by developing alternative microbial platforms that are highly resistant to inhibitors. NCYs often exhibit strong resistance to various stresses, such as heat, acid, and high sugar concentrations, as environmental adaptations. Another key advantage of NCYs is their ability to utilize a wide range of carbon sources [8,19]. Additionally, many NCYs can exist in both haploid and diploid types like *S. cerevisiae*, and sexual reproduction is possible. Therefore, NCY strains with a desired ploidy can be developed through mating depending on the purpose. For example, diploid *P. pastoris* strains having a higher stability than its haploid form were constructed through a well-designed mating process for production of proteins [20], or various auxotrophic *K. marxianus* libraries were constructed using mating and dissection [21]. In this section, the industrial potentials of five promising NCYs are described (Figure 1).

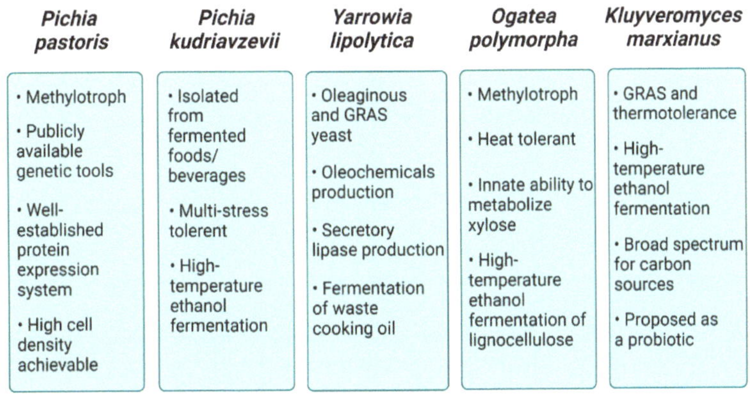

Figure 1. Properties and industrial values of promising NCYs.

2.1. *Pichia pastoris*

P. pastoris is a methylotrophic yeast with a developed peroxisomal system responsible for compartmentalized methanol metabolism [22]. Its gene expression system, together with its secretory system, is well established, which enables easy genetic manipulation using a publicly available commercial kit. *P. pastoris* is mainly used for the production of pharmaceutical proteins and other industrial enzymes, because of simple post-translational modifications process [23,24]. In fact, hyperglycosylation of proteins in *S. cerevisiae* often causes an allergic reaction in human body; therefore, using it as an expression host for producing pharmaceuticals and medicinal proteins is often undesirable. *P. pastoris* exhibits superior growth rate and cell density compared to *S. cerevisiae*, and like *S. cerevisiae*, its protein expression is controlled under strong and tightly regulated methanol inducible promoters (AOX1). Thus, high cell density in addition to a high yield of recombinant proteins, produced either intracellular or extracellular, can be achieved, which may lead to increased biotransformation efficiency of whole cells [23,25].

2.2. *Pichia kudriavzevii*

P. kudriavzevii is a multi-stress tolerant yeast commonly found in fermented foods and beverages, such as *Nuruk*, which is a starter used for making Korean traditional alcoholic beverages and various sub-Saharan African indigenous foods [26–29]. Previous studies have isolated robust strains of *P. kudriavzevii* that can withstand multiple stress factors, such as high salt concentration, high temperature, and low pH. In particular, there are several advantages of using thermotolerant strains as production hosts for the ethanol industry. For *S. cerevisiae*, scaled-up ethanol production through simultaneous saccharification and fermentation is normally performed at 30 °C, above which growth and fermentation are repressed. Ethanol production at a high temperature is beneficial for reducing microbial contamination, as well as energy and water costs required to cool the fermentation system [30–32]. Indeed, thermotolerant *P. kudriavzevii* produce more ethanol at a higher temperature (44 °C) compared with *S. cerevisiae*. Furthermore, considering the information on genome sequence and genetic engineering tools, it is a potent host for various industrial metabolites, such as organic acids (e.g., succinic acid) and bioethanol [30,32–34].

2.3. *Yarrowia lipolytica*

As an oleaginous microorganism that accumulates lipids up to 20% of dry cell weight, *Y. lipolytica* is used for the industrial production of fatty acid-derived products [35]. *Y. lipolytica* is a GRAS organism and assimilates hydrophilic (e.g., glucose, glycerol, alcohols, and acetate) and hydrophobic substrates (e.g., fatty acids, triacylglycerols, and alkanes) [36]. With its high protein secretory capacity and lipophilicity, the organism has also been used for the fermentation of waste cooking oil to achieve bioremediation and waste valorization. During this fermentation process, *Y. lipolytica* produces extracellular lipase, which subsequently generates free fatty acids, utilizable as a carbon source, from waste cooking oil. With the development of metabolic engineering, *Y. lipolytica* can potentially produce several other metabolites, such as organic acids [37], erythritol [38], and flavonoids [39].

2.4. *Ogataea polymorpha*

O. polymorpha is a methylotroph, an organism that utilizes C1 compounds, such as methanol, as its sole carbon source, and is one of the most heat-resistant yeasts. Because native *S. cerevisiae,* the most widely used workhorse for bioethanol production, is incapable of xylose fermentation and engineered *S. cerevisiae* to possess heterologous xylose metabolic pathway may suffer from high metabolic burden, a benefit of *O. polymorpha* is its innate ability to metabolize xylose, the second most abundant sugar of lignocellulosic biomass [40]. Thus, high-temperature (i.e., 45–50 °C) ethanol fermentation of lignocellulose hydrolysate mainly consisting of glucose and xylose is possible. As one of a few methylotrophic yeasts, their key enzymes involved in methanol metabolism are strongly induced by methanol present within membrane-bound peroxisomes, which enables a compartmentalized reaction. Based on this expression machinery, *O. polymorpha* is a useful expression host for producing heterologous and difficult-to-express proteins via establishing expression systems induced by methanol under the control of strong and tightly regulated promoters.

2.5. *Kluyveromyces marxianus*

K. marxianus is a GRAS and thermotolerant ethanol-producing species that can grow at temperatures up to 52 °C [41–43], enabling high-temperature ethanol fermentation. As a Crabtree-negative yeast, this species is also advantageous for synthesizing non-ethanol products. Like *K. lactis* [44], *K. marxianus* has the unique ability to assimilate lactose, which is not feasible with *S. cerevisiae* and other yeasts [44,45]. Owing to its high capacity to grow on a broad spectrum of cheap carbon sources, such as xylose, arabinose, galactose, lactose, pectin, inulin, hemicellulose hydrolysate, cheese whey, and molasses, this species is an excellent microbial source of enzymes, bioethanol, and food ingredients [46] for commercial-scale applications [47–49]. Additionally, *K. marxianus* can produce fructose and fructooligosaccharides, which are industrially pertinent foods and pharmaceutical ingredients through inulinase secretion [50]. Recently, *K. marxianus* has been proposed as a probiotic yeast due to its beneficial roles in the gut [46].

3. Genetic Engineering Tools for NCYs

Although the innate traits of NCYs are beneficial for industrial applications, the productivity is low and must be increased through metabolic engineering techniques. One basis for metabolic engineering is endogenous or heterologous gene expression using a host strain [51]. This can be accomplished through two approaches. The first approach is the development of an episomal plasmid expression system using a self-replicating vector and the second is genome integration [52]. For *S. cerevisiae,* episomal vector systems with high copy numbers have been highly developed through extensive optimization and are widely utilized relative to NCYs [53]. Further, plasmid-based expression systems have also been developed for NCYs [54,55]. For example, an autonomously replicating sequence (ARS)-based episomal vector has been developed for *P. pastoris,* providing higher transformation efficiency and lower interclonal variability compared with the classical integrative plasmid [56]. The lower interclonal variability of transformants obtained from ARS system could be highly related with the frequent occurring non-specific genome integration caused by the integrative vector [51]. However, such an episomal plasmid system suffers from a high cost and inefficiency arising from segregational instability, which requires its maintenance under selective pressure for stable protein expression. Moreover, a centromeric plasmid was engineered, in which various promoters were fused upstream of the centromere sequence to control the function of centromere in *Y. lipolytica.*

Such an approach successfully led to the significant improvement in the plasmid copy number, and in turn, protein expression levels (80%) [54].

From an industrial viewpoint, genome integration is the preferred method because it leads to more homogeneous expression levels and, upon integration, expression cassettes are stably inherited with the respective chromosome. To achieve more sophisticated gene editing and higher integration efficiency, the double-stranded sequence at the desired site must be broken. Nuclease-based tools, such as ZFNs and TALENs, can be used for the genetic engineering of NCYs [57,58]. However, these methods are outdated with the emergence of CRISPR-Cas9 system. The Cre–lox system, which deletes the sequences between direct repeats of two loxP sites by catalyzing their recombination, leaving a single loxP site behind in the genome [59], is also a useful technology for producing knockout mutants without leaving a marker; however, some scars are left in the genome, or unwanted recombination may occur [10]. CRISPR-Cas9 system, established and predominantly used over the past decade, is a powerful gene editing tool that overcomes the problems associated with conventional tools [60].

Contrary to ZFN and TALEN-assisted approaches in which nucleases play a dual role in recognition and cleavage, the CRISPR-Cas9 system relies on designed sgRNAs that are specific to a target gene, and after the recognition, Cas9 cleaves the targeted site. This is an advanced genetic engineering tool that allows a simple-to-design plasmid construction for sgRNA expression, as well as rapid and precise cell programming [10].

4. CRISPR-Cas System-Guided Metabolic Engineering in NCYs

4.1. CRISPR-Cas System: Classification, Components, and Mechanism

The CRISPR-Cas system, comprising a DNA array and associated proteins, is a widely distributed RNA-based adaptive defense mechanism against viruses in bacteria and archaea [61]. Based on the Cas protein effector, CRISPR-Cas systems are mainly categorized into two classes: class I (type I, III, and IV) with a multi-subunit Cas complex and class II (type II, V, and VI) with a single Cas protein [62,63]. Depending on the type and mechanism of Cas endonucleases, each type can be further classified into several subtypes.

Among these, the type II CRISPR-Cas9 system uses a single Cas9 protein and is most commonly used for genetic and metabolic engineering because of its simple structure. In particular, the most widely used CRISPR system is composed of a Cas9 protein derived from *Streptococcus pyogenes* [64]. The CRISPR-Cas9 system consists of two integral components, Cas9 and sgRNA; Cas9 is divided into two domains: a recognition lobe (REC1 and REC2) and nuclease lobe (NUC). Further, the sgRNA is a fusion construct composed of the two RNA types, CRISPR RNA (crRNA) and transactivation crRNA (tracrRNA), connected by a linker sequence (Figure 2) [65]. sgRNA is bound by the REC lobe of Cas9 and REC lobe controls the conformational alteration of the catalytic core of Cas9 (i.e., HNH motif) [66]. The NUC lobe comprises the following three domains: (1) HNH, (2) RuvC cleaving target and nontarget single-stranded DNAs, and (3) protospacer adjacent motif (PAM)-binding domain that scans PAM sequence through weak or transient interactions before forming a stable hybrid between target sequence-sgRNA to perform the specific cleavage.

The process of Cas9-mediated genome editing involves recognition, cleavage, and repair [67]. First, the PAM sequence (i.e., 5′-NGG-3′) is scanned for probing the target for cleavage and subsequently, the target sequence is recognized by the sgRNA bound to REC lobe of Cas9 [64,67]. After the sgRNA forms a duplex with its complementary DNA, the target and nontarget single-stranded DNA sequences located at 3-bp upstream of the PAM sequence are cleaved by HNH and RuvC nucleases, respectively [65]. Finally, the generated double-strand breaks (DSBs) undergo DNA repair; the nucleotide sequence of the cleaved site can be modified through the endogenous DNA repair machinery. Two main repair systems in eukaryotes are based on the homology-directed repair (HDR) and non-homologous end joining (NHEJ) pathways [68]. The predominant pathway in eukaryotes is NHEJ in which random insertions and deletions (indels) of a small number of nucleotides are introduced at the broken ends. Without using template DNA, this process is prone to error and often causes undesirable genetic modifications. In contrast, the HDR pathway uses a homologous DNA template for DSB repair that can be exogenously provided as single- or double-stranded DNA carried as a plasmid or PCR product. The designed nucleotide sequence is integrated into the genome of a host strain via homologous recombination (HR) [69]. This template-based mechanism allows accurate genome editing, modification, and replacement.

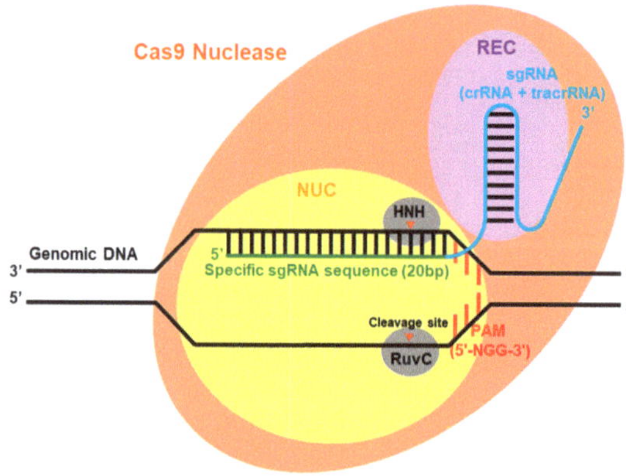

Figure 2. The CRISPR-Cas9 system: Cas9 structure and mechanism. Abbreviations: sgRNA, single guide RNA; crRNA, CRISPR RNA; tracrRNA, transactivation crRNA; PAM, protospacer adjacent motif; REC, recognition lobe; NUC, nuclease lobe.

4.2. Challenges and Strategies of CRISPR-Cas9-Guided Genome Editing in NCYs

As demonstrated in Section 4.1, the CRISPR-Cas9 system has emerged as the most powerful tool for the gene editing of yeast species for biotechnological applications due to its simple design, precise recognition and targeted generation of double strand breaks, and no remnant of antibiotic selection markers [70]. Here, we summarize the CRISPR-Cas9 system applied to NCYs (Table 1).

Table 1. Comparison of CRISPR-Cas9 system for NCYs.

Strain	sgRNA Promoter	Plasmid (Backbone)	Cas9 Promoter	Editing Efficiency [1] (%)	Reference
Y. lipolytica	*SCR1′-tRNA* [2] *TEFin* (Pol II) [3]	pCRISPRyl pCASyl pGGA	*TEF1*	0–68.9	[71–74]
O. polymorpha	*ScSNR52* (Pol III) *ScTDH3* (Pol III)	pCRCT pYTK079	*ScTEF1* *AaTEF*	1–75	[75,76]
P. pastoris	*HTX1* (Pol II) [3] *PFK300* (Pol II) [3] *LAT1*(Pol II) [3]	pPpT4 p414 BB3cH	*HTX1* *GAP*	75–93.8	[77–79]
K. marxianus	*ScTDH3* (Pol II) [3] *RPR1′-tRNA* (Pol III) [2]	pYTK079 pIW601	*AaTEF* *1 ScTEF1*	10–82	[76,80]
P. kudriavzevii	*RPR1* (Pol III) *RPR1′-tRNA* (Pol III) [2]	pRS416 pRS415 pCast	*TEF1*	64	[30,81,82]

[1] Defined as transformation efficiency (i.e., positive colony number/total colony number); [2] synthetic hybrid promoters; [3] self-cleaving ribozyme system was used.

In contrast to the homologous DNA template-based HDR system adopted by *S. cerevisiae*, most NCYs rely on the NHEJ pathway. NHEJ-dominant DNA repair system is the major challenge of CRISPR-Cas9-guided genome editing for NCYs. In particular, NHEJ is an evolutionarily conserved pathway that directly rejoins the broken ends of DNA without a template [83–85]. This recovery system can cause unwanted frameshift indels, which is an obstacle to integrating donor DNA into genomes [17]. To overcome this limitation, a strategy to knockout NHEJ-related genes was devised that was capable of strongly reducing NHEJ, causing remaining recombinants to have a higher percentage of the desired HDR [86]. Through knockout of *Ku70/Ku80*, *Dnl4/Lif1*, *Nej1*, and *Mre11/Rad50/Xrs2* (*MRX*), studies have shown that the NHEJ pathway was inhibited but gene targeting efficiency was increased through the induced HDR pathway [87–89]. Although the stability of strains associated with growth and ultraviolet sensitivity needs to be improved [89], these NHEJ knockout approaches serve as a basis for CRISPR-guided genetic engineering of NCYs. Conversely, there is a strategy to increase genome integration efficiency through overexpression of genes involved in the HDR pathway [90]. For example, RAD51 or 52, a recombinase involved in DSB repair, is the central enzyme of HDR process, and its overexpression is suggested as a methodology to induce HDR pathway. Such an approach has either enhanced or inhibited HDR efficiency in earlier studies [91]. While a study revealed that RAD51 overexpression inhibited HDR process of DSBs in yeast (i.e., *S. cerevisiae*) [91], there are also examples with *P. pastoris* showing that heterologous or homologous expression of RAD52 increases HR efficiency [92]. A similar phenomenon was also observed with *Y. lipolytica* [93].

The CRISPR-Cas9 system requires the optimal expression of sgRNA and Cas9 that are tailored to the individual organism. sgRNA expression is another major challenge, and its efficiency relies on the type of RNA promoter and polymerase. According to a previous study, sgRNA expression was observed under the control of the standard RNA polymerase III promoter in *S. cerevisiae*, but expression of gRNA using the same promoter was at the low level in NCYs, including *P. pastoris* [94]. Furthermore, RNA polymerase II, a polymerase responsible for most protein-coding mRNA synthesis, is not applicable for sgRNA transcription because the transcripts synthesized by this RNA polymerase cause significant alterations, such as polyadenylation, at the transcript termini, which are undesirable for sgRNA expression [95]. To overcome these limitations, Gao et al. developed a chimeric gene system known as ribozyme-sgRNA-ribozyme, through which the RNA transcript is self-cleaved because of the ribozymal nuclease activity to release the sgRNA [95,96]. This system functions well with appropriate promoters using the classical RNA polymerase (e.g., RNA polymerase II), which serves as an efficient tool to increase functional sgRNA expression, even in NCYs (Table 1).

Apart from those mentioned above, challenges, such as off-target effects causing unwanted target cleavage [97] and restriction of NGG PAM motifs [98] remain, and further improvements are needed for industrial application of NCYs. To be specific, variants of Cas9 through protein engineering (for recognition of NAG and NGA PAMs) [99] and PAM sequence can be generated to achieve the more specific target cleavage. Additionally, an off-target effect could be solved by the well-designed sgRNA using *in silico* program [100].

4.3. Biotechnological Application of CRISPR-Cas9-Introduced NCYs

4.3.1. Secondary Plant Metabolites

Plant-derived secondary metabolites are organic compounds with complex molecular structures that exhibit beneficial bioactivity. Flavonoids, terpenes, saponins, alkaloids, and sterols are representative examples. These exhibit antiobesity, anticancer, antioxidant, and antiaging effects [101,102] and can be used in various industries, such as pharmaceuticals, foods, and cosmetics [103,104]. The classical production of these compounds mainly depends on extractive and chemical methods, which are neither environmentally friendly nor economically feasible [105]. Microbial biosynthesis is a promising alternative to overcome the problems associated with current production methods [106]. *Y. lipolytica* is a favorable host strain for the biosynthesis of secondary plant metabolites, especially flavonoids, as it produces large amounts of acetyl-CoA and other flavonoid precursors through the mevalonate (MVA) pathway. With the application of CRISPR-Cas9 technology for genomic integration in *Y. lipolytica*, targeted and marker-free strains have been developed in an easy, simple, and precise manner, thus contributing to the current research [107].

Flavonoids, such as naringenin and resveratrol, are a class of secondary plant metabolites derived from phenylpropanoids [108]; they are biologically synthesized as follows (Figure 3). From phenylalanine and tyrosine, 4-coumaric acid is produced through deamination, which is then converted into 4-coumaroyl CoA by 4-coumaroyl-CoA ligase (4CL) [103]. Next, it is combined with three molecules of malonyl-CoA via polyketide synthase to form flavonoids. In yeasts lacking the aforementioned pathway, the heterologous expression of tyrosine amino lyase (TAL), 4CL, and chalcone synthase (CHS) is used as the primary strategy to produce flavonoid precursors [103,109]. In particular, an engineered *Y. lipolytica* strain in which the xylose metabolism and naringenin biosynthesis pathways were introduced produced 715.3 mg/L naringenin from a mixture of glucose and xylose as the carbon source (Table 2) [109]. Additionally, the exogenous expression of the resveratrol biosynthetic pathway and mutations in 3-deoxy-D-arabino-heptulosonic acid 7-phosphate synthase (ARO4) and chorismate mutase (ARO7) lowered the sensitivity to tyrosine-induced feedback inhibition, thereby increasing resveratrol production to 12.4 g/L [103].

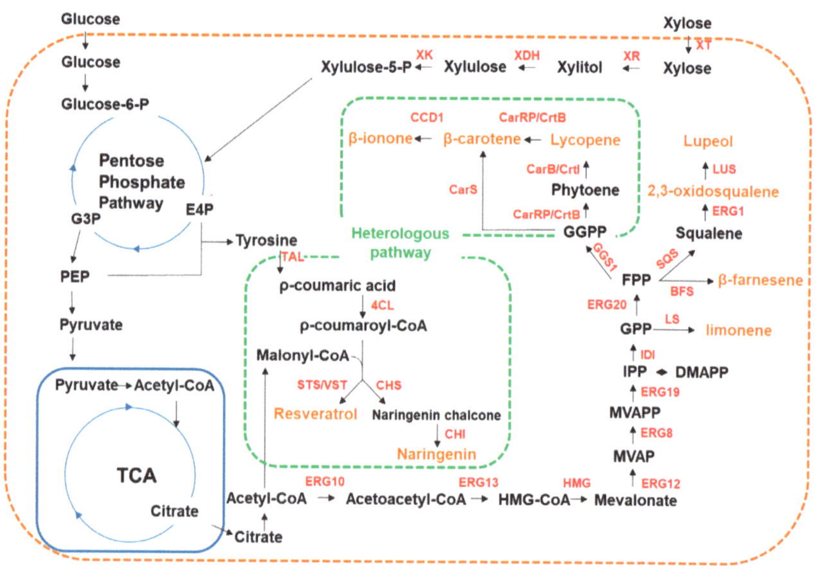

Figure 3. Engineered metabolic pathways for the production of secondary plant metabolites, such as flavonoids and carotenoids, in NCYs. Abbreviations: glucose-6-P, glucose-6-phosphate; xylulose-5-P, xylulose-5-phosphate; E4P, erythrose 4-phosphate; G3P, glyceraldehyde-3-phosphate; PEP, phosphoenolpyruvic acid; acetyl-CoA, acetyl coenzyme A; HMG-CoA, 3-hydroxy-3-methylglutaryl-CoA; MVAP, mevalonate-5-phosphate; MVAPP, mevalonate pyrophosphate; IPP, isopentenyl pyrophosphate; DMAPP, dimethylallyl pyrophosphate; GPP, geranyl pyrophosphate; FPP, farnesyl pyrophosphate; GGPP, geranylgeranyl pyrophosphate; TAL, tyrosine ammonia-lyase; 4CL, 4-coumarate-coA ligase; STS, stilbene synthase; VST, resveratrol synthase; CHS, chalcone synthase; CHI, chalcone isomerase; ERG10, acetyl-CoA C-acetyltransferase; ERG13, HMG-CoA synthase; HMG, HMG-CoA reductase; ERG12, mevalonate kinase; ERG8, phosphomevalonate kinase; ERG19, mevalonate pyrophosphate decarboxylase; IDI, isopentenyl diphosphate isomerase; LS, limonene synthase; ERG20, farnesyl pyrophosphate synthase; BFS, β-farnesene synthase; GGS1, geranylgeranyl diphosphate synthase; SQS, squalene synthase; ERG1, squalene monooxygenase; LUS, lupeol synthase; CarRP/CrtB, phytoene synthase/lycopene cyclase; CarB/CrtI, phytoene dehydrogenase; CarS, multi-functional carotene synthase; CCD1, carotenoid cleavage dioxygenase; XK, xylulokinase; XDH, xylitol dehydrogenase; XR, xylose reductase; XT, xylose transporter.

Carotenoids are tetraterpenes containing 40 carbon atoms and natural antioxidants. Carotenoids derived from vegetable and fruits, called β-carotene, β-cryptoxanthin, and α-carotene, are regarded as great dietary sources of provitamin A, which is converted in the human intestine into vitamin A, demonstrating their high industrial value [110]. *Y. lipolytica* natively produces geranylgeranyl pyrophosphate (GGPP) via the MVA pathway, which can be transformed into carotenoids through the introduction of heterologous genes involved in downstream pathways (Figure 3). In previous studies, lycopene and β-carotenoids were generated from GGPP via the introduction of phytoene dehydrogenase and phytoene synthase/lycopene cyclase from *Mucor circinelloides* or CrtYB and CrtI from *Xanthophyllomyces dendrorhous* [111,112]. Additionally, the production of β-carotene was increased to 4.5 g/L by enhancing the supply of precursors (i.e., GGPP) through the overexpression of MVA metabolic pathway genes, such as 3-hydroxy-3-methylglutaryl-CoA reductase [73].

Industrially useful terpenes, such as monoterpene and sesquiterpene, which are utilized in food, cosmetics, medicine, and next-generation jet fuel, have also been synthesized via the metabolic engineering of NCYs [113].

Table 2. Production of secondary plant metabolites using CRISPR-Cas9 in NCYs.

Strain	Target Genes		Product	Reference
	Endogenous Gene Editing	Heterologous Gene Editing		
Y. lipolytica	*HMG1, GGS1*	*crtE (Pantoea ananatis)*, *crtI (P. ananatis)*, *crtB (P. ananatis)*	Lycopene 3.38 mg/g DCW [2]	[71]
	GGS1	*carB (Mucor circinelloides)*, *carRP (M. circinelloides)*,	β-carotene 4.8 mg/g DCW	[72]
	GGS1, ERG13, HMG	*carB (M. circinelloides)*, *carRP (M. circinelloides)*	β-carotene 4.5 g/L	[73]
	GGS1, HMG1, ERG8, ERG10, ERG12, ERG13, ERG20, ERG19, IDI	*carB (M. circinelloides)*, *carRP (M. circinelloides)*, *CCD1 (Petunia hybrid)*, *PK (Bifidobacterium bifidum)*, *PTA (Bacillus subtilis)*	β-ionone 358.4 mg/L 0.98 g/L (fed-batch)	[112]
	XK, HMG1 [1], *ERG12* [1]	*LS (Agastache rugosa)* [1], *NDPS (Solanum lycopersicum)* [1], *XR (Scheffersomyces stipitis)*, *XDH (S. stipitis)*	Limonene 20.57 mg/L	[113]
	HMG, ERG12, IDI, ERG20, SQS	*BFS (Artemisia annua)*, *LS (Citrus limon)*, *LS (Perilla frutescens)*, *CnVS (Callitropsis nootkatensis)*, *crtI (Xanthophyllomyces dendrorhous)*, *crtYB (X. dendrohous)*, *acs (Salmonella enterica)*	β-farnesene 955 mg/L Limonene 35.9 mg/L Valencene 113.9 mg/L Squalene 402.4 mg/L β-carotene 164 mg/L 2,3-oxidosqualene 22 mg/L	[111]
	HMG1, ERG1, ERG9, OLE1, PAH1, DGK1	*LUS (Ricinus communis)*	Lupeol 441.72 mg/L	[114]
	GGS1	*carS (Schizochytrium* sp.)	β-carotene 0.41 mg/g DCW	[74]
	XT [1], *XR* [1], *XDH* [1], *XKS* [1]	*TAL (Rhodotorula glutinis)*, *4CL (Arabidopsis thaliana)*, *CHS (A. thaliana)*, *CHI (A. thaliana)*	Naringenin 715.3 mg/L	[109]
	ARO4, ARO7	*TAL (Flavobacterium johnsoniae)*, *VST (Vitis vinifera)*, *4CL (A. thaliana)*	Resveratrol 12.4 g/L	[103]
O. polymorpha	-	*TAL (Herpetosiphon aurantiacus)*, *STS (V. vinifera)*, *4CL (A. thaliana)*	Resveratrol 97.23 mg/L	[75]

[1] Plasmid-based engineering; [2] DCW: dry cell weight.

4.3.2. Other Industrial Products

In addition to secondary plant metabolites, various industrially useful products have been synthesized using NCYs engineered with the CRISPR-Cas9 system (Table 3).

For example, itaconic acid is a high value-added compound used to synthesize polymers and chemical intermediates, such as styrene and 2-methyl-1,4-butanediol [115]. Currently, *Aspergillus terreus* is used as the microbial host for the production of itaconic acid with a titer of 160 g/L; however, it has some disadvantages in terms of its pathogenic potential (biosafety level 2), heterogeneous fermentation (as the filamentous fungi), the resultant increase in viscosity of broth, difficult genetic engineering, and high sensitivity to shear stress, etc. [116]. Therefore, there is an increasing demand for an alternative host. Itaconic acid can be produced via cis-aconic acid decarboxylase (CAD), an enzyme that decarbonizes cis-carbonate, which is an intermediate of the tricarboxylic acid (TCA) cycle in yeasts [115]. Studies have focused on the heterologous expression of CAD from *A. terreus* [115,117]. The biosynthesis of itaconic acid using *Y. lipolytica* and *P. kudriavzevi* with biosafety, which have low pH resistance allowing for saving of downstream cost associated with neutralization, fast growth rates, and shear stress resistance, has been conducted [30,116]. One strategy to increase the amount of cis-aconitate secreted by mitochondria and transported into the cytosol is overexpressing the mitochondrial tricarboxylate transporter, which produced up to 22.03 g/L itaconic acid [116].

Lipid-derived oleochemicals have been increasingly synthesized by *Y. lipolytica*, a microbial host with a high potential to produce microbial oils that can replace vegetable oils to increase economic feasibility and reduce environmental pollution [118,119]. A titer of 25 g/L lipids was produced by knocking out phospholipase, an acyl-binding phospholipid hydrolase [119]. The rich malonyl-CoA pool in *Y. lipolytica* is also useful for fatty alcohol production. The heterologous expression of fatty acyl-CoA (FAR) from *Marinobacter aquaeolei* [98] in *Y. lipolytica* led to 5.75 g/L fatty alcohol [120].

K. marxianus can synthesize esters for industrial use as it contains a rich acetyl-CoA pool owing to its high growth rate and relevant endogenous enzymes, such as alcohol acetyltransferase and esterase [121–123]. With the development of CRISPR-based tools, it has become feasible to edit multiple target genes involved in related metabolic pathways. A previous study successfully increased the production of 2-phenylethanol to 850 mg/L in *K. marxianus* by balancing the precursors of shikimate and phenylalanine biosyntheses [48]. Additionally, 150 mg/L ethyl acetate was produced via the TCA cycle and knockdown of electron transport chain-related genes [123].

Table 3. Production of industrially useful products using CRISPR-Cas9 in NCYs.

Strain	Target Genes		Product	Reference
	Endogenous Gene Editing	Heterologous Gene Editing		
Y. lipolytica	-	*CAD* (*Aspergillus terreus*), *mttA* (*A. terreus*) [1]	Itaconic acid 22.03 g/L	[116]
P. kudriavzevii	*ICD*, *mttA* [1]	*CAD* (*A. terreus*) [1]	Itaconic acid 1.23 g/L	[30]
Y. lipolytica	*SCT1*, *OLE1*	*FAR* (*M. aquaeolei*)	Fatty alcohol 5.75 g/L	[120]
Y. lipolytica	PLA_2	-	Lipid 25 g/L	[119]
Y. lipolytica	*AXP*	*celB* (*Pyrococcus furiosus*)	β-glycosidase 187.5 μkat_{oNPGal}/L [2]	[124]
K. marxianus	*ARO1*, *ARO2*, *ARO3*, *ARO4*, *ARO7*, *ARO8*, *ARO9*, *PHA2*, *TAL1*, *TKL1*, *RPE1*, *RKI1*, *LAC4*	*xfpk* (*Bifidobacterium breve*), *ppsA* (*Escherichia coli*), *pta* (*Salmonella enterica*)	2-penylethanol 850 mg/L	[48]
Y. lipolytica	-	*FAP* (*Chlorella variabilis*)	Hydrocarbons 58.7 mg/L	[125]
K. marxianus	*ACO2b*, *SDH2*, *RIP1*, *MSS51*	-	Ethyl acetate 150 mg/L	[123]

[1] Plasmid-based engineering; [2] The amount of enzyme was measured as enzyme activity assayed against *o*-nitrophenyl-β-galactopyranoside (oNPGal).

5. Future Perspectives

The current petroleum-based economy can be replaced by a sustainable bioeconomy with the increase in the development of modern biotechnology based on genetically and metabolically engineered microorganisms. The recent approval of CRISPR-Cas system-mediated genome editing for use in gene therapy clinical trials and relaxed regulation of the commercialization of genetically modified crops by the United States government will accelerate the expansion of applications of genome editing from bench to reality. This can be demonstrated by NCYs with intrinsic commercial value applicable to various industrial sectors. However, to overcome the technical difficulties associated with the genetic engineering of NCYs, careful tailoring of CRISPR-Cas systems in a given organism is required. The continual discovery of novel machinery and development of diverse variants of the CRISPR-Cas system are expected to promote this process. Advances in systems and synthetic biology approaches, in which predicted modeling data are combined with experimental data (i.e., product yield and metabolic flux), will further aid in the selection of suitable strains and optimize engineered pathways more efficiently for NCYs.

6. Conclusions

Each NCY possesses unique characteristics and a high potential for industrial use. Compared with the model yeast *S. cerevisiae*, the genetic and physiological characteristics of NCYs have not been fully elucidated, making it challenging to develop industrial applications using NCYs as microbial hosts. Although the introduction of the revolutionary CRISPR-Cas9 system has accelerated the targeted metabolic engineering of NCYs, error-prone genetic mutations emerging from native DNA repair machinery, depending on the inaccurate NHEJ pathway, are the primary obstacle. To overcome this issue, editing efficiency can be improved using strategies for knocking out NHEJ-related genes or overexpressing HR-related genes. Further improvement can be achieved by customized optimization of CRISPR-Cas9 systems specific to individual organisms.

Author Contributions: Conceptualization, J.P.; writing—original draft preparation, J.P.; writing—review and editing, I.J.K. and S.R.K.; supervision, S.R.K. All authors have read and agreed to the published version of the manuscript. **Funding:** This work was supported by a National Research Foundation of Korea (NRF) grant funded by the Korean government (MSIT) (No. NRF-2022R1A2C1093074). This work was also supported by a National Research Foundation of Korea (NRF) grant funded by the Korean government (MSIT) (No. 2022R1I1A1A01072158).

Conflicts of Interest: The authors declare no conflict of interest.

References

1. Li, W.; Shen, X.; Wang, J.; Sun, X.; Yuan, Q. Engineering microorganisms for the biosynthesis of dicarboxylic acids. *Biotechnol. Adv.* **2021**, *48*, 107710. [CrossRef] [PubMed]
2. Karim, N.; Shishir, M.R.I.; Gowd, V.; Chen, W. Hesperidin-an emerging bioactive compound against metabolic diseases and its potential biosynthesis pathway in microorganism. *Food Rev. Int.* **2021**. [CrossRef]
3. Srinivasan, P.; Smolke, C.D. Biosynthesis of medicinal tropane alkaloids in yeast. *Nature* **2020**, *585*, 614–619. [CrossRef]
4. Raschmanová, H.; Weninger, A.; Glieder, A.; Kovar, K.; Vogl, T. Implementing CRISPR-Cas technologies in conventional and non-conventional yeasts: Current state and future prospects. *Biotechnol. Adv.* **2018**, *36*, 641–665. [CrossRef] [PubMed]
5. Parapouli, M.; Vasileiadis, A.; Afendra, A.-S.; Hatziloukas, E. *Saccharomyces cerevisiae* and its industrial applications. *AIMS Microbiol.* **2020**, *6*, 1. [CrossRef]
6. Rainha, J.; Gomes, D.; Rodrigues, L.R.; Rodrigues, J.L. Synthetic biology approaches to engineer *Saccharomyces cerevisiae* towards the industrial production of valuable polyphenolic compounds. *Life* **2020**, *10*, 56. [CrossRef]
7. Rainha, J.; Rodrigues, J.L.; Rodrigues, L.R. CRISPR-Cas9: A powerful tool to efficiently engineer *Saccharomyces cerevisiae*. *Life* **2020**, *11*, 13. [CrossRef]
8. Kręgiel, D.; Pawlikowska, E.; Antolak, H. Nonconventional yeasts in fermentation processes: Potentialities and limitations. In *Old Yeasts: New Questions*; IntechOpen: London, UK, 2017; p. 87.
9. Bertels, L.-K.; Fernández Murillo, L.; Heinisch, J.J. The pentose phosphate pathway in yeasts-More than a poor cousin of glycolysis. *Biomolecules* **2021**, *11*, 725. [CrossRef]

10. Patra, P.; Das, M.; Kundu, P.; Ghosh, A. Recent advances in systems and synthetic biology approaches for developing novel cell-factories in non-conventional yeasts. *Biotechnol. Adv.* **2021**, *47*, 107695. [CrossRef]
11. Jensen, M.K.; Keasling, J.D. Recent applications of synthetic biology tools for yeast metabolic engineering. *FEMS Yeast Res.* **2015**, *15*, 1–10. [CrossRef]
12. Barrangou, R.; Fremaux, C.; Deveau, H.; Richards, M.; Boyaval, P.; Moineau, S.; Romero, D.A.; Horvath, P. CRISPR provides acquired resistance against viruses in prokaryotes. *Science* **2007**, *315*, 1709–1712. [CrossRef] [PubMed]
13. Walsh, R.M.; Hochedlinger, K. A variant CRISPR-Cas9 system adds versatility to genome engineering. *Proc. Natl. Acad. Sci. USA* **2013**, *110*, 15514–15515. [CrossRef] [PubMed]
14. Levi, O.; Arava, Y. Expanding the CRISPR/Cas9 toolbox for gene engineering in *S. cerevisiae*. *Curr. Microbiol.* **2020**, *77*, 468–478. [CrossRef]
15. Mitsui, R.; Yamada, R.; Ogino, H. CRISPR system in the yeast *Saccharomyces cerevisiae* and its application in the bioproduction of useful chemicals. *World J. Microbiol. Biotechnol.* **2019**, *35*, 111. [CrossRef] [PubMed]
16. Geijer, C.; Ledesma-Amaro, R.; Tomás-Pejó, E. Unraveling the potential of non-conventional yeasts in biotechnology. *FEMS Yeast Res.* **2022**, *22*, foab071. [CrossRef] [PubMed]
17. Cai, P.; Gao, J.; Zhou, Y. CRISPR-mediated genome editing in non-conventional yeasts for biotechnological applications. *Microb. Cell Fact.* **2019**, *18*, 63. [CrossRef] [PubMed]
18. Rebello, S.; Abraham, A.; Madhavan, A.; Sindhu, R.; Binod, P.; Karthika Bahuleyan, A.; Aneesh, E.M.; Pandey, A. Non-conventional yeast cell factories for sustainable bioprocesses. *FEMS Microbiol. Lett.* **2018**, *365*, fny222. [CrossRef] [PubMed]
19. Mukherjee, V.; Radecka, D.; Aerts, G.; Verstrepen, K.J.; Lievens, B.; Thevelein, J.M. Phenotypic landscape of non-conventional yeast species for different stress tolerance traits desirable in bioethanol fermentation. *Biotechnol. Biofuels* **2017**, *10*, 216. [CrossRef]
20. Chen, M.-T.; Lin, S.; Shandil, I.; Andrews, D.; Stadheim, T.A.; Choi, B.-K. Generation of diploid *Pichia pastoris* strains by mating and their application for recombinant protein production. *Microb. Cell Fact.* **2012**, *11*, 91. [CrossRef]
21. Yarimizu, T.; Nonklang, S.; Nakamura, J.; Tokuda, S.; Nakagawa, T.; Lorreungsil, S.; Sutthikhumpha, S.; Pukahuta, C.; Kitagawa, T.; Nakamura, M. Identification of auxotrophic mutants of the yeast *Kluyveromyces marxianus* by non-homologous end joining-mediated integrative transformation with genes from *Saccharomyces cerevisiae*. *Yeast* **2013**, *30*, 485–500. [CrossRef]
22. Kao, Y.-T.; Gonzalez, K.L.; Bartel, B. Peroxisome function, biogenesis, and dynamics in plants. *Plant Physiol.* **2018**, *176*, 162–177. [CrossRef] [PubMed]
23. Ahmad, M.; Hirz, M.; Pichler, H.; Schwab, H. Protein expression in Pichia pastoris: Recent achievements and perspectives for heterologous protein production. *Appl. Microbiol. Biotechnol.* **2014**, *98*, 5301–5317. [CrossRef]
24. Kurtzman, C.P. Biotechnological strains of *Komagataella* (*Pichia*) *pastoris* are *Komagataella phaffii* as determined from multigene sequence analysis. *J. Ind. Microbiol. Biotechnol.* **2009**, *36*, 1435–1438. [CrossRef] [PubMed]
25. Baghban, R.; Farajnia, S.; Ghasemi, Y.; Mortazavi, M.; Zarghami, N.; Samadi, N. New developments in *Pichia pastoris* expression system, review and update. *Curr. Pharm. Biotechnol.* **2018**, *19*, 451–467. [CrossRef] [PubMed]
26. Murata, Y.; Nwuche, C.O.; Nweze, J.E.; Ndubuisi, I.A.; Ogbonna, J.C. Potentials of multi-stress tolerant yeasts, *Saccharomyces cerevisiae* and *Pichia kudriavzevii* for fuel ethanol production from industrial cassava wastes. *Process Biochem.* **2021**, *111*, 305–314. [CrossRef]
27. Greppi, A.; Saubade, F.; Botta, C.; Humblot, C.; Guyot, J.P.; Cocolin, L. Potential probiotic *Pichia kudriavzevii* strains and their ability to enhance folate content of traditional cereal-based African fermented food. *Food Microbiol.* **2017**, *62*, 169–177. [CrossRef]
28. Choi, D.H.; Park, E.H.; Kim, M.D. Isolation of thermotolerant yeast *Pichia kudriavzevii* from *nuruk*. *Food Sci. Biotechnol.* **2017**, *26*, 1357–1362. [CrossRef]
29. Johansen, P.G.; Owusu-Kwarteng, J.; Parkouda, C.; Padonou, S.W.; Jespersen, L. Occurrence and importance of yeasts in indigenous fermented food and beverages produced in sub-Saharan Africa. *Front. Microbiol.* **2019**, *10*, 1789. [CrossRef]
30. Sun, W.; Vila-Santa, A.; Liu, N.; Prozorov, T.; Xie, D.; Faria, N.T.; Ferreira, F.C.; Mira, N.P.; Shao, Z. Metabolic engineering of an acid-tolerant yeast strain *Pichia kudriavzevii* for itaconic acid production. *Metab. Eng. Commun.* **2020**, *10*, e00124. [CrossRef]
31. Li, C.; Liu, Q.; Wang, Y.; Yang, X.; Chen, S.; Zhao, Y.; Wu, Y.; Li, L. Salt stress improves thermotolerance and high-temperature bioethanol production of multi-stress-tolerant *Pichia kudriavzevii* by stimulating intracellular metabolism and inhibiting oxidative damage. *Biotechnol. Biofuels* **2021**, *14*, 222. [CrossRef]
32. Isono, N.; Hayakawa, H.; Usami, A.; Mishima, T.; Hisamatsu, M. A comparative study of ethanol production by *Issatchenkia orientalis* strains under stress conditions. *J. Biosci. Bioeng.* **2012**, *113*, 76–78. [CrossRef] [PubMed]
33. Chan, G.F.; Gan, H.M.; Ling, H.L.; Rashid, N.A.A. Genome sequence of Pichia kudriavzevii M12, a potential producer of bioethanol and phytase. *Eukaryot. Cell* **2012**, *11*, 1300–1301. [CrossRef] [PubMed]
34. Xi, Y.; Zhan, T.; Xu, H.; Chen, J.; Bi, C.; Fan, F.; Zhang, X. Characterization of JEN family carboxylate transporters from the acid-tolerant yeast *Pichia kudriavzevii* and their applications in succinic acid production. *Microb. Biotechnol.* **2021**, *14*, 1130–1147. [CrossRef] [PubMed]
35. Blazeck, J.; Hill, A.; Liu, L.; Knight, R.; Miller, J.; Pan, A.; Otoupal, P.; Alper, H.S. Harnessing *Yarrowia lipolytica* lipogenesis to create a platform for lipid and biofuel production. *Nat. Commun.* **2014**, *5*, 3131. [CrossRef]
36. Liu, H.H.; Ji, X.J.; Huang, H. Biotechnological applications of *Yarrowia lipolytica*: Past, present and future. *Biotechnol. Adv.* **2015**, *33*, 1522–1546. [CrossRef]

37. Cavallo, E.; Charreau, H.; Cerrutti, P.; Foresti, M.L. *Yarrowia lipolytica*: A model yeast for citric acid production. *FEMS Yeast Res.* **2017**, *17*, fox084. [CrossRef]
38. Wang, N.; Chi, P.; Zou, Y.; Xu, Y.; Xu, S.; Bilal, M.; Fickers, P.; Cheng, H. Metabolic engineering of *Yarrowia lipolytica* for thermoresistance and enhanced erythritol productivity. *Biotechnol. Biofuels* **2020**, *13*, 176. [CrossRef]
39. Muhammad, A.; Feng, X.; Rasool, A.; Sun, W.; Li, C. Production of plant natural products through engineered *Yarrowia lipolytica*. *Biotechnol. Adv.* **2020**, *43*, 107555. [CrossRef]
40. Kim, O.C.; Suwannarangsee, S.; Oh, D.B.; Kim, S.; Seo, J.W.; Kim, C.H.; Kang, H.A.; Kim, J.Y.; Kwon, O. Transcriptome analysis of xylose metabolism in the thermotolerant methylotrophic yeast *Hansenula polymorpha*. *Bioprocess Biosyst. Eng.* **2013**, *36*, 1509–1518. [CrossRef]
41. Choudhary, J.; Singh, S.; Nain, L. Thermotolerant fermenting yeasts for simultaneous saccharification fermentation of lignocellulosic biomass. *Electron. J. Biotechnol.* **2016**, *21*, 82–92. [CrossRef]
42. Fonseca, G.G.; Gombert, A.K.; Heinzle, E.; Wittmann, C. Physiology of the yeast *Kluyveromyces marxianus* during batch and chemostat cultures with glucose as the sole carbon source. *FEMS Yeast Res.* **2007**, *7*, 422–435. [CrossRef] [PubMed]
43. Banat, I.M.; Nigam, P.; Marchant, R. Isolation of thermotolerant, fermentative yeasts growing at 52 °C and producing ethanol at 45 °C and 50 °C. *World J. Microbiol. Biotechnol.* **1992**, *8*, 259–263. [CrossRef] [PubMed]
44. Álvarez-Cao, M.E.; Rico-Díaz, A.; Cerdán, M.E.; Becerra, M.; González-Siso, M.I. Valuation of agro-industrial wastes as substrates for heterologous production of α-galactosidase. *Microb. Cell Fact.* **2018**, *17*, 137. [CrossRef] [PubMed]
45. Nurcholis, M.; Lertwattanasakul, N.; Rodrussamee, N.; Kosaka, T.; Murata, M.; Yamada, M. Integration of comprehensive data and biotechnological tools for industrial applications of *Kluyveromyces marxianus*. *Appl. Microbiol. Biotechnol.* **2020**, *104*, 475–488. [CrossRef]
46. Bilal, M.; Ji, L.; Xu, Y.; Xu, S.; Lin, Y.; Iqbal, H.M.N.; Cheng, H. Bioprospecting *Kluyveromyces marxianus* as a robust host for industrial biotechnology. *Front. Bioeng. Biotechnol.* **2022**, *10*, 851768. [CrossRef] [PubMed]
47. Ortiz-Merino, R.A.; Varela, J.A.; Coughlan, A.Y.; Hoshida, H.; Da Silveira, W.B.; Wilde, C.; Kuijpers, N.G.A.; Geertman, J.M.; Wolfe, K.H.; Morrissey, J.P. Ploidy variation in *Kluyveromyces marxianus* separates dairy and non-dairy isolates. *Front. Genet.* **2018**, *9*, 94. [CrossRef] [PubMed]
48. Rajkumar, A.S.; Morrissey, J.P. Rational engineering of *Kluyveromyces marxianus* to create a chassis for the production of aromatic products. *Microb. Cell Fact.* **2020**, *19*, 207. [CrossRef]
49. Cernak, P.; Estrela, R.; Poddar, S.; Skerker, J.M.; Cheng, Y.F.; Carlson, A.K.; Chen, B.; Glynn, V.M.; Furlan, M.; Ryan, O.W.; et al. Engineering *Kluyveromyces marxianus* as a robust synthetic biology platform host. *mBio* **2018**, *9*, e01410–e01418. [CrossRef]
50. Mao, W.; Han, Y.; Wang, X.; Zhao, X.; Chi, Z.; Chi, Z.; Liu, G. A new engineered endo-inulinase with improved activity and thermostability: Application in the production of prebiotic fructo-oligosaccharides from inulin. *Food Chem.* **2019**, *294*, 293–301. [CrossRef] [PubMed]
51. Schwartz, C.; Wheeldon, I. Genome and metabolic engineering in non-conventional yeasts: Current advances and applications. *Synth. Syst. Biotechnol.* **2017**, *2*, 198–207.
52. Da Silva, N.A.; Srikrishnan, S. Introduction and expression of genes for metabolic engineering applications in *Saccharomyces cerevisiae*. *FEMS Yeast Res.* **2012**, *12*, 197–214. [CrossRef]
53. Lee, M.E.; DeLoache, W.C.; Cervantes, B.; Dueber, J.E. A highly characterized yeast toolkit for modular, multipart assembly. *ACS Synth. Biol.* **2015**, *4*, 975–986. [CrossRef]
54. Liu, L.; Otoupal, P.; Pan, A.; Alper, H.S. Increasing expression level and copy number of a *Yarrowia lipolytica* plasmid through regulated centromere function. *FEMS Yeast Res.* **2014**, *14*, 1124–1127.
55. Camattari, A.; Goh, A.; Yip, L.Y.; Tan, A.H.M.; Ng, S.W.; Tran, A.; Liu, G.; Liachko, I.; Dunham, M.J.; Rancati, G. Characterization of a panARS-based episomal vector in the methylotrophic yeast *Pichia pastoris* for recombinant protein production and synthetic biology applications. *Microb. Cell Fact.* **2016**, *15*, 139. [CrossRef]
56. Nakamura, Y.; Nishi, T.; Noguchi, R.; Ito, Y.; Watanabe, T.; Nishiyama, T.; Aikawa, S.; Hasunuma, T.; Ishii, J.; Okubo, Y. A stable, autonomously replicating plasmid vector containing *Pichia pastoris* centromeric DNA. *Appl. Environ. Microbiol.* **2018**, *84*, e02882. [CrossRef]
57. Wang, L.; Lin, J.; Zhang, T.; Xu, K.; Ren, C.; Zhang, Z. Simultaneous screening and validation of effective zinc finger nucleases in yeast. *PLoS ONE* **2013**, *8*, e64687. [CrossRef]
58. Gan, Y.; Lin, Y.; Guo, Y.; Qi, X.; Wang, Q. Metabolic and genomic characterisation of stress-tolerant industrial *Saccharomyces cerevisiae* strains from TALENs-assisted multiplex editing. *FEMS Yeast Res.* **2018**, *18*, foy045. [CrossRef]
59. Song, A.J.; Palmiter, R.D. Detecting and avoiding problems when using the Cre–lox system. *Trends Genet.* **2018**, *34*, 333–340. [CrossRef]
60. Mali, P.; Esvelt, K.M.; Church, G.M. Cas9 as a versatile tool for engineering biology. *Nat. Methods* **2013**, *10*, 957–963. [CrossRef]
61. Mojica, F.J.M.; Montoliu, L. On the origin of CRISPR-Cas technology: From prokaryotes to mammals. *Trends Microbiol.* **2016**, *24*, 811–820. [CrossRef]
62. Makarova, K.S.; Wolf, Y.I.; Alkhnbashi, O.S.; Costa, F.; Shah, S.A.; Saunders, S.J.; Barrangou, R.; Brouns, S.J.; Charpentier, E.; Haft, D.H.; et al. An updated evolutionary classification of CRISPR–Cas systems. *Nat. Rev. Microbiol.* **2015**, *13*, 722–736. [CrossRef]

63. Liu, Z.; Dong, H.; Cui, Y.; Cong, L.; Zhang, D. Application of different types of CRISPR/Cas-based systems in bacteria. *Microb. Cell Fact.* **2020**, *19*, 172. [CrossRef]
64. Anders, C.; Niewoehner, O.; Duerst, A.; Jinek, M. Structural basis of PAM-dependent target DNA recognition by the Cas9 endonuclease. *Nature* **2014**, *513*, 569–573. [CrossRef]
65. Asmamaw, M.; Zawdie, B. Mechanism and applications of CRISPR/Cas-9-mediated genome editing. *Biologics* **2021**, *15*, 353–361.
66. Palermo, G.; Chen, J.S.; Ricci, C.G.; Rivalta, I.; Jinek, M.; Batista, V.S.; Doudna, J.A.; McCammon, J.A. Key role of the REC lobe during CRISPR–Cas9 activation by 'sensing','regulating', and 'locking' the catalytic HNH domain. *Q. Rev. Biophys.* **2018**, *51*, e91. [CrossRef]
67. Ming, S.H.; Tian-Rui, X.U.; Ce-Shi, C.H. The big bang of genome editing technology: Development and application of the CRISPR/Cas9 system in disease animal models. *Zool. Res.* **2016**, *37*, 191–204.
68. Daley, J.M.; Palmbos, P.L.; Wu, D.; Wilson, T.E. Nonhomologous end joining in yeast. *Annu. Rev. Genet.* **2005**, *39*, 431–451. [CrossRef]
69. Brandsma, I.; Gent, D.C. Pathway choice in DNA double strand break repair: Observations of a balancing act. *Genome Integr.* **2012**, *3*, 9. [CrossRef]
70. Meng, J.; Qiu, Y.; Shi, S. CRISPR/Cas9 systems for the development of *Saccharomyces cerevisiae* cell factories. *Front. Bioeng. Biotechnol.* **2020**, *8*, 594347. [CrossRef]
71. Schwartz, C.; Shabbir-Hussain, M.; Frogue, K.; Blenner, M.; Wheeldon, I. Standardized markerless gene integration for pathway engineering in *Yarrowia lipolytica*. *ACS Synth. Biol.* **2017**, *6*, 402–409. [CrossRef]
72. de Souza, C.P.; Ribeiro, B.D.; Zarur Coelho, M.A.; Almeida, R.V.; Nicaud, J.M. Construction of wild-type *Yarrowia lipolytica* IMUFRJ 50682 auxotrophic mutants using dual CRISPR/Cas9 strategy for novel biotechnological approaches. *Enzyme Microb. Technol.* **2020**, *140*, 109621. [CrossRef]
73. Zhang, X.K.; Wang, D.N.; Chen, J.; Liu, Z.J.; Wei, L.J.; Hua, Q. Metabolic engineering of β-carotene biosynthesis in *Yarrowia lipolytica*. *Biotechnol. Lett.* **2020**, *42*, 945–956. [CrossRef]
74. Gao, S.; Tong, Y.; Zhu, L.; Ge, M.; Jiang, Y.; Chen, D.; Yang, S. Production of β-carotene by expressing a heterologous multifunctional carotene synthase in *Yarrowia lipolytica*. *Biotechnol. Lett.* **2017**, *39*, 921–927. [CrossRef]
75. Wang, L.; Deng, A.; Zhang, Y.; Liu, S.; Liang, Y.; Bai, H.; Cui, D.; Qiu, Q.; Shang, X.; Yang, Z.; et al. Efficient CRISPR–Cas9 mediated multiplex genome editing in yeasts. *Biotechnol. Biofuels* **2018**, *11*, 277. [CrossRef]
76. Juergens, H.; Varela, J.A.; Gorter de Vries, A.R.; Perli, T.; Gast, V.J.; Gyurchev, N.Y.; Rajkumar, A.S.; Mans, R.; Pronk, J.T.; Morrissey, J.P. Genome editing in *Kluyveromyces* and *Ogataea* yeasts using a broad-host-range Cas9/gRNA co-expression plasmid. *FEMS Yeast Res.* **2018**, *18*, foy012. [CrossRef]
77. Siripong, W.; Angela, C.; Tanapongpipat, S.; Runguphan, W. Metabolic engineering of *Pichia pastoris* for production of isopentanol (3-methyl-1-butanol). *Enzyme Microb. Technol.* **2020**, *138*, 109557. [CrossRef]
78. Liu, Q.; Shi, X.; Song, L.; Liu, H.; Zhou, X.; Wang, Q.; Zhang, Y.; Cai, M. CRISPR–Cas9-mediated genomic multiloci integration in *Pichia pastoris*. *Microb. Cell Fact.* **2019**, *18*, 144. [CrossRef]
79. Gassler, T.; Heistinger, L.; Mattanovich, D.; Gasser, B.; Prielhofer, R. CRISPR/Cas9-mediated homology-directed genome editing in Pichia pastoris. In *Recombinant Protein Production in Yeast*; Springer: Berlin, Germany, 2019; pp. 211–225.
80. Löbs, A.-K.; Engel, R.; Schwartz, C.; Flores, A.; Wheeldon, I. CRISPR–Cas9-enabled genetic disruptions for understanding ethanol and ethyl acetate biosynthesis in *Kluyveromyces marxianus*. *Biotechnol. Biofuels* **2017**, *10*, 164. [CrossRef]
81. Tran, V.G.; Cao, M.; Fatma, Z.; Song, X.; Zhao, H. Development of a CRISPR/Cas9-based tool for gene deletion in *Issatchenkia orientalis*. *mSphere* **2019**, *4*, e00345-19. [CrossRef]
82. Lee, Y.G.; Kim, C.; Kuanyshev, N.; Kang, N.K.; Fatma, Z.; Wu, Z.Y.; Cheng, M.H.; Singh, V.; Yoshikuni, Y.; Zhao, H.; et al. Cas9-based metabolic engineering of *Issatchenkia orientalis* for enhanced utilization of cellulosic hydrolysates. *J. Agric. Food Chem.* **2022**, *70*, 12085–12094. [CrossRef]
83. Emerson, C.H.; Bertuch, A.A. Consider the workhorse: Nonhomologous end-joining in budding yeast. *Biochem. Cell Biol.* **2016**, *94*, 396–406. [CrossRef]
84. Yang, H.; Matsumoto, Y.; Trujillo, K.M.; Lees-Miller, S.P.; Osley, M.A.; Tomkinson, A.E. Role of the yeast DNA repair protein Nej1 in end processing during the repair of DNA double strand breaks by non-homologous end joining. *DNA Repair* **2015**, *31*, 1–10. [CrossRef]
85. Zhang, Y.; Hefferin, M.L.; Chen, L.; Shim, E.Y.; Tseng, H.M.; Kwon, Y.; Sung, P.; Lee, S.E.; Tomkinson, A.E. Role of Dnl4–Lif1 in nonhomologous end-joining repair complex assembly and suppression of homologous recombination. *Nat. Struct. Mol. Biol.* **2007**, *14*, 639–646. [CrossRef]
86. Utomo, J.C.; Hodgins, C.L.; Ro, D.K. Multiplex Genome Editing in Yeast by CRISPR/Cas9–A Potent and Agile Tool to Reconstruct Complex Metabolic Pathways. *Front. Plant Sci.* **2021**, *12*, 719148. [CrossRef]
87. Yan, Y.; Finnigan, G.C. Development of a multi-locus CRISPR gene drive system in budding yeast. *Sci. Rep.* **2018**, *8*, 17277. [CrossRef]
88. Rajkumar, A.S.; Varela, J.A.; Juergens, H.; Daran, J.M.G.; Morrissey, J.P. Biological parts for *Kluyveromyces marxianus* synthetic biology. *Front. Bioeng. Biotechnol.* **2019**, *7*, 97. [CrossRef] [PubMed]
89. Näätsaari, L.; Mistlberger, B.; Ruth, C.; Hajek, T.; Hartner, F.S.; Glieder, A. Deletion of the *Pichia pastoris* KU70 homologue facilitates platform strain generation for gene expression and synthetic biology. *PLoS ONE* **2012**, *7*, e39720. [CrossRef]

90. Zhang, K.; Duan, X.; Cai, P.; Gao, L.; Wu, X.; Yao, L.; Zhou, Y.J. Fusing an exonuclease with Cas9 enhances homologous recombination in *Pichia pastoris*. *Microb. Cell Fact.* **2022**, *21*, 182. [CrossRef]
91. Paffett, K.S.; Clikeman, J.A.; Palmer, S.; Nickoloff, J.A. Overexpression of Rad51 inhibits double-strand break-induced homologous recombination but does not affect gene conversion tract lengths. *DNA Repair* **2005**, *4*, 687–698. [CrossRef]
92. Cai, P.; Duan, X.; Wu, X.; Gao, L.; Ye, M.; Zhou, Y.J. Recombination machinery engineering facilitates metabolic engineering of the industrial yeast *Pichia pastoris*. *Nucleic Acids Res.* **2021**, *49*, 7791–7805. [CrossRef]
93. Ji, Q.; Mai, J.; Ding, Y.; Wei, Y.; Ledesma-Amaro, R.; Ji, X.-J. Improving the homologous recombination efficiency of *Yarrowia lipolytica* by grafting heterologous component from *Saccharomyces cerevisiae*. *Metab. Eng. Commun.* **2020**, *11*, e00152. [CrossRef]
94. Weninger, A.; Hatzl, A.M.; Schmid, C.; Vogl, T.; Glieder, A. Combinatorial optimization of CRISPR/Cas9 expression enables precision genome engineering in the methylotrophic yeast *Pichia pastoris*. *J. Biotechnol.* **2016**, *235*, 139–149. [CrossRef]
95. Gao, Y.; Zhao, Y. Self-processing of ribozyme-flanked RNAs into guide RNAs in vitro and in vivo for CRISPR-mediated genome editing. *J. Integr. Plant Biol.* **2014**, *56*, 343–349. [CrossRef] [PubMed]
96. Ryan, O.W.; Skerker, J.M.; Maurer, M.J.; Li, X.; Tsai, J.C.; Poddar, S.; Lee, M.E.; DeLoache, W.; Dueber, J.E.; Arkin, A.P.; et al. Selection of chromosomal DNA libraries using a multiplex CRISPR system. *eLife* **2014**, *3*, e03703. [CrossRef] [PubMed]
97. Naito, Y.; Hino, K.; Bono, H.; Ui-Tei, K. CRISPRdirect: Software for designing CRISPR/Cas guide RNA with reduced off-target sites. *Bioinformatics* **2015**, *31*, 1120–1123. [CrossRef] [PubMed]
98. Kleinstiver, B.P.; Prew, M.S.; Tsai, S.Q.; Nguyen, N.T.; Topkar, V.V.; Zheng, Z.; Joung, J.K. Broadening the targeting range of *Staphylococcus aureus* CRISPR-Cas9 by modifying PAM recognition. *Nat. Biotechnol.* **2015**, *33*, 1293–1298. [CrossRef]
99. Rees, H.A.; Yeh, W.-H.; Liu, D.R. Development of hRad51–Cas9 nickase fusions that mediate HDR without double-stranded breaks. *Nat. Commun.* **2019**, *10*, 2212. [CrossRef]
100. Schusterbauer, V.; Fischer, J.E.; Gangl, S.; Schenzle, L.; Rinnofner, C.; Geier, M.; Sailer, C.; Glieder, A.; Thallinger, G.G. Whole Genome Sequencing Analysis of Effects of CRISPR/Cas9 in *Komagataella phaffii*: A Budding Yeast in Distress. *J. Fungi* **2022**, *8*, 992. [CrossRef]
101. Butelli, E.; Titta, L.; Giorgio, M.; Mock, H.P.; Matros, A.; Peterek, S.; Schijlen, E.G.; Hall, R.D.; Bovy, A.G.; Luo, J.; et al. Enrichment of tomato fruit with health-promoting anthocyanins by expression of select transcription factors. *Nat. Biotechnol.* **2008**, *26*, 1301–1308. [CrossRef]
102. Kawser Hossain, M.; Abdal Dayem, A.; Han, J.; Yin, Y.; Kim, K.; Kumar Saha, S.; Yang, G.M.; Choi, H.Y.; Cho, S.G. Molecular mechanisms of the anti-obesity and anti-diabetic properties of flavonoids. *Int. J. Mol. Sci.* **2016**, *17*, 569. [CrossRef]
103. Sáez-Sáez, J.; Wang, G.; Marella, E.R.; Sudarsan, S.; Cernuda Pastor, M.; Borodina, I. Engineering the oleaginous yeast *Yarrowia lipolytica* for high-level resveratrol production. *Metab. Eng.* **2020**, *62*, 51–61. [CrossRef] [PubMed]
104. Pyne, M.E.; Narcross, L.; Martin, V.J.J. Engineering plant secondary metabolism in microbial systems. *Plant Physiol.* **2019**, *179*, 844–861. [CrossRef] [PubMed]
105. Zha, J.; Wu, X.; Gong, G.; Koffas, M.A.G. Pathway enzyme engineering for flavonoid production in recombinant microbes. *Metab. Eng. Commun.* **2019**, *9*, e00104. [CrossRef] [PubMed]
106. Sheng, H.; Sun, X.; Yan, Y.; Yuan, Q.; Wang, J.; Shen, X. Metabolic engineering of microorganisms for the production of flavonoids. *Front. Bioeng. Biotechnol.* **2020**, *8*, 589069. [CrossRef]
107. Soreanu, I.; Hendler, A.; Dahan, D.; Dovrat, D.; Aharoni, A. Marker-free genetic manipulations in yeast using CRISPR/CAS9 system. *Curr. Genet.* **2018**, *64*, 1129–1139. [CrossRef]
108. Lv, Y.; Marsafari, M.; Koffas, M.; Zhou, J.; Xu, P. Optimizing oleaginous yeast cell factories for flavonoids and hydroxylated flavonoids biosynthesis. *ACS Synth. Biol.* **2019**, *8*, 2514–2523. [CrossRef] [PubMed]
109. Wei, W.; Zhang, P.; Shang, Y.; Zhou, Y.; Ye, B.C. Metabolically engineering of *Yarrowia lipolytica* for the biosynthesis of naringenin from a mixture of glucose and xylose. *Bioresour. Technol.* **2020**, *314*, 123726. [CrossRef]
110. Toti, E.; Chen, C.Y.O.; Palmery, M.; Villaño Valencia, D.; Peluso, I. Non-provitamin A and provitamin A carotenoids as immunomodulators: Recommended dietary allowance, therapeutic index, or personalized nutrition? *Oxid. Med. Cell. Longev.* **2018**, *2018*. [CrossRef] [PubMed]
111. Arnesen, J.A.; Kildegaard, K.R.; Cernuda Pastor, M.; Jayachandran, S.; Kristensen, M.; Borodina, I. *Yarrowia lipolytica* strains engineered for the production of terpenoids. *Front. Bioeng. Biotechnol.* **2020**, *8*, 945. [CrossRef]
112. Lu, Y.; Yang, Q.; Lin, Z.; Yang, X. A modular pathway engineering strategy for the high-level production of β-ionone in *Yarrowia lipolytica*. *Microb. Cell Fact.* **2020**, *19*, 49. [CrossRef]
113. Yao, F.; Liu, S.C.; Wang, D.N.; Liu, Z.J.; Hua, Q.; Wei, L.J. Engineering oleaginous yeast *Yarrowia lipolytica* for enhanced limonene production from xylose and lignocellulosic hydrolysate. *FEMS Yeast Res.* **2020**, *20*, foaa046. [CrossRef]
114. Zhang, J.L.; Bai, Q.Y.; Peng, Y.Z., Fan, J., Jin, C.C.; Cao, Y.X.; Yuan, Y.J. High production of triterpenoids in *Yarrowia lipolytica* through manipulation of lipid components. *Biotechnol. Biofuels* **2020**, *13*, 133. [CrossRef] [PubMed]
115. Zhao, M.; Lu, X.; Zong, H.; Li, J.; Zhuge, B. Itaconic acid production in microorganisms. *Biotechnol. Lett.* **2018**, *40*, 455–464. [CrossRef] [PubMed]
116. Zhao, C.; Cui, Z.; Zhao, X.; Zhang, J.; Zhang, L.; Tian, Y.; Qi, Q.; Liu, J. Enhanced itaconic acid production in *Yarrowia lipolytica* via heterologous expression of a mitochondrial transporter MTT. *Appl. Microbiol. Biotechnol.* **2019**, *103*, 2181–2192. [CrossRef]
117. Blazeck, J.; Miller, J.; Pan, A.; Gengler, J.; Holden, C.; Jamoussi, M.; Alper, H.S. Metabolic engineering of *Saccharomyces cerevisiae* for itaconic acid production. *Appl. Microbiol. Biotechnol.* **2014**, *98*, 8155–8164. [CrossRef] [PubMed]

118. Yook, S.D.; Kim, J.; Gong, G.; Ko, J.K.; Um, Y.; Han, S.O.; Lee, S.M. High-yield lipid production from lignocellulosic biomass using engineered xylose-utilizing *Yarrowia lipolytica*. *Glob. Change Biol. Bioenergy* **2020**, *12*, 670–679. [CrossRef]
119. Li, J.X.; Xu, J.; Ruan, J.C.; Meng, H.M.; Su, H.; Han, X.F.; Lu, M.; Li, F.L.; Wang, S.A. Disrupting a phospholipase A2 gene increasing lipid accumulation in the oleaginous yeast *Yarrowia lipolytica*. *J. Appl. Microbiol.* **2021**, *130*, 100–108. [CrossRef]
120. Zhang, J.L.; Cao, Y.X.; Peng, Y.Z.; Jin, C.C.; Bai, Q.Y.; Zhang, R.S.; Liu, D.; Yuan, Y.J. High production of fatty alcohols in *Yarrowia lipolytica* by coordination with glycolysis. *Sci. China Chem.* **2019**, *62*, 1007–1016. [CrossRef]
121. Reyes-Sánchez, F.J.; Páez-Lerma, J.B.; Rojas-Contreras, J.A.; López-Miranda, J.; Soto-Cruz, N.Ó.; Reinhart-Kirchmayr, M. Study of the enzymatic capacity of *Kluyveromyces marxianus* for the synthesis of esters. *J. Mol. Microbiol. Biotechnol.* **2019**, *29*, 1–9. [CrossRef] [PubMed]
122. Li, M.; Lang, X.; Moran Cabrera, M.; De Keyser, S.; Sun, X.; Da Silva, N.; Wheeldon, I. CRISPR-mediated multigene integration enables Shikimate pathway refactoring for enhanced 2-phenylethanol biosynthesis in *Kluyveromyces marxianus*. *Biotechnol. Biofuels* **2021**, *14*, 3. [CrossRef]
123. Löbs, A.K.; Schwartz, C.; Thorwall, S.; Wheeldon, I. Highly multiplexed CRISPRi repression of respiratory functions enhances mitochondrial localized ethyl acetate biosynthesis in *Kluyveromyces marxianus*. *ACS Synth. Biol.* **2018**, *7*, 2647–2655. [CrossRef] [PubMed]
124. Swietalski, P.; Hetzel, F.; Seitl, I.; Fischer, L. Secretion of a low and high molecular weight β-glycosidase by *Yarrowia lipolytica*. *Microb. Cell Fact.* **2020**, *19*, 100. [CrossRef] [PubMed]
125. Bruder, S.; Moldenhauer, E.J.; Lemke, R.D.; Ledesma-Amaro, R.; Kabisch, J. Drop-in biofuel production using fatty acid photodecarboxylase from Chlorella variabilis in the oleaginous yeast *Yarrowia lipolytica*. *Biotechnol. Biofuels* **2019**, *12*, 202. [CrossRef] [PubMed]

fermentation

MDPI

Article

Effects of Diamond Nanoparticles Immobilisation on the Surface of Yeast Cells: A Phenomenological Study

Yuri Dekhtyar [1], Dagnis Abols [1], Liga Avotina [2], Anita Stoppel [3], Sascha Balakin [3], Galina Khroustalyova [1,4], Joerg Opitz [3], Hermanis Sorokins [1], Natalia Beshchasna [3], Patricija Tamane [1] and Alexander Rapoport [4,*]

1 Institute of Biomedical Engineering and Nanotechnologies, Riga Technical University, Kipsalas Str. 6B, LV-1048 Riga, Latvia
2 Institute of Chemical Physics, University of Latvia, Jelgavas Str. 1, LV-1004 Riga, Latvia
3 Fraunhofer Institute for Ceramic Technologies and Systems IKTS, Maria-Reiche-Straße 2, 01109 Dresden, Germany
4 Laboratory of Cell Biology, Institute of Microbiology and Biotechnology, University of Latvia, Jelgavas Str. 1, LV-1004 Riga, Latvia
* Correspondence: rapoport@mail.eunet.lv

Abstract: An interesting development of biotechnology has linked microbial cell immobilisation with nanoparticles. The main task of our research was to reveal the possible influences of differently electrically charged diamond nanoparticles upon physiological characteristics of the yeast *Saccharomyces cerevisiae*. It was revealed that the adverse impact of these nanoparticles can manifest not only against prokaryotes, but also against eukaryotic yeast cells. However, the obtained results also indicate that it is possible to reduce and, most likely, completely eliminate the dangerous effects of nanoparticles to cells by using special physical approaches. Comparison of non-arylated and arylated nanoparticles showed that in terms of changes in the physiological activity of cells, which are important to biotechnology and biomedicine, the selection of certain nanoparticles (non-arylated or arylated) may be necessary in each specific case, depending on the purpose of their use.

Keywords: yeast; immobilisation; nanoparticles; electric charge; viability

Citation: Dekhtyar, Y.; Abols, D.; Avotina, L.; Stoppel, A.; Balakin, S.; Khroustalyova, G.; Opitz, J.; Sorokins, H.; Beshchasna, N.; Tamane, P.; et al. Effects of Diamond Nanoparticles Immobilisation on the Surface of Yeast Cells: A Phenomenological Study. *Fermentation* **2023**, *9*, 162. https://doi.org/10.3390/fermentation9020162

Academic Editor: Ronnie Willaert

Received: 24 December 2022
Revised: 1 February 2023
Accepted: 7 February 2023
Published: 10 February 2023

Copyright: © 2023 by the authors. Licensee MDPI, Basel, Switzerland. This article is an open access article distributed under the terms and conditions of the Creative Commons Attribution (CC BY) license (https://creativecommons.org/licenses/by/4.0/).

1. Introduction

Immobilisation of microorganisms on the surface or inside various carriers is a widely used approach in different biotechnological processes. There are a lot of advantages of immobilised cell application compared to free cells. It was shown that microbial cell immobilisation may lead to enhancement of fermentation productivity, increased stability of cells, as well as decreased overall costs of recovery, regeneration, recycling and downstream processes [1]. Application of immobilised systems allows the possibility of reusing them in the repeated cycles of various biotechnological processes. Besides these advantages, it was shown that, for example, immobilisation of yeast cells may change their physiological characteristics and improve their biotechnological properties, such as ethanol productivity and heavy metal sorption for the purification of the environment, among others [2–5]. Immobilisation of yeast cells improves their resistance to desiccation, which is important for the storage of industrially important strains in plant collections [6].

An interesting development in this industrial microbiology direction is linked with microbial cell immobilisation with nanoparticles. The systems that are obtained as a result of this process are called either immobilised microbial nanoparticles, microorganisms functionalised by nanoparticles, microorganisms immobilised on nanoparticles or microorganisms decorated with nanoparticles [5,7–11]. We suppose that they may also be referred to as nanoparticles immobilised on the surface of microorganisms or nanoparticles immobilised on or inside the microorganisms. Most of these studies are linked to the application of magnetic nanoparticles for immobilisation and further use of these systems for the biosorption and removal of pollutants and purification of wastewaters, soils and aquatic environments [5]. At the same time, this approach has also been applied to other biotechnological, medical and pharmacological goals over the last few years [11–15]. Another new and important direction of these studies is linked to nanodiamond (ND) particle application in biotechnological and medical fields. It was found that these nanoparticles could enter cells without causing any damage to the cell walls or the cell. This allows for the possibility of using them as efficient potential drug delivery carriers. It is supposed that they are non-toxic and non-reactive [16]. Recent studies revealed the possibility of using nanodiamond particles in treatments for cancers, including brain tumours and breast cancers, autoimmune diseases and cardiovascular affections [16–20].

At the same time, there are different results of studies of microbial sensitivity to diamond nanoparticles. The authors of one study concluded that these nanoparticles have antimicrobial activity in relation to *Escherichia coli* and *Bacillus subtilis*. They revealed interaction of diamond nanoparticles with bacterial surfaces and supposed their influence upon permeability of the bacterial cell wall and/or membrane [21]. Other researchers, who used yeast in their studies, have not found any negative effects of diamond nanoparticles on these cells [22,23].

The surface charge of a substrate has a strong influence on immobilisation of cells and their physiological activity [3]. However, the influence of the surface charge value of the nanoparticles on their connection with the cell and the physiological activity of the latter are unknown.

Diamond nanoparticles are unique dielectric materials in which electric charge may be easily induced.

The yeast *Saccharomyces cerevisiae* have been chosen for the present research. *S. cerevisiae* cells have a strong cell wall that the nanoparticle likely may not deform and, because of this, they influence cell physiology.

The main task of our research was to reveal possible influences of differently electrically charged diamond nanoparticles on physiological characteristics of *S. cerevisiae*. We took into account, on the one hand, the practical importance of yeast in various areas of the food industry and biotechnology and that, on the other hand, their cells are unique and a perfect model of any eukaryotic cell that allows them to be used in various molecular biology, medical and pharmacological studies.

2. Materials and Methods

2.1. Nanoparticles and Their Functionalisation

Commercial native Raw Detonation Nanodiamonds (nanoparticles manufactured Adamasnano, Raleigh, NC, USA, average particle size 4–5 nm, aggregated in ~170 nm particles, 2% ash content) were used. To functionalise nanoparticles (NP), physical and chemical approaches were employed.

Physically, the electrical functionalisation of NP (deposition of the electric charge) was provided with ultraviolet radiation (UV). The UV lamp Lightningcure LC5 (Hamamatsu, Japan) was used. The intensity of the ultraviolet light lamp was 3500 W/cm^2 with an output wavelength of 365 nm. Following the manufacturer's certificate of the lamp, the radiation stability was within ±5%. The distance between the light source and the substrates surface was 30 cm. The exposure time was 30 and 90 min. It was verified that the substrate was not heated in such conditions.

To identify the deposited surface electric charge, near-threshold photoelectron emission spectroscopy was employed [24]. The electron work function (ϕ) is directly proportional to the electric charge density (I) at the surface emitting electrons. To measure the electron work function, the NP were radiated with UV light and the energy of the photons was varied from 4.2–6.0 eV. Alongside electron emission current was measured. The electron work function was identified at the energy of the photon when I (hυ) = 0 [24]. To achieve this, the I (hv) line was extrapolated to I = 0. Measurements were taken in vacuum conditions ~10^{-5} Torr.

To functionalise ND particles chemically, arylation functionalisation was used. First, 1 M borane–tetrahydrofuran (BH3•THF) (Sigma-Aldrich Chemie GmbH, Taufkirchen, Germany), dry tetrahydrofuran (THF) (Life Technologies GmbH, Darmstadt, Germany) and 4-aminobenzoic acid (Sigma-Aldrich Chemie GmbH, Taufkirchen, Germany) were employed.

Arylation of ND was performed by adding 10 g of aminobenzoic acid and 5 mL of amyl nitrite to 10 mL of a 150 mg/mL aqueous ND suspension under stirring at 80 °C for 15 h. Amyl nitrite was used for the in situ generation of diazonium salts [25]. It was conjugated with the NDs via the click chemistry method [26]. Five washing cycles in ddH20 and 15 in acetone were required to obtain a transparent supernatant after centrifugation and to ensure that excess chemicals were removed. All ND samples were dried overnight using a vacuum desiccator after surface modification.

Fourier transmission infrared spectroscopy (FTIR) (FTIR, VERTEX 70v, Bruker) was used to characterise the ND surface. The infrared absorption spectra were recorded in the range from 450 cm^{-1} to 4000 cm^{-1} with four scan cycles and a resolution of 2 cm^{-1}. Three measurements were performed for each sample and the mean value was determined for further analysis.

Nanoparticle tracking analysis (NTA) (ZetaView PMX120, Particle Metrix GmbH, Inning am Ammersee, Germany) was used to determine the particle size distribution. Then, 1 mg of ND was mixed with ddH_2O and sufficiently diluted. Measurements were recorded at RT and a neutral pH = 7. The wavelength of the laser was 488 nm. Prior to the measurement, nanodiamond suspensions were sonicated using ultrasound bath for 10 min. The final ND suspension was stable for at least 10 h.

Depending on the interplay of electrostatic and van der Waals interactions, ND can aggregate in aqueous media or exist as homogeneous dispersions [27]. The stability of aggregates was tested in water. The mean diameter of arylated nanodiamonds particles was ~134 nm.

2.2. Strains, Nutrient Medium and Culture Conditions

Yeast strain *Saccharomyces cerevisiae 77* from Microbial Strain Collection of Latvia was used in this investigation. The strain was maintained on YPD agar medium (gL^{-1}): yeast extract—10, peptone—20, glucose—20 and agar—20 at 4 °C. The yeast was cultivated at 30 °C in a nutrient-rich, complex YPD liquid medium (gL^{-1}): yeast extract—10, peptone—20 and glucose—20. Culture density was monitored at a wavelength of 595 nm using a Tecan's Sunrise Absorbance Microplate Reader (Switzerland). Mean values of optical density were plotted against incubation times for each sample, as well as the control. At least three independent cultivation experiments were performed during this study.

2.3. Preparation of Yeast Suspension and Immobilisation Procedure

Yeast biomass was collected from 10 mL suspension of *S. cerevisiae 77* after 24 h of growth by centrifugation. The biomass was washed with distilled water. The supernatant was discarded after centrifugation, and the washed biomass was diluted with distilled water up to 80 mL in a 100 mL volumetric cylinder to obtain an optical density of $OD_{595} \approx 0.5$ using a Tecan's Sunrise Absorbance Microplate Reader (Switzerland). If necessary, the suspension was additionally diluted with distilled water until the optimum suspension density was reached. The prepared yeast suspension was added to a certain amount of aqueous suspension of nanodiamond particles in a ratio of 1:1. The samples were placed on an orbital shaker and rotated at 50 rpm for 1 h at 25 °C.

2.4. Immobilisation Evaluation

The nanoparticles' connection with the cells was evaluated using visual observation under the microscope and the segregation pace of the particle–cell aggregates.

The visual observation was performed in phase-contrast mode, which allowed for the possibility of evaluating the nanoparticle immobilisation level rather quickly. Images of the cells with immobilised nanoparticles were obtained with a computer-controlled

Olympus BX51 fluorescence microscope equipped with a 100-W mercury lamp, attached to a Peltier-cooled CCD camera DP71 (Olympus, Tokio, Japan). Microscope magnification 1000 was used to visualise cells in phase-contrast mode. Image acquisition was performed using Cell^F imaging software (Olympus, Tokio, Japan).

The segregation pace of the particle–cell aggregates was indemnified using spectrophotometry (UV-VIS Thermo Spectronic Helios Gamma, Zagreb, Croatia).

The NPs were mixed with cells in the distilled water. The optical absorbance of the solutions at 595 nm was around 1. The absorbance (A) time-dependent relaxation was measured. The time (t) constant (τ) (Equation (1)) of relaxation was identified at the beginning of relaxation when the latter had the highest pace (t < 4 min).

$$\frac{1}{\tau} \sim -\frac{\ln(A)}{t}, \tag{1}$$

Influence of UV exposure to native and chemically processed NP-cell aggregates on time constant was evaluated.

2.5. Cell Viability Determination

The classical method of colony forming units (CFU) and fluorescent microscopy with the application of fluorochrome primuline (mol. wt. 475.5) [28] were used for viability determination. Fluorescent images were taken after 500 ms of exposure with a U-MNIBA2 mirror unit (470–490 nm excitation/510–550 nm emission). About 1000 cells were counted in microscope for the determination of the viability of each sample using the fluorescent method. At least 3 parallel experiments were evaluated for each sample.

2.6. Determinations of Lag Phase

Next, 10 μL of suspension of the immobilised system of nanodiamond particles on and in yeast cells was taken in 10–12 parallel repetitions from each sample. These aliquots were used to inoculate YPG media (190 μL) in a 96-well microplate. Cells were grown aerobically with shaking under optimal temperature conditions (30 °C). The optical density at a wavelength of 595 nm was then monitored in a Tecan's Sunrise Absorbance Microplate Reader (Switzerland) at 30 °C for 24 h. The mean values of optical density were plotted against incubation times for each treated sample, as well as for the control.

2.7. Statistical Methods

In microbiological studies, at least 3 independent experiments (each in triplicate or 10–12 parallel repetitions for growth curves) were performed. Results were expressed as the mean ± standard deviation. The results were analysed through a one-way analysis of variance (ANOVA test). In this way, significant differences among samples and sample groups were calculated. A p-value < 0.05 was considered statistically significant.

3. Results and Discussion

3.1. ND Surface Characterisation and Charge Deposition

Following the near-threshold photoelectron emission measurements (Figure 1), the native and chemically functionalised particles were characterised with the values of ϕ (5.10 ± 0.10 and 4.90 ± 0.07 eV, respectively).

The result demonstrates that chemical functionalisation tended to deposit more positive charge on the ND particles surface. This could be because the native ND particles are coated with a layer of atmospheric water consisting of O-H dipoles; however, the chemically functionalised particles covered with polymeric film have C-H dipoles (Figure 2). Because the electron emits via the electric field created by the dipole, the latter has an influence on ϕ. The electric field of O-H is around four times stronger compared to the field created by C-H, because their dipole moments are equal to 1.500 D and 0.339 D, respectively [29,30].

FTIR analyses demonstrated the presence of C-H (wavenumber = 3300 cm^{-1}) against O-H (wavenumber = 3420 cm^{-1}) couples on the chemically processed NPs (Figure 2).

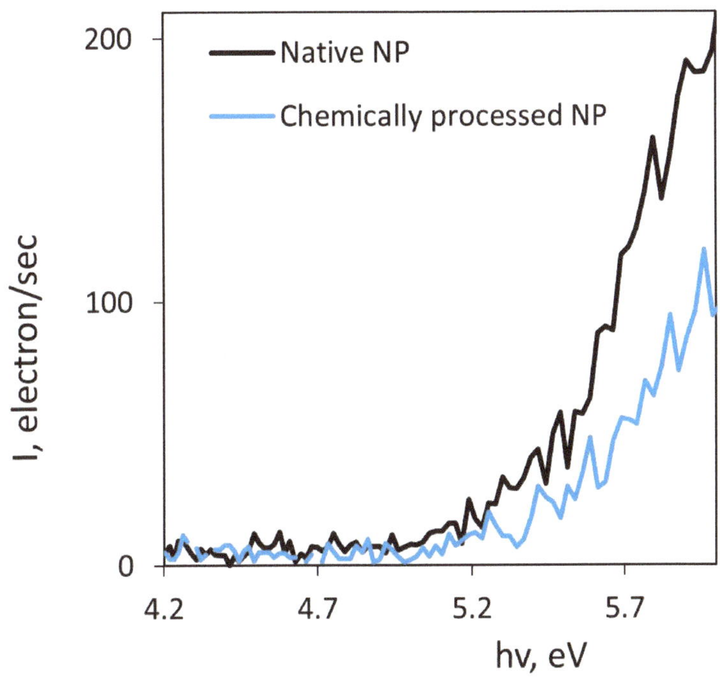

Figure 1. Photoemission spectra (I—electric charge density, electron/sec; hυ—photon energy, eV).

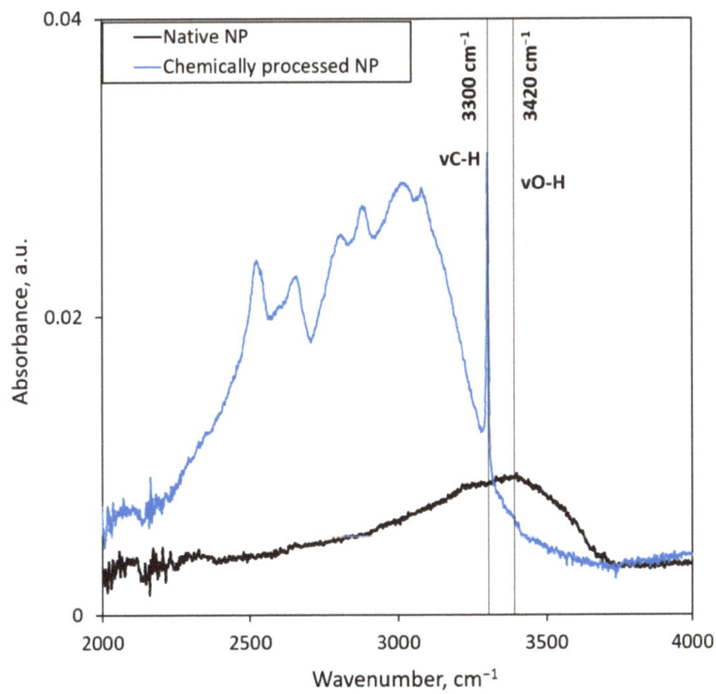

Figure 2. FTIR spectra of the native and chemically processed ND particles.

To explore the possibility of controlling ϕ of the chemically processed ND particles, they were irradiated with UV alongside the native particles. Figure 3 demonstrates the increment (Δϕ) of ϕ under UV.

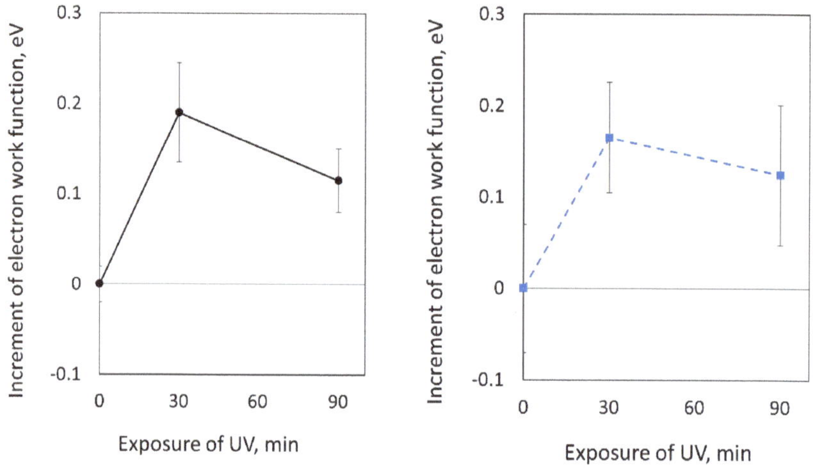

(a) (b)

Figure 3. Influence of UV exposure on increment (Δϕ) of electron work function ϕ: (**a**) native ND particles; (**b**) chemically processed ND particles.

Following the results in Figure 3, the UV irradiation increases the ϕ of both native and chemically processed ND particles. This means the UV deposits the negative charge onto the ND particles.

3.2. *Viability of Yeast Cells with Immobilised Nanodiamond Particles*

The ND particles were immobilised on/in the yeast cells using a traditional method of joint incubation under non-growing conditions. In this case, the water suspension of ND particles was mixed for a certain time (1 h in this study) with the suspension of the yeast taken from the stationary growth phase of the culture in water. Figure 4 shows the results of visual observation of the quality of native ND particles immobilisation on/in yeast cells using light microscopy.

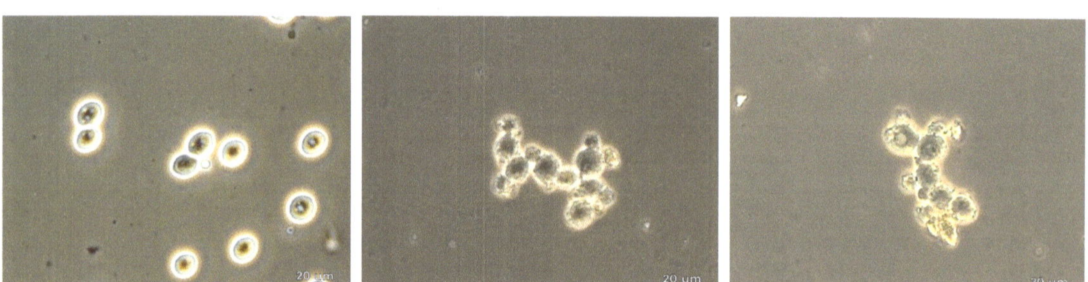

(a) (b) (c)

Figure 4. Phase-contrast microscopy images of UV-non-pretreated and UV- pretreated native ND particles immobilised on/in the *S. cerevisiae 77* cells. The total magnification—1000, scale bar 20 μm: (**a**) free yeast cells; (**b**) yeast cells with ND particles without UV treatment; (**c**) yeast cells with ND particles that were treated with UV during 90 min.

As it seen in Figure 4, yeast cells with immobilised non-treated and UV-treated native ND particles show a tendency to aggregate. It should be noted here that the aqueous suspension of these ND particles is unstable. Irregular conglomerates formed quite quickly, and they remained after sufficiently hard shaking of the suspension at Vortex. Another situation was observed for yeast cells with chemically processed ND (ArND) immobilised on their surface. In this case, the shape and size of nanodiamond agglomerates were more regular, and the formation of cell clusters was not observed (Figure 5).

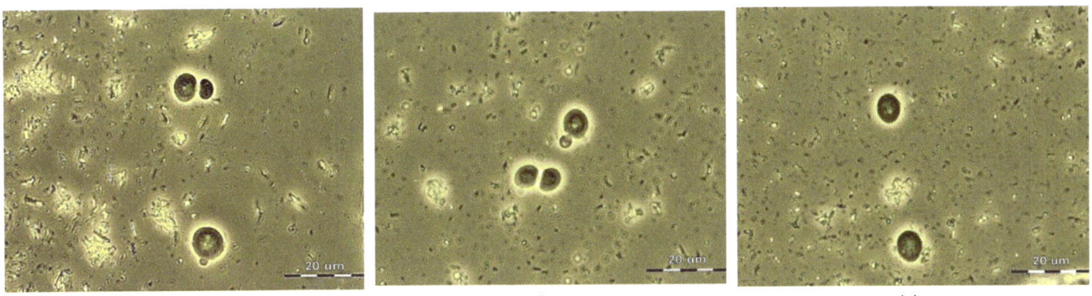

(**a**) (**b**) (**c**)

Figure 5. Phase-contrast microscopy images of UV-pretreated chemically processed ND (ArND immobilised by the *Saccharomyces cerevisiae 77* yeast cells. The total magnification—1000, scale bar 20 µm: (**a**) yeast cells with ArND particles without UV treatment; (**b**) yeast cells with ArND that were treated with UV during 30 min; (**c**) yeast cells with ArND that were treated with UV during 90 min.

Spectrophotometric measurements were applied to evaluate NP-cell aggregates segregation pace. Figure 6 demonstrates that time constant τ decreases with UV exposure in both native and chemically processed NP-cell aggregates.

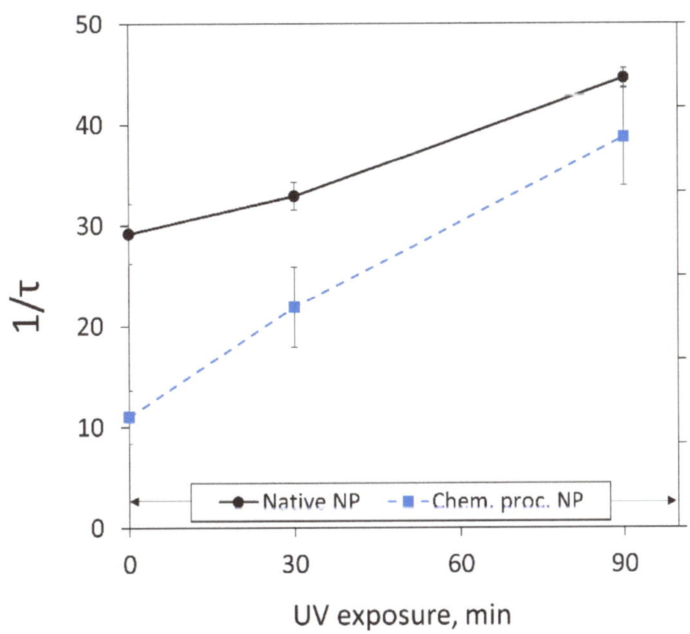

Figure 6. Influence of UV on $1/\tau$.

Figure 6 gives the sense that NP-cell aggregates segregate faster if UV exposure increases. Because of this, one might assume that the size (i.e., weight) of the NP-cell aggregates is controlled by UV: heavier aggregates "fall down" faster.

At the beginning of the study of the physiological state of yeast, after immobilisation on/in diamond nanoparticles, we determined the viability of cells. Two different methodological approaches were used. The first one was based on the use of fluorescent microscopy and fluorochrome primulin (mol. wt. 475.5). This fluorochrome does not penetrate inside viable cells and the only wall of the fluorochrome-stained cell fluoresces. In severely damaged dead cells, this fluorochrome binds to cytoplasm, resulting in a bright yellow-green fluorescence of the whole yeast cell. These experiments showed that immobilisation of ND particles on/in yeast cells led to the loss of viability of some part of the cell population. Viability of the cells with immobilised ND particles that were non-treated with UV was 62 ± 3%. A short pretreatment (for 30 min) of these particles with UV mitigated the negative effect of their immobilisation, and the viability of cells in this case was 75 ± 4%. All these differences are statistically significant ($p < 0.05$). However, as previously noted, these particles in aqueous suspension of yeast cells have irregular shapes and sizes and have a tendency to form conglomerates that make the determination of cell viability via the fluorescence method more difficult. Therefore, these results were also checked using the classical method of colony forming units (CFU) (Figure 7).

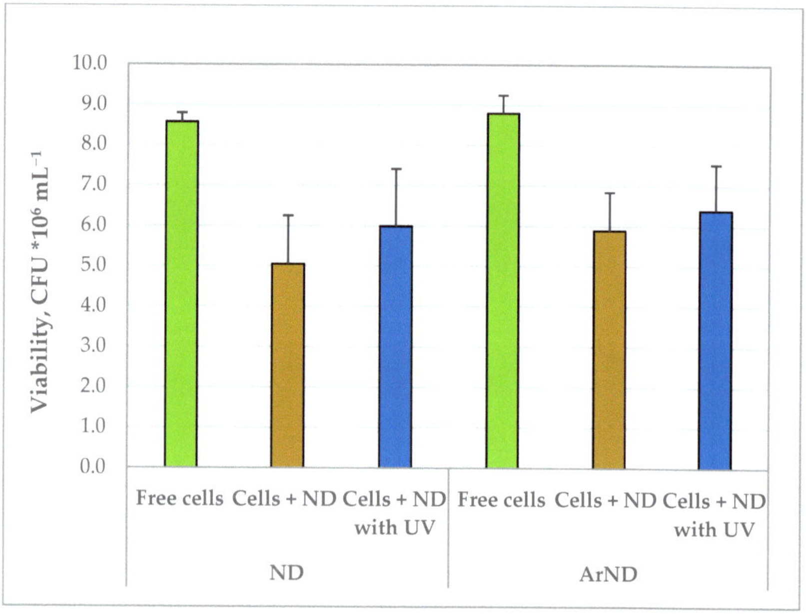

Figure 7. Viability of yeast cells with immobilised on them ND particles without UV treatment or after UV treatment for 30 min.

In this case, similar results were obtained, as shown above. Viability determinations using the fluorescent microscopy method were also performed for arylated ND particles. The obtained results were somewhat better than those obtained for the yeast cells with non-arylated ND particles. The viability of the cells with immobilised ArND particles that were non-treated with UV was 72 ± 2%. In this case, a short pretreatment of ArND particles with UV also mitigated the negative effect of their immobilisation, and the viability of cells was 82 ± 3%. All these differences are statistically significant ($p < 0.05$). We suppose that first of all, these data evidence that the immobilisation of diamond nanoparticles on the surface or in yeast cells is at least not always non-damaging. At the same time, the highest part of the cell's population still maintains viability. Besides that, these results evidence that arylation of ND particles before their immobilisation on/in yeast reduces the decrease in cell viability compared with non-arylated ND particles. It is very interesting and important that pretreatment of these nanoparticles with UV significantly mitigates their negative effect on yeast cells. We suppose that this means that it would most likely be possible to determine the conditions for the additional or complete mitigation of these negative effects of diamond nanoparticles on yeast cells, if necessary.

3.3. Growth of Yeast Cells with Immobilised Nanodiamond Particles

In further experiments, we checked the growth of yeast with immobilised nanoparticles, comparing it with growth of free cells taken as a control (Figures 8 and 9). As can be seen from Figure 8, immobilisation of ND particles without UV treatment or with short UV-treatment (30 min) does not significantly change the duration of the growth lag phase compared with the culture of free cells. In all these cases, its duration is around 5 h. At the same time, longer exposure of ND particles by UV (90 min) before their immobilisation on/in the cells leads to an essential decrease in yeast culture with immobilised ND lag-phase (2.5–3 h). During the lag phase, the cell metabolism aims to increase the amount of proteins, RNA and the synthesis of enzymes for cell multiplication in a specific environment. Therefore, such results may indicate that this UV treatment not only mitigates some negative effects of ND immobilisation, but somehow activates these cells' physiological state. Such a conclusion can also be drawn when comparing the biomass yield after 20 h of cultivation of yeast cell cultures with immobilised particles non-treated with UV and treated with ND. In this case, it can be seen that immobilisation of non-treated with UV NDs decreases the yield of accumulated biomass, comparing it to both the control-free cell cultures and the cell cultures with immobilised ND particles treated with UV. Figure 8 reveals the differences between the effects of immobilisation on/in the cells of native ND particles and chemically processed NP.

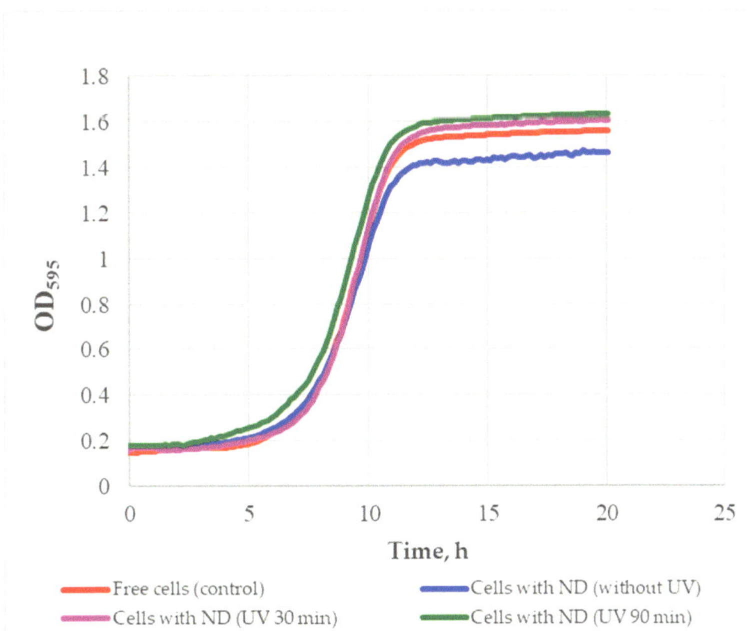

Figure 8. Growth kinetics of *S. cerevisiae 77* with immobilised native ND particles in YPG medium.

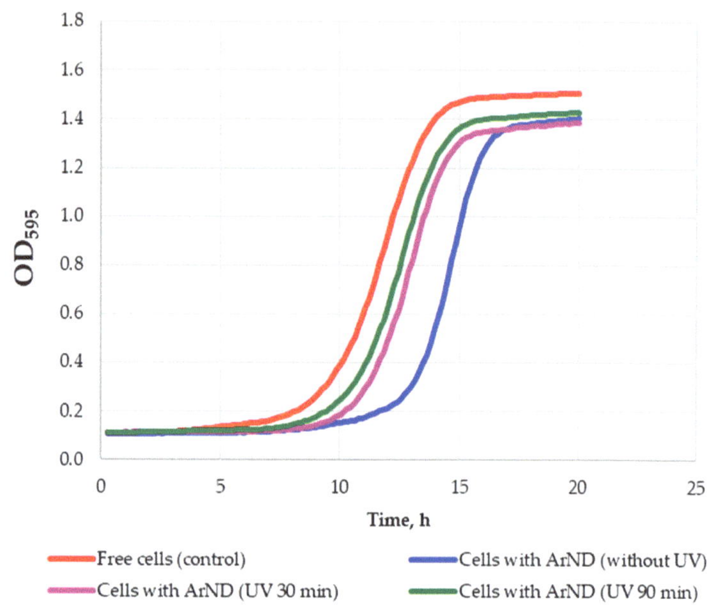

Figure 9. Growth kinetics of *S. cerevisiae 77* with immobilised ArND particles in YPG media.

In the latter case, cells with immobilised ArND particles demonstrated worse results compared to the free cells. The duration of the lag phase of any culture with immobilised ArND particles was essentially longer than for the free cells, for which it was about 5 h. Additionally, in these experiments, preliminary (to immobilisation) UV treatment of ArND particles decreased the duration of a yeast culture's lag phase. It changed from 9 h (for the yeast culture non-treated with UV particles) to about 7 h (for the cultures that were immobilised with ArND particles that were UV treated for 90 min). When discussing these results, it is necessary to mention that lag phase duration can be indicative, or independent, of cellular stress [31]. Additionally, the total yield of biomass after 20 h of cultivation in these experiments was the highest for the free cell cultures.

Summarising these results, we can conclude that immobilisation on/in yeast cells of non-arylated ND particles is preferential if we require a shorter lag phase duration and higher yield of the biomass. At the same time, a lower yield of the biomass may be linked to higher synthesis of some valuable compounds, which are the targets of biotechnological processes. Additionally, in the case of various biomedical applications of ND particles, it is not clear if the lag phase duration and higher cells proliferation are always better for the achievements of the goals of ND particles use. So, we may conclude that the results of this study showed that the selection of certain ND particles (non-arylated or arylated) may be necessary in each specific case, depending on the purpose of their use.

4. Conclusions

As previously mentioned, one of the important directions of current microbiological studies may be linked to nanodiamond particle application in biotechnological and medical fields. The surface of industrially produced nanodiamonds is polyfunctional [32]. There are various chemically active functional groups (carboxylic, hydroxyl, carbonyl, etc.) on the surface of the nanosized particles. Such properties of nanodiamonds are formed during their production, the technology of which is associated with a directed explosion and subsequent chemical purification of impurities. The physicochemical properties of this material and the chemically active surface of the nanoparticles determine their high sorption properties with respect to biological macromolecules and various chemical compounds. Thus, nanodiamonds find practical application in different fields of biology and for medical purposes. Various applications of ND particles for biomedical and biotechnological goals were described during the last few years. In medicine, nanodiamonds may be used as tissue scaffolds and surgical implants, as fluorescent biomarkers, for photoacoustic microscopy and for MRI applications [33]. The possibility of great success in cancer therapy may be associated with the use of nanodiamonds as carriers for drug delivery. Such drug delivery systems may combine therapeutic and diagnostic agents on a single platform [34]. ND particles may also be used for antibacterial and antiviral treatments, as well as for the construction of medical devices such as nanorobots. ND particles may be used also for biotechnological purposes; for example, for nanocatalysis, nanoseparation and as nanosensors [35,36]. Additionally, there is another important problem linked with nanoparticles. The rapid development of nanotechnology leads to a wide distribution of nanoparticles in the environment. Correspondingly, they may enter inside various tissues of humans and animals through their inhalation and direct contact with the skin [37]. However, there is currently no unambiguous assessment of the results of the use of nanomaterials; the mechanisms of their possible toxicity and potential health risks have not been sufficiently studied. Additionally, in the case of microorganisms, there is no clarity on the possible damaging effects or safety of nanoparticles, including nanodiamond particles, as previously mentioned [21–23].

One of the objectives of this study was to analyse the possibility of using the native and electrically charged uncoated and arylated nanodiamonds for their immobilisation on/in yeast cells and to obtain the first phenomenological results regarding the physiological activity of yeast cells with immobilised nanoparticles for their possible use in biotechnology and biomedicine.

Our work confirmed the ambiguity of the results in relation to the possible damaging effects of nanodiamond particles on microorganisms. It revealed that some adverse impact of these nanoparticles can manifest not only against prokaryotes, but also against eukaryotic yeast cells. However, the obtained results also indicate that it is possible to reduce and, most likely, completely eliminate the damage of nanoparticles to cells using special physical approaches. In our case, this is a change in the electric charge of the surface of nanoparticles as a result of their exposure to UV. At the same time, it has been shown that in terms of changes in the physiological activity of cells, which are important for the further possible use of these effects in biotechnology and biomedicine, the use of arylated nanoparticles for this purpose is of greater interest. It should also be noted that our work aimed only to detect the first phenomenological characteristics of the effect of immobilisation of diamond nanoparticles on yeast cells; in subsequent studies, it will be necessary to explore the mechanisms of the detected phenomena.

Author Contributions: Conceptualisation, Y.D., N.B. and A.R.; experimental work, methodology G.K., D.A., P.T., L.A., S.B., J.O., H.S. and A.S.; result evaluation, G.K., P.T., A.R., S.B., A.S. and Y.D.; writing—original draft, G.K., A.R., A.S. and Y.D.; writing—review and editing, A.R., S.B., N.B., J.O. and Y.D.; supervision, Y.D., S.B., N.B., J.O. and A.R.; funding acquisition, N.B. and Y.D.; project administration, Y.D. and N.B. All authors have read and agreed to the published version of the manuscript.

Funding: This research was funded by HORIZON, ERA-NET RUS Plus project #RUS_ST2019-126 BIOACTIVE CARDIOVASCULAR STENT FOR ANTIATHEROSCLEROSIS TREATMENT AND REDUCED RESTENOSIS" (German project part 01DJ21012 was funded by Federal Ministry of Education and Research (BMBF). Latvian project part was funded by State Education Development Agency of the Republic of Latvia, Project No. ES RTD/2021/4).

Institutional Review Board Statement: Not applicable.

Informed Consent Statement: Not applicable.

Data Availability Statement: The data supporting the results in this study are available on request from the corresponding author.

Conflicts of Interest: The authors declare no conflict of interest.

References

1. Kourkoutas, Y.; Bekatorou, A.; Banat, I.M.; Marchant, R.; Koutinas, A.A. Immobilization technologies and support materials suitable in alcohol beverages production: A review. *Food Microbiol.* **2004**, *21*, 377–397. [CrossRef]
2. Peng, Q.; Liu, Y.; Zeng, G.; Xu, W.; Yang, C.; Zhang, J. Biosorption of copper (II) by immobilizing Saccharomyces cerevisiae on the surface of chitosan-coated magnetic nanoparticles from aqueous solution. *J. Haz. Mat.* **2010**, *177*, 676–682. [CrossRef]
3. Rapoport, A.; Borovikova, D.; Kokina, A.; Patmalnieks, A.; Polyak, N.; Pavlovska, I.; Mezinskis, G.; Dekhtyar, Y. Immobilisation of yeast cells on the surface of hydroxyapatite ceramics. *Proc. Biochem.* **2011**, *46*, 665–670. [CrossRef]
4. Borovikova, D.; Scherbaka, R.; Patmalnieks, A.; Rapoport, A. Effects of yeast immobilisation on bioethanol production. *Biotechnol. Appl. Biochem.* **2014**, *61*, 33–39. [CrossRef] [PubMed]
5. Giese, E.C.; Silva, D.D.V.; Costa, A.F.M.; Almeida, S.G.C.; Dussán, K.J. Immobilized microbial nanoparticles for biosorption. *Crit. Rev. Biotechnol.* **2020**, *40*, 653–666. [CrossRef] [PubMed]
6. Borovikova, D.; Rozenfelde, L.; Pavlovska, I.; Rapoport, A. Immobilisation increases yeast cells resistance to dehydration-rehydration treatment. *J. Biotechnol.* **2014**, *184*, 169–171. [CrossRef]
7. Dotto, G.L.; Lima, E.C.; Pinto, L.A. Biosorption of food dyes onto Spirulina platensis nanoparticles: Equilibrium isotherm and thermodynamic analysis. *Biores. Technol.* **2012**, *103*, 123–130. [CrossRef]
8. Azeez, R.A.; Al-Zuhairi, F.K.I. Biosorption of dye by immobilized yeast cells on the surface of magnetic nanoparticles. *Alexandr. Engineer. J.* **2022**, *61*, 5213–5222. [CrossRef]
9. Li, X.; Ming, Q.Y.; Cai, R.; Yue, T.; Yuan, Y.; Gao, Z.; Wang, Z. Biosorption of Cd2+ and Pb2+ from apple juice by the magnetic nanoparticles functionalized lactic acid bacteria cells. *Food Control.* **2020**, *109*, 106916. [CrossRef]
10. Khatoon, H.; Rai, J. Augmentation of atrazine biodegradation by two Bacilli immobilized on a-Fe2O3 magnetic nanoparticles. *Sci. Rep.* **2018**, *8*, 17831. [CrossRef]
11. Taghizadeh, S.-M.; Ebrahiminezhad, A.; Ghoshoon, M.B.; Dehshahri, A.; Berenjian, A.; Ghasemi, Y. Magnetic immobilization of Pichia pastoris cells for the production of recombinant human serum albumin. *Nanomaterials* **2020**, *10*, 111. [CrossRef] [PubMed]
12. Ebrahiminezhad, A.; Varma, V.; Yang, S.; Berenjian, A. Magnetic immobilization of Bacillus subtilis natto cells for menaquinone-7 fermentation. *Appl. Microbiol. Biotechnol.* **2016**, *100*, 173–180. [CrossRef] [PubMed]
13. Ebrahiminezhad, A.; Varma, V.; Yang, S.; Ghasemi, Y.; Berenjian, A. Synthesis and application of amine functionalized iron oxide nanoparticles on menaquinone-7 fermentation: A step towards process intensification. *Nanomaterials* **2015**, *6*, 1. [CrossRef]
14. Raee, M.J.; Ebrahiminezhad, A.; Gholami, A.; Ghoshoon, M.B.; Ghasemi, Y. Magnetic immobilization of recombinant *E. coli* producing extracellular asparaginase: An effective way to intensify downstream process. *Sep. Sci. Technol.* **2018**, *53*, 1397–1404. [CrossRef]
15. Taghizadeh, S.-M.; Jafari, S.; Ahmad-Kiadaliri, T.; Mobasher, M.A.; Lal, N.; Raee, M.J.; Berenjian, A.; Ghasemi, Y.; Ebrahiminezhad, A. Magnetic immobilisation as a promising approach against bacteriophage infection. *Mater. Res. Express* **2020**, *6*, 1250a8. [CrossRef]
16. Gupta, C.; Prakash, D.; Gupta, S. Cancer treatment with nano-diamonds. *Front. Biosci.-Schol.* **2017**, *9*, 62–70. [CrossRef]
17. Wang, M.; Thanou, M. Targeting nanoparticles to cancer. *Pharmacol. Res.* **2010**, *62*, 90–99. [CrossRef]
18. Li, C. A targeted approach to cancer imaging and therapy. *Nat. Mater.* **2014**, *13*, 110–115. [CrossRef]
19. Getts, D.R.; Terry, R.L.; Getts, M.T.; Deffrasnes, C.; Müller, M.; van Vreden, C.; Ashhurst, T.M.; Chami, B.; McCarthy, D.; Wu, H.; et al. Therapeutic inflammatory monocyte modulation using immune-modifying microparticles. *Sci. Transl. Med.* **2014**, *6*, 219ra7. [CrossRef]
20. Gupta, A.S. Nanomedicine approaches in vascular disease: A review. *Nanomed. Nanotechnol. Biol. Med.* **2011**, *7*, 763–779. [CrossRef]
21. Beranova, J.; Seydlova, G.; Kozak, H.; Benada, O.; Fiser, R.; Artemenko, A.; Konopasek, I.; Kromka, A. Sensitivity of bacteria to diamond nanoparticles of various size differs in gram-positive and gram-negative cells. *FEMS Microbiol. Lett.* **2014**, *351*, 179–186. [CrossRef] [PubMed]
22. Hemelaar, S.R.; van der Laan, K.J.; Hinterding, S.R.; Koot, M.V.; Ellermann, E.; Perona-Martinez, F.P.; Roig, D.; Hommelet, S.; Novarina, D.; Takahashi, H.; et al. Generally applicable transformation protocols for fluorescent nanodiamond internalization into cells. *Sci. Rep.* **2017**, *7*, 5862. [CrossRef] [PubMed]
23. Morita, A.; Hamoh, T.; Sigaeva, A.; Norouzi, N.; Nagl, M.A.; van der Laan, K.J.; Evans, E.P.P.; Schirhagl, R. Targeting nanodiamonds to the nucleus in yeast cells. *Nanomaterials* **2020**, *10*, 1962. [CrossRef]
24. Baltacis, K.; Bystrov, V.; Bystrova, A.; Dekhtyar, Y.; Freivalds, T.; Raines, J.; Rozenberga, K.; Sorokins, H.; Zeidaks, M. Physical Fundamentals of Biomaterials Surface Electrical Functionalization. *Materials* **2020**, *13*, 4575. [CrossRef] [PubMed]
25. Liang, X.; Jacobs, R.; Martens, W.; Hu, Y.; Adriaensens, P.; Quirynen, M.; Lambrichts, I. Macro- and micro-anatomical, histological and computed tomography scan characterization of the nasopalatine canal. *J Clin. Periodontol.* **2009**, *36*, 598–603. [CrossRef]

26. Poonthiyil, V.; Lindhorst, T.K.; Golovko, V.B.; Fairbanks, A.J. Recent applications of click chemistry for the functionalization of gold nanoparticles and their conversion to glyco-gold nanoparticles. *Beilstein. J. Org. Chem.* **2018**, *3*, 11–24. [CrossRef] [PubMed]
27. Thess, A.; Lee, R.; Nikolaev, P.; Dai, H.; Petit, P.; Robert, J.; Xu, C.; Lee, Y.H.; Kim, S.G.; Rinzler, A.G.; et al. Crystalline Ropes of Metallic Carbon Nanotubes. *Science* **1996**, *26*, 483–487. [CrossRef] [PubMed]
28. Rapoport, A.I.; Meysel, M.N. Survival rates of yeast organisms after dehydration as determined by fluorescence microscopy. *Microbiology* **1985**, *54*, 53–55.
29. Chemistry LibreTexts. Available online: https://chem.libretexts.org/Bookshelves/Physical_and_Theoretical_Chemistry_Textbook_Maps/Supplemental_Modules_(Physical_and_Theoretical_Chemistry)/Physical_Properties_of_Matter/Atomic_and_Molecular_Properties/Dipole_Moments (accessed on 5 December 2022).
30. Lazzeretti, P.; Zanasi, R. On the CH bond dipole moment in alkanes. *J. Chem. Phys.* **1987**, *87*, 1681. [CrossRef]
31. Hamill, P.G.; Stevenson, A.; McMullan, P.E.; Williams, J.P.; Lewis, A.D.R.; Sudharsan, S.; Stevenson, K.E.; Farnsworth, K.D.; Khroustalyova, G.; Takemoto, J.Y.; et al. Microbial lag phase can be indicative of, or independent from, cellular stress. *Sci. Rep.* **2020**, *10*, 5948. [CrossRef]
32. Schrand, A.M.; Ciftan Hens, S.A.; Shenderova, O.A. Nanodiamond particles: Properties and perspectives for bioapplications. *Crit. Rev. Solid State Mater. Sci.* **2009**, *34*, 18–74. [CrossRef]
33. Garifo, S.; Stanicki, D.; Ayata, G.; Muller, R.N.; Laurent, S. Nanodiamonds as nanomaterial for biomedical field. *Front. Mater. Sci.* **2021**, *15*, 334–351. [CrossRef]
34. Lai, H.; Stenzel, M.H.; Xiao, P. Surface engineering and applications of nanodiamonds in cancer treatment and imaging. *Int. Mater. Rev.* **2020**, *65*, 189–225. [CrossRef]
35. Yadav, A.K.; Shukla, R.; Flora, S.J.S. Nanodiamonds: A Versatile Drug-Delivery System in the Recent Therapeutics Scenario. *Critic. Rev. Therap. Drug Carrier Syst.* **2021**, *38*, 39–78. [CrossRef]
36. Bilal, M.; Cheng, H.; Gonzalez-Gonzalez, R.B.; Parra-Saldıvar, R.; Iqbal, H.M.N. Bio-applications and biotechnological applications of nanodiamonds. *J. Mater. Res. Technol.* **2021**, *15*, 6175–6189. [CrossRef]
37. Oberdörster, G.; Oberdörster, E.; Oberdörster, J. Nanotoxicology: An emerging discipline evolving from studies of ultrafine particles. *Environ. Health Perspect.* **2005**, *113*, 823–839. [CrossRef]

Disclaimer/Publisher's Note: The statements, opinions and data contained in all publications are solely those of the individual author(s) and contributor(s) and not of MDPI and/or the editor(s). MDPI and/or the editor(s) disclaim responsibility for any injury to people or property resulting from any ideas, methods, instructions or products referred to in the content.

fermentation

Article

Assessment of Starters of Lactic Acid Bacteria and Killer Yeasts: Selected Strains in Lab-Scale Fermentations of Table Olives (*Olea europaea* L.) cv. *Leccino*

Grazia Federica Bencresciuto [1], Claudio Mandalà [1], Carmela Anna Migliori [1], Giovanna Cortellino [2], Maristella Vanoli [2] and Laura Bardi [1,*]

[1] Research Centre for Engineering and Agro-Food Processing, CREA Council for Agricultural Research and Economics, 10135 Turin, Italy
[2] Research Centre for Engineering and Agro-Food Processing, CREA Council for Agricultural Research and Economics, 20133 Milan, Italy
* Correspondence: laura.bardi@crea.gov.it

Abstract: Olives debittering, organoleptic quality and safety can be improved with yeasts and lactic acid bacteria (LABs) selected strain starters, that allow for better fermentation control with respect to natural fermentation. Two selected killer yeasts (*Wickerhamomyces anomalus* and *Saccharomyces cerevisiae*) and *Lactobacillus plantarum* strains were tested for olive (cv. *Leccino*) fermentation to compare different starter combinations and strategies; the aim was to assess their potential in avoiding pretreatments and the use of excessive salt in the brines and preservatives. Lactobacilli, yeasts, molds, *Enterobacteriaceae* and total aerobic bacteria were detected, as well as pH, soluble sugars, alcohols, organic acids, phenolic compounds, and rheological properties of olives. Sugars were rapidly consumed in the brines and olives; the pH dropped quickly, then rose until neutrality after six months. The oleuropein final levels in olives were unaffected by the treatments. The use of starters did not improve the LABs' growth nor prevent the growth of *Enterobacteriaceae* and molds. The growth of undesirable microorganisms could have been induced by the availability of selective carbon source such as mannitol, whose concentration in olive trees rise under drought stress. The possible role of climate change on the quality and safety of fermented foods should be furtherly investigated. The improvement of olives' nutraceutical value can be induced by yeasts and LABs starters due to the higher production of hydroxytyrosol.

Keywords: table olives; fermentation; starter cultures; killer yeasts; lactic acid bacteria

Citation: Bencresciuto, G.F.; Mandalà, C.; Migliori, C.A.; Cortellino, G.; Vanoli, M.; Bardi, L. Assessment of Starters of Lactic Acid Bacteria and Killer Yeasts: Selected Strains in Lab-Scale Fermentations of Table Olives (*Olea europaea* L.) cv. *Leccino*. *Fermentation* **2023**, *9*, 182. https://doi.org/10.3390/fermentation9020182

Academic Editor: Ronnie Willaert

Received: 30 January 2023
Revised: 14 February 2023
Accepted: 15 February 2023
Published: 17 February 2023

Copyright: © 2023 by the authors. Licensee MDPI, Basel, Switzerland. This article is an open access article distributed under the terms and conditions of the Creative Commons Attribution (CC BY) license (https://creativecommons.org/licenses/by/4.0/).

1. Introduction

Table olives are a processed food widespread in the Mediterranean area obtained from fruits of *Olea europaea* L. Fruit processing is necessary to reduce their natural bitterness due to phenolic compounds, mainly oleuropein [1]. Phenolic compounds are among the most important components of olive drupes that contribute to taste and texture [2] and to their antioxidant, anti-inflammatory, antibacterial and cytotoxic properties [3]. Oleuropein has been extensively studied for its many beneficial properties due to antioxidant, scavenging of radicals, antimicrobial, antihypertensive and anticancer activities. During the ripening and fermentation of olives, oleuropein and ligstroside, another important phenolic compound, are hydrolyzed in a process involving the action of β-glucosidases that break the glycosidic bond with the formation of the aglycone forms of oleuropein and ligstroside, and esterases, which release elenoic acid, hydroxytyrosol (3,4 DHPEA) from oleuropein aglycone and tyrosol (p-HPEA) from ligstroside aglycone [4]. Tyrosol and hydroxytyrosol also act as scavengers of free radicals and as chelators of metals; furthermore, they have cardioprotective, anticancer and neuroprotective properties, and antimicrobial effects [5].

The abundance of phenolic compounds in table olives provides a high nutraceutical value to this fermented food.

Olive debittering can be reached by fermentation; the most commonly used processes are "Spanish style", in which green olives are soaked in diluted lye solution before fermentation in the brines, "Californian-style" in which olives are darkened by oxidation in lye solutions and air bubbling, and natural fermentation, also known as "Greek style", in which olives are immersed in 6–10% NaCl brines and fermentation takes place lasting 8–10 months due to the growth of autochthonous microorganisms of olives and plants [6–9]. Yeasts and lactic acid bacteria (LABs) are the main agents of natural fermentations due to their ability to grow in high NaCl concentrations: they rapidly acidify the brines so that high salinity and low pH prevent the growth of spoilage or pathogenic microorganisms. Moreover, LABs and yeasts can significantly improve the sensory properties of processed olives producing secondary metabolites [9–11]. LABs have been widely assessed and featured for their ability to improve the olives quality and safety through brine acidification and release of bacteriocines, whereas yeasts can play a double role, as they may also be agents of the deterioration of the fruits, gas-pocket formation, softening of the olive flesh, production of off-flavors and odors [12], or cause brines clouding or package swelling [13].

Furthermore, yeasts can contribute to debittering, promote LAB growth, improve flavor, taste and texture of final product, inhibit the growth of spoilage yeasts and molds by producing killer toxin and inhibit several adverse microorganisms. Due to these positive actions, lesser use of salt and additives could be allowed [9,13–16]. Moreover, yeasts can make up for LABs when their activity is inhibited by high phenols and/or salt concentrations or excessively low pH [17].

The industrial production of black and green olives by spontaneous fermentation is subject to autochthonous microflora, physicochemical conditions, olive varieties, fermentable substrates and brine salt content; the results can be unpredictable [18]. The growth of undesirable strains can lead to abnormal fermentation and a non-compliant final product [6]. Better control of the fermentation can be reached with a starter culture of selected LABs and yeast strains. Selected strains are exploited to inhibit spoilage due to undesirable microorganisms or pathogens, reduce debittering times, and obtain a better final product. Starters could also be useful for producing probiotic table olives [19].

Among the species of LABs, *Lactobacillus plantarum* and *Lactobacillus pentosus* strains have been mainly characterized, while the yeasts most used as starters are *Wickerhamomyces anomalus, Saccharomyces cerevisiae* and *Candida boidinii* [20].

In recent years, several studies focused on killer yeasts as they can inhibit the growth of spoilage yeasts by producing killer toxins in fermented foods, including olives [9,14–16]. Among them, *W. anomalus* and *S. cerevisiae* killer strains were selected for olive fermentation [14]. The mechanisms of action can be variable and influenced by various factors; a killer toxin can inhibit DNA replication, induce membrane permeability changes and arrest the cell cycle [21–23].

In the present work, the use of two selected yeast strains for olives fermentation was studied to evaluate their potential to obtain a good final product that can avoid the pretreatments of olives, excessive salt in the brines and further additives and preservatives. *W. anomalous* and *S. cerevisiae* strains previously characterized for killer activity and debittering capability were chosen. Olives cv. *Leccino* were fermented in the brines with the inocula of starter cultures of these yeast strains, alone or associated with a selected strain of *L. plantarum*, in comparison to spontaneous fermentation.

2. Materials and Methods

2.1. Olive Samples

Natural black olives cv. *Leccino* were provided by Romeo Ficacci s.r.l., a table olive industry located in Castelmadama (Rome, Italy). Black olives (70 kg) were collected at the BBCH (Biologische Bundesanstalt, Bundessortenamt and Chemical Industry) phenological stage 8, maturity of fruit [24] and washed with tap water to eliminate plant materials (residues of leaves, branches) and superficial contaminants.

2.2. Microbial Strains and Starter Production

Two selected oleuropeinolytic killer yeast strains (collection of the University of Sassari, Italy), *Wickerhamomyces anomalus* (Wa1) and *Saccharomyces cerevisiae* (Sc24), and one selected bacterial strain, *Lactobacillus plantarum* (B51) (collection CREA-IT, Pescara, Italy) were used as starters.

Bacterial strain precultures were obtained in 100 mL MRS (Man Rogosa and Sharpe: Peptone 10 g/L, Beef Extract 10 g/L, Yeast Extract 5 g/L, Glucose 20 g/L, Di-potassium Hydrogen Phosphate 2 g/L, Sodium Acetate 5 g/L, Di-amonium Citrate 2 g/L, Magnesium Sulphate 0.2 g/L, Manganous Sulphate 0.05 g/L and Tween® 80 1 g/L) broth incubated at 30 °C for 24 h. Bacterial cells were then adapted to the saline environment of the brine by incubation in 100 mL MRS broth supplemented with 3% NaCl at 30 °C for 24 h, followed a further incubation in 100 mL MRS broth supplemented with 6% NaCl at 30 °C for 24 h.

S. cerevisiae and *W. anomalus* precultures were produced in 200 mL of YEPD (10 g/L Yeast Extract, 20 g/L Peptone, 20 g/L Dextrose) broth at 28 °C for 48 h. Afterward, the yeasts were adapted to the salinity of the brines by sequential incubation at 28 °C for 48 h in YEPD broth supplemented with 3% NaCl, then in YEPD broth with 6% NaCl.

Yeast and bacteria cells were collected by centrifugation (3500 rpm at 15 °C for 15 min), then inoculated in 8% NaCl sterile brine to achieve an initial population of 10^6 CFU/mL in each jar according to the experimental setup.

2.3. Experimental Set Up

Lab-scale fermentations were carried out in 5 L capacity glass jars according to the Greek-style method. 2.5 kg of olives were placed in each jar, then filled with 2 L of 8% NaCl (*w*/*v*) previously sterilized at 121 °C × 20 min.

Seven different fermentation conditions were compared: (i) spontaneous fermentation (SP), (ii) fermentation inoculated with *L. plantarum* B51 (LP), (iii) fermentation inoculated with *W. anomalus* (WA), (iv) fermentation inoculated with mixed *L. plantarum* and *S. cerevisiae* (MIX1), (v) fermentation with sequential inoculation of *L. plantarum* followed one month later by *S. cerevisiae* (LY1), (vi) fermentation inoculated with mixed *L. plantarum* and *W. anomalus* (MIX2), (vii) fermentation with sequential inoculation of *L. plantarum* followed one month later by *W. anomalus* (LY2).

All treatments were performed in triplicate (three fermentation jars per treatment) at room temperature (between 19 °C and 25 °C) for 188 days. Brine samples (10 mL) and olives (20 drupes) were collected aseptically and subjected to microbiological and chemical analysis throughout the fermentations.

2.4. Microbiological Analysis

Brines were analyzed to determine the content of Lactobacilli, Yeasts, Molds, *Enterobacteriaceae* and Total Aerobic Bacteria.

The samples (1 mL) were aseptically transferred to 9 mL of sterile saline solution. Serial dilutions in the same sterile saline were prepared, and 1 mL or 0.1 mL samples of the appropriate dilutions were spread or mixed on the following media: Man Rogosa and Sharpe Agar plates for the detection of Lactobacilli, 3M™ Petrifilm™ Aerobic Count Plates for the detection of Total Aerobic Bacteria, 3M™ Petrifilm™ Enterobacteriaceae Count Plates for the detection of *Enterobacteriaceae*, and 3M™ Petrifilm™ Yeast and Mold Count Plates for the detection of yeast and mold.

The 3M™ Petrifilm™ Plate is a sample-ready culture medium system that contains nutrients, a cold-water-soluble gelling agent, and an indicator system that facilitates microorganism enumeration. The 3M™ Petrifilm™ formulation are (i) Violet Red Bile with Glucose nutrients, cold-water soluble gel, and tetrazolium indicator (3M™ Petrifilm™ Enterobacteriaceae Count Plates, (6420/6421), Manufactured at Brookings, SD, USA (ISO

9001:2015, FM 14552)), (ii) Standard Methods nutrients, cold-water soluble gel, and tetrazolium indicator (3M™ Petrifilm™ Aerobic Count Plates, (6400/6406/6442), Manufactured at Brookings, SD, USA (ISO 9001:2015, FM 14552)), (iii) Nutrients supplemented with antibiotics, a cold-water-soluble gel, and indicator (3M™ Petrifilm™ Yeast and Mold Count Plates, (6407/6417/6445), Manufactured at Brookings, SD, USA (ISO 9001:2015, FM 14552)).

Incubation was carried out at 28 °C for 5 days for Lactobacilli, at 37 °C for 24 h for *Enterobacteriaceae*, at 28 °C for 5 days for yeasts and mold and at 37 °C for 48 h for Total Aerobic Bacteria. Results were expressed as colony-forming units per milliliters (CFU/mL) of brines.

Fungal isolation was performed directly from emerging colonies that developed on 3M™ Petrifilm™ plates by detaching them diligently with a sterilized bacterial loop and transferring them on PDA (4 g/L Potato extract, 20 g/L Dextrose, 15 g/L Agar) plates. Monoconidial cultures were produced for each isolate to obtain pure fungal colonies. Each monoconidial culture was incubated at 25 °C for 48 h; at the end of the incubation period, colony characteristics (color, mycelium, colony type and shape) were observed. Dimensions of available conidia and hyphal features (color, shape, presence or absence of chlamydospores) were recorded.

2.5. Physico-Chemical Analysis

2.5.1. Sugars and Alcohols

Sugars and alcohols were extracted from olives following the method described by Fibiani et al. [25]. Brine samples and extract from olives pulp were analyzed by HPLC after filtration through 0.45 μm Nylon filters. Sucrose, glucose, fructose, mannitol, sorbitol, ethanol and glycerin were analyzed by chromatographic analysis with HPLC (Jasco PU-980, Jasco, Tokyo, Japan) equipped with a RI530 refractive index detector, combined with a Chromnav data acquisition system version 3.32, in a Benson 87C carbohydrate column (300 mm × 8 mm) maintained at 80 °C. The mobile phase was ultrapure water (HPLC grade), 0.7 mL/min flow. Commercial reference standards were used to create calibration curves [25].

2.5.2. Organic Acids

Brine samples were analyzed after filtration through 0.45 μm Nylon filters. An Agilent 1200 Series HPLC was used for organic acid analysis coupled to a UV/Vis detector with detection at 214 nm. The column used was a Repromer H + column (300 mm × 8 mm), maintained at 63 °C. The mobile phase was H_2SO_4 25 nM, 0.5 mL/min flow. Commercial reference standards of citric, malic, pyruvic, succinic, lactic and acetic acids were used to create calibration curves [25].

2.5.3. pH

All pH measurements were made in a Crison BASIC 20 + pH-meter (Crison Instruments S.A., Alella, Spain) fitted with a Crison electrode. The calibration was made with standard buffers at pH 4.00 and 7.00. The determinations were executed in triplicate.

2.5.4. Phenolic Compounds

The brines were filtered through 0.45 μm Nylon filters and diluted 1:1 in a methanol/acidified water mixture (80/20). Extraction from destoned olives was carried out following the method of Ambra et al. [2], except for the use of frozen samples homogenized by an Ultraturrax IKA T18 Basic (IKA Works Inc., Wilmington, NC, USA) for 30 s in the presence of the extraction mixture and treatment of the homogenized sample in an ultrasonic bath (Elmasonic S, Singen, Germany) for 30 min instead of liquid nitrogen treatment. The samples were then filtered with 0.45 μm membranes to remove any residue. Subsequently, olives extracts and brine samples were analyzed by HPLC technique following the method described in Bleve et al. [26], using a Jasco BS-997 HPLC equipped with a Supelco ODS80Ts RPC18 column (250 mm × 4.6 mm, 5 μm). All analyses were performed in triplicate. Phenol compounds were quantified using calibration curves of authentic phenolic standards, with concentrations between 250 ppm and 1000 ppm for oleuropein, between 125 ppm and 750 ppm for tyrosol and between 250 ppm and 1000 ppm for hydroxytyrosol.

2.6. Rheological Properties of Olive

The rheological properties of olives were measured by texture profile analysis (TPA) and a puncture test. Both analyses were performed by means of a TA.XT Plus texture analyzer (Stable Micro System Ltd., Godalmig, UK) and data acquisition and integration by using Texture Exponent 32-bit software. All analyses were done at room temperature. For both tests, twenty olives were analyzed for each jar and sixty for each treatment. The texture was evaluated only at the end of treatment. Each olive was centered horizontally under the probe for measurement.

2.6.1. Texture Profile Analysis (TPA)

The olives were subjected to a double compression cycle using a cylindrical flat probe (10 cm diameter) at a test speed of 1 mm/s and a deformation of 15% of the initial size. By the acquired force (g)/displacement (mm) curve, multiple textural parameters were acquired:

- Hardness: Maximum force of the first compression.
- Springiness: Ratio between the distance traveled during the first compression cycle and the distance during the second cycle. It is the rate at which a deformed sample returns to its original size.
- Cohesiveness: Ratio between the area (work) during the second compression and the area (work) during the first compression. It is the degree to which a product can be deformed before it breaks.
- Gumminess: Hardness × cohesiveness. It is the energy required to disintegrate the product to the state ready for swallowing.
- Chewiness: Hardness × cohesiveness × springiness. It is the number of chews needed to masticate the product until it is ready for swallowing.

2.6.2. Puncture Test

A puncture test was carried out using a needle probe (2 mm diameter) with a crosshead speed of 6.67 mm/s. From the force curve, the following parameters were extracted: firmness corresponding to the maximum force (firmness, g), rigidity index corresponding to the slope of the last part of the force curve (dF/dS, g/mm) and work required to puncture the olive (g × mm).

2.7. Chemicals

Reference standards of HPLC-grade oleuropein, tyrosol and hydroxytyrosol were purchased from Cayman Chemicals (Ann Arbor, MI, USA). References standards (HPLC grades) of sugars, alcohols and organic acids; HPLC-grade solvents (methanol, acetonitrile, hexane) and acetic acid 99–100% were purchased from Sigma-Aldrich (Milan, Italy). HPLC-grade Ultrapure water was purchased from Carlo Erba reagents (Milan, Italy), standard buffers at pH 4.00 and 7.00 from CRISON (Crison Instruments S.A., Alella, Spain). MRS Broth with tween 80, BIOLIFE, Milan, Italy; Sodium Chloride, SACCO s.r.l., Cadorago (Como), Italy; YEPD, Sigma-Aldrich, Milan, Italy; MRSCM0361, Oxoid, Hampshire, United Kingdom; 3M™ Petrifilm™ count plates specific for the detection of yeast, mold, *Enterobacteriaceae* and total aerobic bacteria were purchased from 3M Italy, Pioltello, Milan (Italy).

2.8. Statistical Analyses

The data were subjected to analysis of variance (ANOVA) and post-hoc LSD test, using $p \leq 0.05$ as the cut-off level of significance. ORIGIN software (OriginPro, Version 2022. OriginLab Corporation, Northampton, MA, USA) was used to check for significant differences among treatments at each sampling time.

3. Results

3.1. Microbial Analysis

Among the checked microbial populations (Figure 1), Lactobacilli showed the fastest and highest growth at the early phase of fermentation; then, their cell concentrations showed a decreasing trend (Figure 1A). On the 40th day, they reached cell concentrations between 3.6×10^7 CFU/mL in LY1 and 9.4×10^7 CFU/mL in MIX2 treatments; moreover, cell concentrations were highest where *W. anomalus* selected strain was inoculated in the mix with *L. plantarum* (40th day) or alone (63rd day), while they were lowest when *L. plantarum* was associated to *S. cerevisiae*. The total aerobic bacteria grown at 37 °C showed a trend similar to LABs, but at slightly lower concentrations (Figure 1C); during the third month of fermentation, they were significantly lower in treatments inoculated with both yeasts and LAB selected strains. *Enterobacteriaceae* followed an inverse trend throughout the fermentation: their cell concentrations rose from the 77th to the 140th day, then dropped sharply on the 188th day (Figure 1D). On the 40th day, yeast growth (Figure 1B) was highest in the SP treatment; later, the trend changed, showing that, in general, cell concentration rose until the 77th day, followed by a subsequent descent.

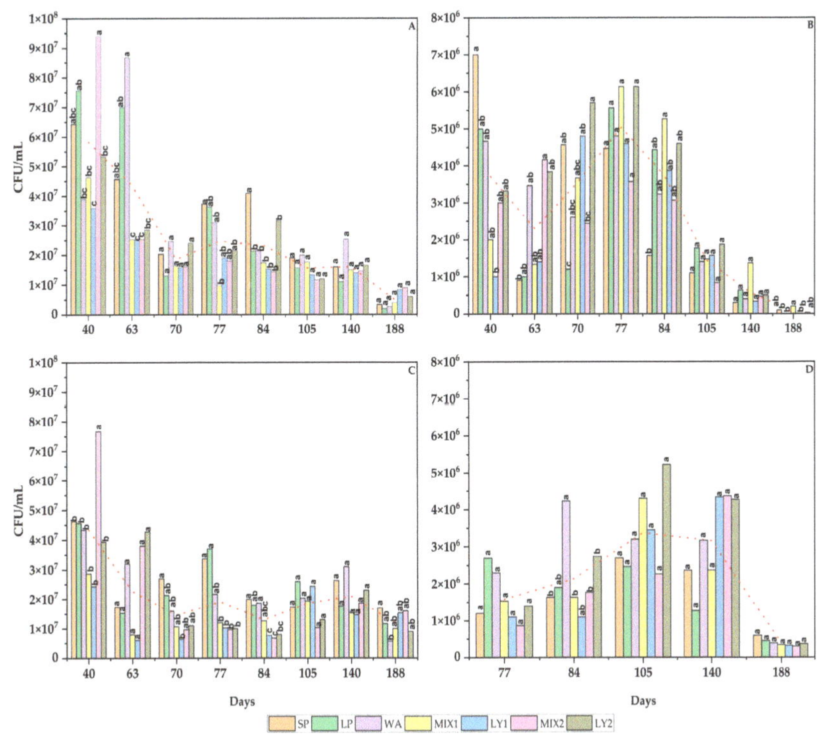

Figure 1. Lactobacilli (**A**), yeasts (**B**), total aerobic bacteria (**C**) and *Enterobacteriaceae* (**D**) in the brines of the different treatments. The data are expressed as means of the triplicate measurements followed by ANOVA tests. Significant differences are indicated by different letters ($p \leq 0.05$). The red dotted line represents the average microbial counts for each sampling time.

Maximum yeast cell concentration was reached on the 70th day in SP and LY, on the 77th day in LP, WA, MIX1, MIX2 and LY2 treatments; it sharply dropped on the 84th day in SP and on the 105th day in the other treatments. From the 63rd day onward, the presence of molds was observed, reaching significant abundance during the late phase of fermentation. High variability detected in terms of quantification did not allow significant differences to be evidenced among treatments, with the exception of the last sampling time (188th), where the molds were lowest in MIX1, which was significantly lower than in LP treatment. From monoconidial cultures obtained from isolated colonies, it was possible to identify the *Fusarium* spp. genera as the prevalent mold in all treatments except for the MIX1 treatment. In particular, the microscopic observations of the morphological characteristics allowed the identification of *Fusarium solani* as the prevalent species. Specifically, the *F. solani* traits are reported in the Supplementary Materials (Figure S1).

3.2. *Physico-Chemical Analysis*

3.2.1. Sugars and Alcohols

The compositions of the sugars and alcohols of the raw olive pulp are reported in Table 1. The most abundant sugar was glucose (9477.74 ppm ± 203.12) followed by sucrose (703.93 ppm ± 125.43) and fructose (615.80 ppm ± 26.43). Among sugar alcohols, mannitol showed the highest concentration (3619.97 ppm ± 570.04), while the sorbitol concentration was 92.41 ppm ± 66.39, with the latter showing very high variability (CV = 72%).

Table 1. Sugar, alcohol and polyphenol content in the raw olive pulp. Data are the average from the three replicates ± standard error.

Sugars and Alcohols	**ppm**
Sucrose	703.93 ± 125.43
Glucose	9477.74 ± 203.12
Fructose	615.80 ± 26.43
Mannitol	3619.97 ± 570.04
Sorbitol	92.41 ± 66.39
Phenolic Compounds	**ppm**
Oleuropein	682.09 ± 230.58
Hydroxytyrosol	1632.23 ± 675.70
Tyrosol	1816.50 ± 545.35

During fermentation, the olives' sugar content rapidly decreased, with glucose being the most rapidly consumed (Table 2). At the 23rd day, about 76% of the glucose was consumed on average, followed by sucrose (45%) and fructose (38%), whereas among sugar alcohols (Table 3), sorbitol strongly decreased (62%), while the decrease in mannitol was only 8%. Glucose and sucrose were more rapidly consumed in SP, but fructose was more rapidly consumed in LY2. On the 40th day of fermentation, glucose and sucrose were almost completely consumed (96% and 94% respectively on average), while 26% fructose residue was still present, with the lowest concentrations found in MIX1, LY1, MIX 2 and LY2; also, mannitol showed a strong decrease from 23rd to 40th day of fermentation, when its concentration dropped to 15% mean residue, highest in LP, lowest in SP. All sugars and sugar alcohols were almost zeroed on the 105th day in all treatments; only traces of fructose were found in treatments LP, MIX2 and LY2, and glucose in treatment SP. At the same sampling time, a high ethanol concentration was detected in olives, ranging from 468 ppm to 852 ppm in SP and WA treatments, respectively. Sucrose, after zeroing on the 105th day, was detected at 73 ppm mean concentration on the 140th day and 30 ppm on the 188th day. Sorbitol and glycerol concentrations showed the same trend during fermentations in olive pulp, rising throughout the fermentations to the highest values on the 140th day (mean values 424 ppm and 1450 ppm, respectively) (Table 3).

Table 2. Change in concentrations of sugars in the pulp of *Leccino* olives during spontaneous (SP) or starter-driven fermentation, measured by HPLC analysis. Different letters indicate significant differences among the different treatments ($p \leq 0.05$) at the same sampling time. Data are averaged from the three replicates ± standard error.

Day	23	40	105	140	188
Treatments			**Sucrose (ppm)**		
SP	305.49 ± 4.85 [e]	52.76 ± 3.88 [b]	2.16 ± 0.25 [a]	80.71 ± 1.73 [a]	0.00 ± 0.00 [d]
LP	426.14 ± 6.52 [ab]	124.19 ± 7.67 [a]	0.00 ± 0.00 [b]	103.55 ± 6.64 [a]	0.00 ± 0.00 [d]
WA	410.69 ± 14.47 [abc]	141.17 ± 17.57 [a]	0.00 ± 0.00 [b]	94.39 ± 3.51 [a]	11.78 ± 8.98 [cd]
MIX1	386.85 ± 19.64 [c]	0.00 ± 0.00 [c]	0.00 ± 0.00 [b]	93.32 ± 3.03 [a]	133.98 ± 12.17 [a]
LY1	397.47 ± 5.29 [bc]	0.00 ± 0.00 [c]	0.00 ± 0.00 [b]	31.69 ± 2.33 [c]	40.14 ± 2.31 [b]
MIX2	340.79 ± 7.95 [d]	0.00 ± 0.00 [c]	0.00 ± 0.00 [b]	72.33 ± 6.96 [ab]	23.25 ± 8.05 [bc]
LY2	439.57 ± 11.71 [a]	0.00 ± 0.00 [c]	0.00 ± 0.00 [b]	34.10 ± 9.14 [bc]	0.00 ± 0.00 [d]
			Glucose (ppm)		
SP	1336.53 ± 30.22 [f]	264.96 ± 5.08 [c]	5.14 ± 4.06 [ab]	225.35 ± 64.80 [a]	0.00 ± 0.00 [b]
LP	2678.28 ± 89.35 [ab]	589.65 ± 138.75 [b]	0.73 ± 0.73 [b]	20.60 ± 1.27 [b]	0.00 ± 0.00 [b]
WA	2370.15 ± 74.86 [cd]	886.36 ± 209.44 [a]	0.00 ± 0.00 [b]	18.90 ± 3.25 [b]	17.17 ± 7.52 [a]
MIX1	2091.05 ± 104.84 [e]	231.74 ± 17.91 [c]	0.00 ± 0.00 [b]	12.64 ± 0.94 [b]	9.27 ± 2.12 [ab]
LY1	2730.10 ± 38.10 [a]	172.82 ± 7.96 [c]	0.00 ± 0.00 [b]	4.93 ± 0.72 [b]	3.80 ± 0.55 [b]
MIX2	2472. 06 ± 34.42 [bc]	217.64 ± 18.48 [c]	1.38 ± 0.20 [b]	0.00 ± 0.00 [b]	0.00 ± 0.00 [b]
LY2	2204.15 ± 108.27 [de]	212.25 ± 3.30 [c]	10.17 ± 4.36 [a]	2.54 ± 0.37 [b]	0.00 ± 0.00 [b]
			Fructose (ppm)		
SP	425.31 ± 11.99 [ab]	234.09 ± 3.06 [a]	1.05 ± 0.87 [b]	37.69 ± 14.64 [a]	0.00 ± 0.00 [c]
LP	366. 63 ± 33.94 [ab]	377.97 ± 52.56 [a]	5.62 ± 0.87 [b]	1.92 ± 0.56 [b]	0.00 ± 0.00 [c]
WA	379.57 ± 13.07 [ab]	248.65 ± 114.49 [a]	0.00 ± 0.00 [b]	0.00 ± 0.00 [b]	17.52 ± 2.73 [b]
MIX1	468.63 ± 59.45 [a]	60.56 ± 7.77 [b]	0.00 ± 0.00 [b]	3.29 ± 0.73 [b]	44.27 ± 3.28 [a]
LY1	374.99 ± 7.21 [ab]	83.75 ± 9.93 [b]	0.00 ± 0.00 [b]	2.91 ± 0.69 [b]	6.24 ± 2.15 [c]
MIX2	359.85 ± 12.16 [ab]	42.41 ± 20.16 [b]	21.93 ± 6.53 [b]	12.01 ± 4.14 [b]	0.00 ± 0.00 [c]
LY2	287.43 ± 118.51 [b]	53.84 ± 8.28 [b]	63.09 ± 19.98 [a]	20.09 ± 10.27 [ab]	0.00 ± 0.00 [c]

Table 3. Change in concentration of alcohols in the pulp of *Leccino* olives during spontaneous (SP) or starter-driven fermentation, measured by HPLC analysis. Different letters indicate significant differences among the different treatments ($p \leq 0.05$) at the same sampling time. Data are average from three replicates ± standard error.

Day	23	40	105	140	188
Treatments			**Mannitol (ppm)**		
SP	2841.82 ± 184.23 [bc]	330. 47 ± 15.47 [b]	1.03 ± 0.52 [a]	107.09 ± 13.69 [b]	66.47 ± 20.02 [b]
LP	2721.22 ± 443.95 [c]	881.01 ± 406.13 [a]	0.00 ± 0.00 [b]	0.00 ± 0.00 [d]	666.53 ± 20.87 [a]
WA	4248.15 ± 117.43 [a]	533.65 ± 12.23 [ab]	0.00 ± 0.00 [b]	0.00 ± 0.00 [d]	608.07 ± 111.80 [a]
MIX1	4205.16 ± 192.19 [a]	613.97 ± 89.47 [ab]	0.00 ± 0.00 [b]	99.60 ± 10.39 [b]	123.15 ± 16.39 [b]
LY1	3505.10 ± 184.46 [ab]	460.93 ± 31.92 [ab]	0.00 ± 0.00 [b]	66.89 ± 5.74 [c]	103.76 ± 19.62 [b]
MIX2	3146.29 ± 29.75 [bc]	445.93 ± 51.91 [ab]	0.00 ± 0.00 [b]	166.02 ± 6.45 [a]	0.00 ± 0.00 [b]
LY2	2539.49 ± 361.20 [c]	466.11 ± 72.69 [ab]	0.00 ± 0.00 [b]	183.03 ± 9.68 [a]	43.10 ± 7.92 [b]

Table 3. *Cont.*

	Day	23	40	105	140	188
				Sorbitol (ppm)		
SP		37.29 ± 13.06 [ab]	125.09 ± 3.96 [bcd]	0.13 ± 0.08 [a]	808.75 ± 39.18 [a]	90.20 ± 9.68 [c]
LP		27.99 ± 5.01 [ab]	97.59 ± 0.44 [cd]	0.00 ± 0.00 [b]	321.37 ± 41.25 [bc]	46.20 ± 5.41 [d]
WA		36.42 ± 5.99 [ab]	59.67 ± 27.49 [d]	0.00 ± 0.00 [b]	402.61 ± 54.24 [bc]	113.06 ± 0.20 [b]
MIX1		66.90 ± 16.76 [a]	211.87 ± 39.55 [b]	0.00 ± 0.00 [b]	235.62 ± 75.17 [c]	198.39 ± 8.32 [a]
LY1		18.58 ± 7.61 [b]	201.85 ± 10.38 [bc]	0.00 ± 0.00 [b]	432.52 ± 46.01 [b]	0.00 ± 0.00 [e]
MIX2		16.25 ± 2.69 [b]	337.44 ± 84.51 [a]	0.00 ± 0.00 [b]	290.22 ± 88.32 [bc]	16.42 ± 4.77 [e]
LY2		41.03 ± 26.42 [ab]	177.08 ± 8.37 [bc]	0.00 ± 0.00 [b]	479.53 ± 86.09 [b]	10.22 ± 4.86 [e]
				Ethanol (ppm)		
SP		0.00 ± 0.00	0.00 ± 0.00	467.75 ± 132.03 [d]	0.00 ± 0.00	0.00 ± 0.00 [b]
LP		0.00 ± 0.00	0.00 ± 0.00	822.96 ± 32.33 [ab]	0.00 ± 0.00	0.00 ± 0.00 [b]
WA		0.00 ± 0.00	0.00 ± 0.00	851.71 ± 78.11 [a]	0.00 ± 0.00	0.00 ± 0.00 [b]
MIX1		0.00 ± 0.00	0.00 ± 0.00	598.26 ± 50.76 [bcd]	0.00 ± 0.00	0.00 ± 0.00 [b]
LY1		0.00 ± 0.00	0.00 ± 0.00	773.19 ± 40.86 [abc]	0.00 ± 0.00	0.00 ± 0.00 [b]
MIX2		0.00 ± 0.00	0.00 ± 0.00	576.26 ± 23.88 [cd]	0.00 ± 0.00	41.49 ± 5.33 [a]
LY2		0.00 ± 0.00	0.00 ± 0.00	535.13 ± 102.12 [d]	0.00 ± 0.00	0.00 ± 0.00 [b]
				Glycerol (ppm)		
SP		0.00 ± 0.00	0.00 ± 0.00 [c]	0.00 ± 0.00	1342.62 ± 287.81 [a]	2362.26 ± 496.59 [a]
LP		0.00 ± 0.00	0.00 ± 0.00 [c]	0.00 ± 0.00	1331.19 ± 52.87 [a]	1487.78 ± 150.80 [bc]
WA		0.00 ± 0.00	470.96 ± 156.54 [b]	0.00 ± 0.00	1353.50 ± 156.54 [a]	950.54 ± 137.67 [cd]
MIX1		0.00 ± 0.00	740.64 ± 228.03 [b]	0.00 ± 0.00	1726.35 ± 49.04 [a]	1792.01 ± 77.12 [ab]
LY1		0.00 ± 0.00	880.76 ± 74.01 [ab]	0.00 ± 0.00	1681.73 ± 207.55 [a]	450.85 ± 32.96 [de]
MIX2		0.00 ± 0.00	1221.38 ± 274.93 [a]	0.00 ± 0.00	1345.29 ± 137.99 [a]	0.00 ± 0.00 [e]
LY2		0.00 ± 0.00	858.80 ± 73.42 [ab]	0.00 ± 0.00	1368.59 ± 205.04 [a]	0.00 ± 0.00 [e]

In the brines, only traces of sugars were found at early sampling times and throughout the fermentations (Figure 2A,B). Sucrose was not detected. The highest concentrations at the last sampling time were 278.40 ppm glucose (treatment MIX2) and 187.20 ppm fructose (treatment LY2). Instead, sugar alcohols were detected at higher concentrations (Figure 2C,D): mean values of mannitol were 2405.53 ppm on the 16th day and 1521.34 ppm on the 105th day, and sorbitol appeared on the 188th day at an average concentration of 594.94 ppm. Ethanol concentration rose, reaching mean concentrations of 45,873.69 ppm on the 16th and 3822.06 ppm on the 140th day, then a more rapid increase was observed until the 180,581.91 ppm mean concentration on the 188th day (Figure 3D). Glycerol was not detected in the brines.

3.2.2. Organic Acids

Lactic acid, acetic acid and pyruvic acid reached the highest concentration in the brines very rapidly (Figure 3A–C). On the 16th day, lactic acid, acetic acid, and pyruvic acid were highest in WA (4100.25 ppm, 32,819.01 ppm and 956.21 ppm, respectively) and lowest in MIX2 (1207.10 ppm, 5875.03 ppm and 394.55 ppm, respectively). Then, a decrease was observed in pyruvic acid in all treatments, lactic acid in all treatments except MIX2, and acetic acid in all treatments except MIX2 and LY2.

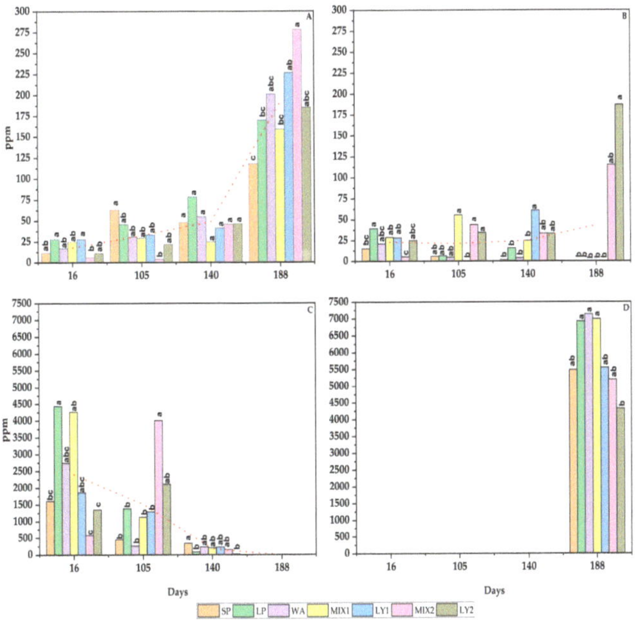

Figure 2. Glucose (**A**), fructose (**B**), mannitol (**C**) and sorbitol (**D**) evolution of the brines during fermentation. The data are expressed as means of triplicate measurements. Significant differences (LSD) are indicated by different letters ($p \leq 0.05$). The red dotted line represents the average values at each sampling time.

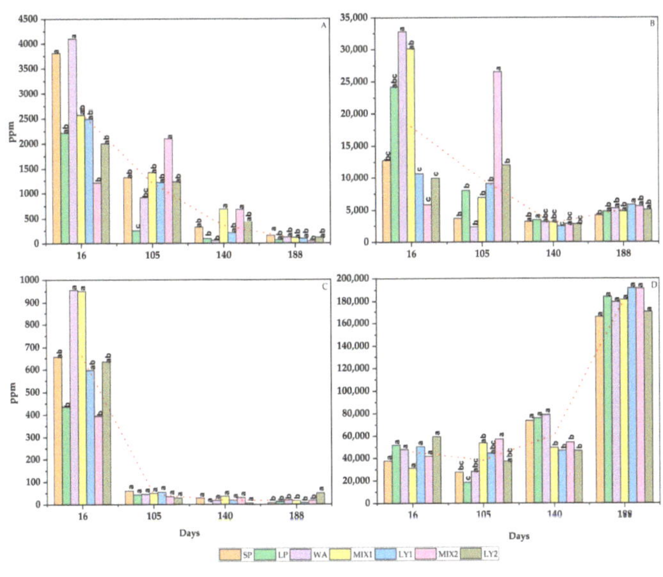

Figure 3. Lactic acid (**A**), acetic acid (**B**) and pyruvic acid (**C**) and ethanol (**D**) values in the brines during fermentations. The data are expressed as averages of triplicate measurements. Significant differences (LSD) are indicated by different letters ($p \leq 0.05$). The red dotted line represents the average values at each sampling time.

Citric acid concentrations were very low in the brines throughout the fermentations, but an increase was observed at the last sampling time on the 188th day when it reached a mean concentration of 294.17 ppm (Figure 4A). Malic acid was detected in the brines already on the 16th day at concentrations between 1075.4 ppm in WA and 415.07 ppm in MIX2, then it decreased to almost disappearing on the 140th day; from the 16th to 105th day, it strongly decreased in all treatments except for the MIX2 (Figure 4B). The succinic acid concentration in the brines was very low on the 16th day of fermentation, then an increase was observed. On the 105th and 140th days, the mean concentrations were 214.72 ppm and 199.03 ppm, respectively, while they further decreased on the 188th day except for in the LY2 treatment (Figure 4C).

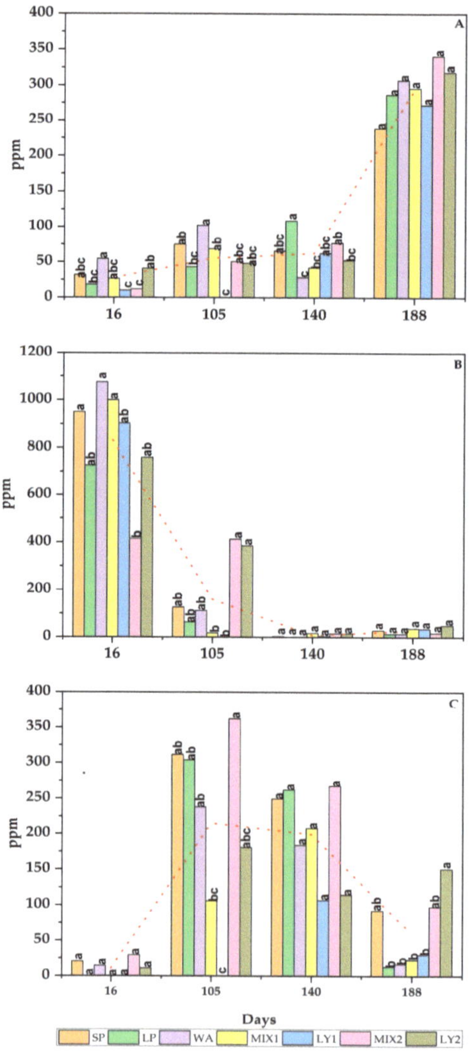

Figure 4. Citric acid (**A**), malic acid (**B**) and succinic acid (**C**) in the brines. The data are expressed as averages of the triplicate measurements. Significant differences (LSD) are indicated by different letters ($p \leq 0.05$). Red dotted line represents the average values at each sampling time.

3.2.3. pH

pH levels quickly dropped to values between 4.37 (LY1) and 4.43 (SP) in the second week of fermentation (Figure 5). A slow increase was observed starting from the 40th day, reaching a 5.02 mean value on the 70th day and almost neutrality (6.76 mean value) at the end of fermentation. Significant differences were found between the treatments from the start, with lower pH in the inoculated treatments compared to spontaneous fermentation; these differences became gradually more pronounced, evidencing lower pH in treatments with the inocula of both yeasts and LAB strains (MIX1, LY1, MIX2, LY2), with the lowest values detected in the MIX2 and LY2 treatments at the last sampling time.

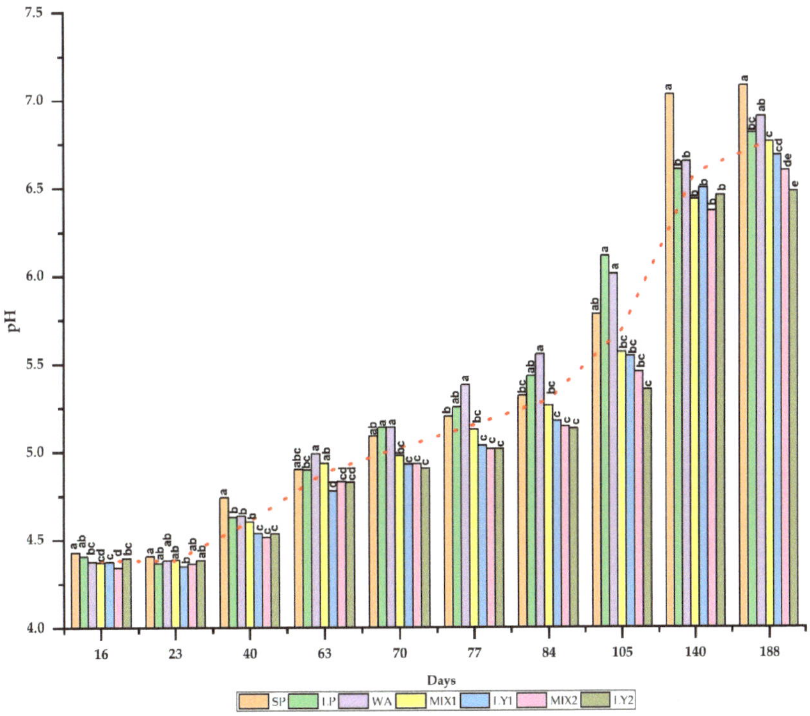

Figure 5. Brine pH values during fermentation. The data are expressed as averages of the triplicate measurements. Significant differences (LSD) are indicated by different letters ($p \leq 0.05$). The red dotted line represents the average pH values at each sampling time.

3.2.4. Phenolic Compounds

Oleuropein aglycone, hydroxytyrosol and tyrosol were quantified in raw fruits (Table 1) and during fermentation both in the brines (Figure 6A–C) and olive pulp (Table 4) during fermentations.

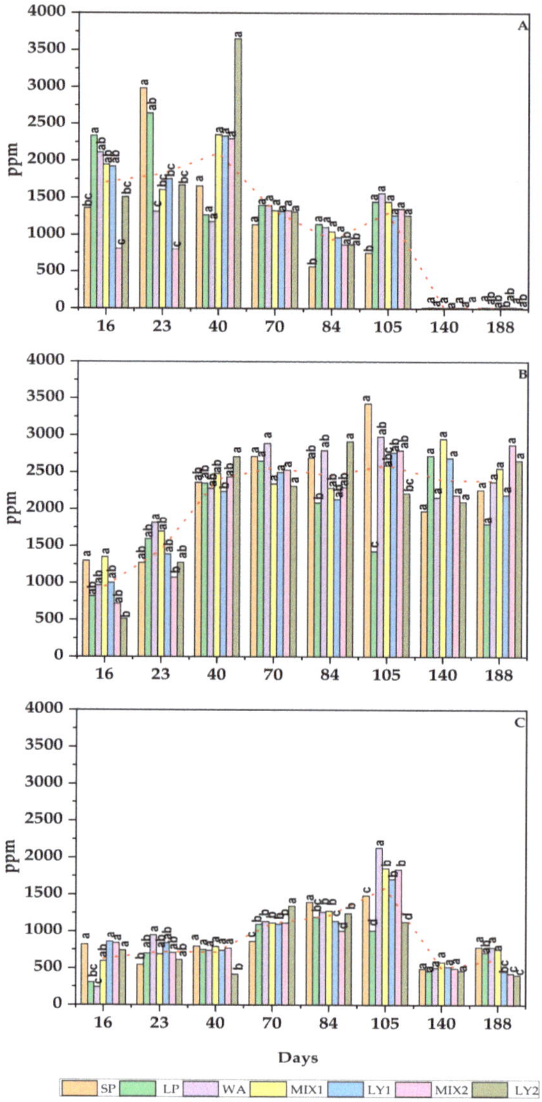

Figure 6. Polyphenol compound concentrations, oleuropein aglycone (**A**), hydroxytyrosol (**B**) and tyrosol (**C**) in the brines during fermentation. The data are expressed as averages of the triplicate measurements. Significant differences (LSD) are indicated by different letters ($p \leq 0.05$). The red dotted line represents the average of a single molecule for each sampling time.

Table 4. Change in concentration of oleuropein, hydroxytyrosol and tyrosol in the pulp of olives during fermentations, measured by HPLC analysis. Different letters indicate significant differences (LSD) among the different treatments ($p \leq 0.05$) within the same sampling time. Data are the averages from three replicates ± standard error.

Day	23	40	105	140	188
Treatments			**Oleuropein (ppm)**		
SP	2368.83 ± 1194.84 [bc]	555.58 ± 502.88 [a]	258.69 ± 26.71 [a]	724.3 ± 680.23 [a]	435.61 ± 63.41 [a]
LP	4709.37 ± 455.77 [a]	342.6 ± 37.11 [a]	228.39 ± 51.85 [a]	313.01 ± 198.86 [a]	456.20 ± 159.71 [a]
WA	4281.64 ± 647.67 [a]	1887.59 ± 2613.48 [a]	242.86 ± 53.21 [a]	421.81 ± 144.07 [a]	328.18 ± 53.63 [a]
MIX1	3821.5 ± 159.55 [ab]	398.43 ± 146.34 [a]	379.31 ± 173.46 [a]	410.80 ± 108.86 [a]	545.42 ± 368.49 [a]
LY1	3969.62 ± 1211.10 [a]	374.93 ± 32.43 [a]	283.02 ± 97.63 [a]	415.54 ± 81.80 [a]	337.83 ± 193.36 [a]
MIX2	2470.98 ± 1182.53 [bc]	358.59 ± 30.75 [a]	270.93 ± 49.17 [a]	450.19 ± 117.43 [a]	318.80 ± 98.89 [a]
LY2	1367.09 ± 394.85 [c]	384.06 ± 61.36 [a]	230.36 ± 63.59 [a]	466.44 ± 77.55 [a]	290.03 ± 21.01 [a]
			Hydroxytyrosol (ppm)		
SP	2009.22 ± 1752.86 [a]	2130.88 ± 823.28 [b]	2428.56 ± 174.15 [a]	1940.845 ± 860.32 [a]	1917.57 ± 879.92 [ab]
LP	2881.68 ± 78.045 [a]	3567.21 ± 257.33 [a]	2381.63 ± 399.71 [a]	1675.35 ± 397.48 [a]	1725.02 ± 865.91 [ab]
WA	3574.43 ± 251.95 [a]	4034.87 ± 90.911 [a]	2494.15 ± 357.76 [a]	1506.91 ± 995.07 [a]	1877.16 ± 184.35 [ab]
MIX1	1991.57 ± 1059.90 [a]	3878.65 ± 398.48 [a]	2379.66 ± 932.88 [a]	2143.85 ± 315.19 [a]	1015.06 ± 105.44 [b]
LY1	2292.77 ± 852.84 [a]	3531.42 ± 544.91 [a]	1768.14 ± 1134.18 [ab]	1813.44 ± 582.48 [a]	1258.71 ± 666.4 [ab]
MIX2	3397.36 ± 479.51 [a]	3741.07 ± 214.48 [a]	1131.05 ± 1794.00 [b]	2407.07 ± 1195.88 [a]	1275.51 ± 239.39 [ab]
LY2	2822.89 ± 828.89 [a]	3033.28 ± 1070.70 [ab]	1411.71 ± 897.04 [b]	2276.69 ± 573.87 [a]	2254.53 ± 439.46 [a]
			Tyrosol (ppm)		
SP	719.41 ± 221.50 [b]	379.89 ± 260.59 [a]	163.25 ± 6.44 [ab]	413.38 ± 103.69 [a]	277.7 ± 40.36 [a]
LP	767.18 ± 55.87 [b]	192.77 ± 9.35 [b]	150.4 ± 13.28 [b]	357.33 ± 21.34 [a]	304.11 ± 32.64 [a]
WA	1069.78 ± 170.11 [ab]	219.16 ± 5.20 [ab]	161.03 ± 29.77 [ab]	365.74 ± 80.85 [a]	255.81 ± 24.90 [a]
MIX1	913.75 ± 65.43 [ab]	211.62 ± 13.81 [ab]	219.47 ± 56.96 [a]	359.77 ± 36.99 [a]	310.21 ± 75.51 [a]
LY1	978.33 ± 345.91 [ab]	205.96 ± 20.49 [ab]	190.66 ± 58.56 [ab]	398.12 ± 37.70 [a]	244.64 ± 75.15 [a]
MIX2	1187.25 ± 379.71 [a]	216.53 ± 21.95 [ab]	188.65 ± 25.85 [ab]	391.93 ± 71.35 [a]	284.12 ± 55.74 [a]
LY2	759.83 ± 99.78 [b]	200.88 ± 21.34 [b]	192.52 ± 36.58 [ab]	323.54 ± 84.63 [a]	261.95 ± 35.40 [a]

Oleuropein in the brines (Figure 6A) was detected starting from the first sampling time (16th day), with starting values within a range of 800 ppm and 2300 ppm, with significant differences among treatments. The highest value was recorded in LP (2336.57 ppm). Oleuropein values gradually decreased during the six months of fermentations, with significant differences among treatments in almost all sampling times. The highest percentage of oleuropein depletion (between 98.26% and 99.14%) was recorded for all treatments between the 105th and 140th day after the beginning of the test.

After 188 days, all the samples reached contents of oleuropein aglycone ranging between 12 ppm and 18 ppm, with slight but significant differences among treatments. MIX1 showed the lowest content (12.52 ppm), while SP had the highest level of oleuropein in the brine at the end of the process, with a value similar to MIX2.

In contrast with the trend of oleuropein, hydroxytyrosol in the brines (Figure 6B) gradually increased in almost all the treatments up to the 105th day. At early fermentation, after 16 days, there was a great variability of the concentration of this molecule among treatments, with values ranging between 513.00 ppm (LY2) and 1349.04 ppm (MIX1). Between 105 and 140 days, a decrease was noticed in SP, WA and MIX2 theses, while for LP and MIX1, a significant increase was recorded. At the end of fermentation, the values were all similar (between 1800 ppm and 2800 ppm) for all treatments without significant differences.

Following the same trend of hydroxytyrosol, tyrosol (Figure 6C) in the brines showed a gradual growth up to 105th days, where the highest value was showed by WA treatment (2134.88 ppm) followed by a decrease (54–69%) on the 140th days. Between 140 and 188 days, SP, LP, WA and MIX1 increased their content in tyrosol, while in the other theses, the values were quite constant. During the process, SP was the treatment with lower concentrations of tyrosol except for the sampling on the 16th, 40th and 84th days. Instead, at the end of fermentation, SP, WA, LP and MIX1 showed higher values ranging between 708.80 ppm (LP) and 781.64 ppm (SP), while the other treatments ranged between 406.93 ppm (LY2) and 457.40 ppm (LY1).

The quantifications of the main phenolic compounds identified in olive pulp are reported in Table 4. The data highlighted the different rates of hydrolysis and/or solubilization of oleuropein between the treatments only at the beginning of the fermentation, where the concentrations of oleuropein ranged between 1367.09 ppm to 4709.37 ppm, for LY2 and LP respectively. At the end of fermentation, oleuropein in olives ranged between 290 ppm (LY2) and 545 ppm (MIX 1) without significant differences among the treatments.

At the beginning of the trial, hydroxytyrosol in olives contents were between 1991.57 ppm (MIX1) and 3574.43 ppm (WA). During fermentation, it showed an increase only between the first and the second sampling for all treatments, except for SP, which showed constant values until the end of the experiment (~2000 ppm). After the third sampling time, there was a linear decrease in the hydroxytyrosol levels for all treatments except for MIX2 and LY2, which showed a higher content in the last sampling with respect to the previous one. On the 188th day, the hydroxytyrosol values ranged between 1015.06 ppm (MIX1) and 2254.53 ppm (LY2).

The quantification of tyrosol in olives showed a decrease in all treatments between the first and the second sampling, with a depletion ranging between 47% (SP) and 81% (MIX2). Except for SP, showing a deep decrease between the second and the third sampling, the other treatments maintained constant values until the 140th day, which showed a slight increase. Moving from the initial values ranging between 719 ppm (SP) and 1187 ppm (MIX2) to values between 244 ppm (LY1) and 310 ppm (MIX1), it has been shown that this polyphenol is able to pass into brine in all treatments with no significant differences among them.

3.3. Textural Proprieties

The puncture test, due to the probe morphology, mainly evaluates the texture properties of the fruit peel. Treatment significantly affected firmness and rigidity index. The sample MIX2 was characterized by the highest firmness and rigidity index and the LY1 by the softest and least rigid peel. The other samples showed quite similar peel characteristics (Table 5).

Table 5. Puncture test of final products. Different letters indicate significant differences among the treatment ($p \leq 0.05$). Data are average from three replicates ± standard error.

Treatments	Firmness (g)	Area (g × mm)	Rigidity Index (g/mm)
SP	37.59 ± 1.01^{a}	30.73 ± 1.24^{a}	20.68 ± 0.56^{ab}
LP	36.21 ± 0.91^{ab}	28.46 ± 1.01^{a}	21.01 ± 0.71^{ab}
WA	36.05 ± 0.83^{ab}	29.30 ± 1.03^{a}	19.36 ± 0.36^{ab}
MIX1	36.44 ± 0.90^{ab}	28.26 ± 0.89^{a}	20.69 ± 0.55^{ab}
LY1	33.28 ± 0.96^{b}	26.68 ± 1.16^{a}	18.90 ± 0.48^{b}
MIX2	38.23 ± 1.27^{a}	30.05 ± 1.27^{a}	21.75 ± 0.63^{a}
LY2	36.81 ± 1.04^{b}	28.04 ± 1.14^{a}	22.00 ± 0.78^{a}

The TPA test allowed for the evaluation of the texture of the olive pulp. Except for cohesiveness, all the other parameters were significantly influenced by treatment. The overall data showed that LP, WA, MIX1 and LY1 samples differed from the control SP, for their highest value of hardness, gumminess and chewiness, while MIX2 and LY2 samples seemed to be more similar to the control, even though these differences were not always statistically significant (Table 6).

Table 6. Texture Profile Analysis (TPA) of final products. Different letters indicate significant differences among the treatments ($p \leq 0.05$). Data are average from three replicates ± standard error.

Treatments	Hardness (g)	Springiness (mm)	Cohesiveness	Gumminess (g)	Chewiness (g)
SP	286.35 ± 11.47 ^{ab}	0.66 ± 0.01 ^{ab}	0.50 ± 0.01 ^{a}	142.46 ± 5.45 ^{abc}	96.79 ± 4.79 ^{bc}
LP	315.20 ± 17.71 ^{ab}	0.67 ± 0.01 ^{ab}	0.50 ± 0.01 ^{a}	155.40 ± 8.30 ^{ab}	126.90 ± 6.43 ^{a}
WA	334.29 ± 12.26 ^{ab}	0.66 ± 0.01 ^{ab}	0.48 ± 0.01 ^{a}	158.83 ± 5.65 ^{ab}	107.31 ± 4.31 ^{abc}
MIX1	337.11 ± 10.55 ^{a}	0.69 ± 0.01 ^{a}	0.48 ± 0.01 ^{a}	162.80 ± 5.14 ^{a}	116.23 ± 3.90 ^{ab}
LY1	321.51 ± 13.74 ^{ab}	0.69 ± 0.01 ^{a}	0.48 ± 0.01 ^{a}	155.81 ± 6.29 ^{abc}	115.08 ± 4.54 ^{ab}
MIX2	277.62 ± 11.61 ^{b}	0.61 ± 0.01 ^{c}	0.48 ± 0.01 ^{a}	130.37 ± 5.39 ^{c}	81.52 ± 4.04 ^{c}
LY2	314.41 ± 23.30 ^{ab}	0.65 ± 0.01 ^{cb}	0.47 ± 0.01 ^{a}	134.42 ± 5.75 ^{bc}	101.06 ± 6.45 ^{abc}

The MIX1 sample was characterized by the highest value of hardness, springiness and gumminess. On the other hand, the MIX2 sample showed the lowest values for all parameters.

4. Discussion

Yeast growth during the early phases of fermentation showed a high variability and higher cell concentrations (Figure 1B) compared to what was previously reported for spontaneous fermentations of *Leccino* [27], which is probably due to higher fermentation temperature. Less growth was observed on the 40th day in treatments in which there were no yeast inoculations (statistically significant only for LY1); this could be ascribed to an inhibition of the growth of autochthonous yeasts from killer selected strains, whose growth, due to lower adaptability to the brines with respect to autochthonous yeasts, was lower at the start, which is likely due to a longer lag phase. In particular, this was observed in LY1, where *S. cerevisiae* was added just 10 days before the analysis, and to a lesser extent, also in MIX1, where *S. cerevisiae* was added simultaneously with *L. plantarum* 40 days before the analysis. Greater growth in yeasts was observed where *W. anomalus* was added (WA, MIX2 and LY2); this could denote a major adaptation ability of *W. anomalus* to brines, or its lower killer activity in terms of *S. cerevisiae*. Yeast growth continued until the 77th day, then yeasts' cell concentrations decreased. The maximum cell concentration was reached on the 70th day for LY1 and SP, and on the 77th day for LP, WA, MIX1, MIX2 and LY2. Even if the differences observed were not statistically significant, in general, growth was higher for *W. anomalus* when added in sequential inoculations, and for *S. cerevisiae* when added in mixed inoculations with *L. plantarum*. This could indicate a better synergy between *S. cerevisiae* and LABs, while *W. anomalous* growth could benefit from slowing of the LABs' activity. In the SP treatment, yeasts dropped rapidly on the 84th day; in the other treatments, the decline was slower. MIX1 and LY2 showed the highest yeast cell concentrations.

One possible advantage of the use of starters cultures of yeasts in olive fermentation is the improvement of lactic acid bacterial growth and lactic acid production [9,13]; this effect was not observed in our assays. LABs developed early and declined after the second month (Figure 1A). The addition of starter cultures did not improve their growth with respect to spontaneous fermentations, as previously described by Lanza et al. [12]; however, a positive effect of *W. anomalus*, inoculated alone or mixed with *L. plantarum*, was observed on the 40th and 63rd days, which is statistically significant in terms of the other inoculation strategies involving yeast starters, but not for spontaneous fermentations and inoculations with the *L. plantarum* selected strain. Bleve et al. [27] reported that LABs were undetectable until the 135th day in spontaneous fermentations of *Leccino*, and they reached 1.4×10^5 CFU/mL maximum cell concentration on day 180, while in our assay, they reached much higher cell concentrations with mean values of 2–3 $\times 10^7$ CFU/mL from the 70th to the 140th day of fermentation; as for yeasts, this could be due to higher fermentation temperatures.

The growth of total aerobic bacteria (Figure 1C) reduced in the third month of fermentation when inocula of both yeasts and LABs were added, but these differences decreased during late fermentation. On the 40th day, their growth was significantly higher in MIX2, as was also observed for LAB growth. *Enterobacteriaceae* are considered undesired microorganisms, and their presence in processed foods is considered a signal for possible safety risks. In olive fermentation, they can be present during early phases, but they usually disappear soon after the growth of LABs and yeasts [27]. In our assay, their cell concentrations were higher than expected and rose until the fourth month (Figure 1D). Their growth was not prevented by LABs and/or yeasts inoculation. It could have been favored by pH rising and by the availability of mannitol as selective energy and carbon source [28].

Sugars are the main soluble components in olive tissues. During fermentation, they provide energy and carbon sources to microorganisms producing secondary metabolites correlated to the flavor of the product [11]. Raw olive flesh contains both simple sugars, mainly represented by glucose, fructose and sucrose and polyols, like mannitol and sorbitol [29]. Significant differences in sugars and polyols concentrations occur among cultivars, and they are likely related to the genotype and to different climatic and environmental conditions [11].

In our assays, sugars were rapidly consumed in the brines (Figure 2) and their concentration decreased rapidly in olives (Table 2). Among sugar alcohols, mannitol in olives on the 23rd day was unchanged with respect to the raw olives. In the brines, on the 16th day, it was present at concentrations higher than the glucose and fructose, indicating poor consumption from microorganisms. Mannitol is scarcely soluble in water: only 18% *w/v* at 25 °C, then low amounts can dissolve from olives to brines. It was still present on the 40th day in olives and on the 105th day in the brines. In olives on the 105th day, when mannitol disappeared, a significant presence of ethanol was detected (Table 3). Mannitol can be used as an energy and carbon source by several microorganisms, including microorganisms usually found in olives and brines, such as several *Enterobacteriaceae* (i.e., *Klebsiella*, *Serratia*, *Proteus*, *Escherichia coli*) [30], homofermentative LABs (while heterofermentative LABs can produce, but not consume, mannitol), yeasts (including *Saccharomyces*), fungi (including *Fusarium* spp.), and other bacteria (including *Staphylococcus aureus*) [31,32]. The simultaneous disappearance of mannitol and rise of ethanol observed on the 105th day (Table 3) could be ascribed to the development of mannitol-consuming microorganisms inside the olives.

In our assays, mannitol was about 25% of the total sugars and sugar alcohols in the olives, the most abundant after glucose (65%) (Table 3). The growth of *Enterobacteriaceae* (Figure 1D), and the presence of abundant molds, mainly *F. solani*, could be ascribed to high mannitol content, as these microorganisms can use mannitol as a carbon source. *Enterobateriaceae* usually grow early when fermentation starts, then rapidly decrease when the yeasts and LAB become dominant [9], while in our assays they were higher than usually reported for olive fermentations. In the olive tree, mannitol plays a major role in osmoregulation, and it is accumulated as a response to drought stress [33]; therefore, increased drought stress due to climate change could induce an increase in mannitol in olives in the coming years, and consequently, it could modify the equilibrium among microbial populations during fermentation in favor of mannitol-catabolizing microorganisms. Significant attention should be paid to these factors, as an increased risk of potentially pathogenic microorganisms, such as *S. aureus* or *E. coli*, or other spoilage microorganisms, such as molds [34], could arise in olives fermentation. Another sugar alcohol, sorbitol, was also present, but only in traces in raw olives (Table 1), while it rose during fermentations in olives (Table 3) and reached high concentrations in the brines only at late fermentation (Figure 2D). It is known that several yeasts can produce sorbitol, including yeasts that usually grow during olive fermentation, such as *Candida boidinii* and *S. cerevisiae*, whereas LABs do not produce sorbitol, even if some of them can catabolize it (i.e., *L. plantarum*) [31]. Therefore, sorbitol increase in the brines at late fermentation could likely be ascribed to its accumulation in yeast cells as a form of protection from osmotic stress, followed by release during cell lysis. In addition, sorbitol present in the olives could be released in the brines during late fermentation due to lysis induced by hydrolytic enzymes produced, i.e., by molds. Glycerol in olives followed the same trend as sorbitol, but it was not found in the brines (Table 3). Textural property deterioration (Tables 5 and 6) that indicates the destructuring of cell walls and membranes might result in the decompartmentalization of cellular enzymes, the release of products of degradation (i.e., glycerol from triacylglycerol hydrolysis, glucose from hydrolysis of residual oleuropein), and absorption of compounds from brine (Tables 2–4).

Organic acids in the brines originate from the olives and from microbial metabolism. Citric, malic and succinic acids are typically found as constituents in olives [35]. Citrate can also be produced by molds [36]: this could explain its progressive increase (Figure 4A). Succinic acid as well can be produced by microorganisms usually found in olive fermentation, such as yeasts, *Enterobacteriaceae* and *Propionibacteriaceae* [9,37,38]: it showed a highly variable, bell-shaped curve, which is likely due to the interplay between metabolisms of different microbial populations (Figure 4C). The rapid disappearance of malic acid (Figure 4B) can be expected, as it is easily metabolized by LABs. Lactic, acetic and pyruvic acids, along with ethanol, are the main products of sugars fermentation from heterofermentative LABs. Lactic, acetic and pyruvic acids were found early at high concentrations on the 16th day, then a decrease was observed (Figure 3). The decrease in lactic acid is a significant problem in olive fermentations due to its role in maintaining a low pH. In our assays, on the 140th day, the concentrations of lactic acid, and organic acids in general, were very low, concurrently with a very high pH (Figure 5). This could be ascribed to the appearance of molds from the 63rd day, when the pH rise started, even if also propionibacteria can oxidize lactic acid to acetic acid.

Both lactic and acetic acid played an important role in lowering pH: on the 105th day, the highest pH was found in LP, where the lactic acid concentration was the lowest, and the lowest pH was found in MIX2, where the lactic acid concentration was the highest, and acetic acid was lowest in SP and WA, where pH is high, even if less than in LP. MIX2, where the pH was the lowest, showed the highest concentrations of lactic and acetic acids. Even if the pH gradually and considerably raised until the end of fermentation, its rise was reduced in treatments where both yeasts and LABs were inoculated. The high pH, probably due to mold growth, might have favored the growth of *Enterobacteriaceae*.

The total absence of *Fusarium* spp. on MIX1, the mold responsible for olive softening as *Penicillium* spp. and *Aspergillus niger* [39], may explain the highest values of hardness and gumminess found on MIX1 treatment. Molds and *Enterobacteriaceae* could be responsible for fermented olives softening in all treatments except in MIX1, as shown by the textural proprieties measured by TPA. This result is likely due to the presence of *L. plantarum* and *S. cerevisiae* as inoculum.

These microorganisms could also be responsible for very high ethanol concentrations in the brines (Figure 3D). Ethanol was also found in olives on the 105th day (Table 3), corresponding to the highest development of molds and *Enterobacteriaceae* (Figure 1D). In addition, the slight glucose increase detected in the brines at the end of the assay, difficult to be explained, could be ascribed to cell lytic processes.

The main final purpose of fermentation is to debitter the olives; as expected in the final product, a deep decrease in oleuropein content occurred both in the brines and in the olive pulp. The detected values of oleuropein aglycone, hydroxytyrosol and tyrosol in raw olives were similar to the data reported by Servilli et al. [40] for cv. *Leccino* in terms of oleuropein, while the values of hydroxytyrosol and tyrosol were quite higher than data showed in the same paper. Ranalli et al. [41] recorded similar levels of oleuropein for *Leccino* olives at the same ripening stage (1.33 g/kg of olives).In olive pulp, the different treatments seemed not to influence the final level of oleuropein; in the brine samples, at the end of the experiment, SP and MIX2 showed the highest content of oleuropein. The lowest value of oleuropein was reached in MIX1, so it can be assumed that the use of *S. cerevisiae* had a positive effect on the oleuropein hydrolysis. This result is in good accordance with other papers, where starter-driven fermentations gave the brines with lower content of oleuropein in comparison with spontaneous fermentation [18,42,43]. As hypothesized in several works, in co-inoculated fermentation, a synergic effect occurs between the action of LABs and yeasts [44,45].

In terms of hydroxytyrosol and tyrosol, scarce data are available in the literature for raw drupes of the tested cultivar. As reported by Lanza et al. [4], the hydroxytyrosol content in debittered olives pulp is the consequence of an equilibrium between the solubilization of this molecule from pulp to brine and the enzymatic hydrolysis of oleuropein carried out by LABs. Despite a decrease in hydroxytyrosol during fermentation, the residual quantity of this molecule, ranging between 1000 ppm and 2254 ppm, provides significant nutraceutical potential for the final product and correlates to its antioxidant and antiradical activity. This property is due to the electron-donating ability of the hydroxyl groups in the ortho position, and the subsequent formation of stable intramolecular hydrogen bonds with the phenoxylic radical [46]. In other studies [47], a correlation between the use of olive mill wastewater extracts (rich in hydroxytyrosol) and reduced activities related to tumor cell behavior in lung cancer cell lines was found. The treatment LY2 showed the highest content of hydroxytyrosol, highlighting a synergic effect of LABs and *W. anomalus* inocula in the retention of this precious molecule.

5. Conclusions

Starter cultures of selected strains of killer yeasts and LABs assessed as a sustainable way to control olive fermentation by avoiding pretreatments and the use of excessive salt in the brines provide unexpected results. Even if the fermentations started regularly, reaching rapidly low pH and a good level of debittering, an excessive growth of undesirable microorganisms was observed, such as *Enterobacteriaceae* and molds, which induced progressive pH growth, olives softening, and anomalous release in the brines of the olive compounds.

The addition of microbial starters partially reduced the pH rise, in particular when both LABs and yeasts were inoculated, but after six months of fermentation, the brines reached almost neutral pH. The excessive growth of undesirable microorganisms could have been induced by high winter temperatures and the availability of selective carbon and energy sources such as mannitol, whose concentration in olives can be increased by tree drought stress during cultivation. The possible role of climate change on the quality and safety of fermented foods should be investigated further and garner more attention. A possible positive effect on the nutraceutical value of olives due to yeasts and LAB inocula can be signaled with significant production of hydroxytyrosol. Due to the recent attribution of neuroprotective activity to hydroxytyrosol and its several potential therapeutic effects against degenerative and cardiovascular diseases, preserving this compound in the final product is important. The higher presence of this molecule in starter-driven fermented olives suggests that an enhancement of the nutraceutical value of this product can be obtained with further studies involving different combinations of strains of yeasts and LABs.

Supplementary Materials: The following supporting information can be downloaded at: https://www.mdpi.com/article/10.3390/fermentation9020182/s1, Figure S1: Morphological description of *Fusarium solani*. (A) Arrow indicates long monophialides; (B) Aerial mycelium presentation: arrow indicates false heads (microconidia in situ); (C) chlamydospores produced in PDA; (D) Macroconidia from sporodochia. Scale bars: 25 μm.

Author Contributions: Conceptualization, L.B.; Formal analysis, G.F.B., C.A.M., C.M., L.B., G.C. and M.V.; Investigation, G.F.B., C.M., C.A.M. and M.V.; Supervision, L.B.; Visualization, G.F.B. and C.M.; Writing—original draft, L.B., G.C. and C.A.M.; Writing—review and editing, G.F.B. and C.M. All authors have read and agreed to the published version of the manuscript.

Funding: This research was funded by the Italian Ministry of Agricultural, Food, and Forestry Policies (MiPAAF), D.M. 93882/2017, D.M. n. 30654/2019 and D.M. 35902/2019, Project DEAOLIVA, CUP C54I17000020006. The APC was offered by MDPI.

Institutional Review Board Statement: Not applicable.

Informed Consent Statement: Not applicable.

Data Availability Statement: The data presented in this study are available in this article.

Acknowledgments: The authors acknowledge Ilaria Mannazzu for her kind willingness to provide yeast strains and operational and conceptual support; Roberto Lo Scalzo for his support in chemical analysis; Alessia Cerullo for her assistance with the formal analysis during her curricular internship, and our colleague Mauro Solomita, in honor of his well-deserved retirement.

Conflicts of Interest: The authors declare no conflict of interest.

References

1. Marsilio, V.; Lanza, B.; Pozzi, N. Progress in table olive debittering: Degradation in vitro of oleuropein and its derivatives by *Lactobacillus plantarum*. *JAOCS J. Am. Oil Chem. Soc.* **1996**, *73*, 593–597. [CrossRef]
2. Ambra, R.; Natella, F.; Bello, C.; Lucchetti, S.; Forte, V.; Pastore, G. Phenolics fate in table olives (*Olea europaea* L. cv. *Nocellara del Belice*) debittered using the Spanish and Castelvetrano methods. *Food Res. Int.* **2017**, *100*, 369–376. [CrossRef] [PubMed]
3. Xu, F.; Li, Y.; Zheng, M.; Xi, X.; Zhang, X.; Han, C. Structure properties, acquisition protocols, and biological activities of oleuropein aglycone. *Front. Chem.* **2018**, *6*, 239. [CrossRef] [PubMed]
4. Lanza, B.; Zago, M.; Di Marco, S.; Di Loreto, G.; Cellini, M.; Tidona, F.; Bonvini, B.; Bacceli, M.; Simone, N. Single and multiple inoculum of *lactiplantibacillus plantarum* strains in table olive lab-scale fermentations. *Fermentation* **2020**, *6*, 126. [CrossRef]
5. Marković, A.K.; Torić, J.; Barbarić, M.; Brala, C.J. Hydroxytyrosol, tyrosol and derivatives and their potential effects on human health. *Molecules* **2019**, *24*, 2001. [CrossRef]
6. Panagou, E.Z.; Schillinger, U.; Franz, C.M.A.P.; Nychas, G.J.E. Microbiological and biochemical profile of cv. Conservolea naturally black olives during controlled fermentation with selected strains of lactic acid bacteria. *Food Microbiol.* **2008**, *25*, 348–358. [CrossRef]
7. Charoenprasert, S.; Mitchell, A. Factors influencing phenolic compounds in table olives (*Olea europaea*). *J. Agric. Food Chem.* **2012**, *60*, 7081–7095. [CrossRef]
8. Grounta, A.; Doulgeraki, A.I.; Nychas, G.J.E.; Panagou, E.Z. Biofilm formation on *Conservolea* natural black olives during single and combined inoculation with a functional *Lactobacillus pentosus* starter culture. *Food Microbiol.* **2016**, *56*, 35–44. [CrossRef]
9. Campus, M.; Degirmencioglu, N.; Comunian, R. Technologies and trends to improve table olive quality and safety. *Front. Microbiol.* **2018**, *9*, 617. [CrossRef]
10. Bonatsou, S.; Paramithiotis, S.; Panagou, E.Z. Evolution of yeast consortia during the fermentation of *Kalamata* natural black olives upon two initial acidification treatments. *Front. Microbiol.* **2018**, *8*, 2673. [CrossRef]
11. Marsilio, V.; Campestre, C.; Lanza, B.; De Angelis, M. Sugar and polyol compositions of some European olive fruit varieties (*Olea europaea* L.) suitable for table olive purposes. *Food Chem.* **2001**, *72*, 485–490. [CrossRef]
12. Lanza, B.; Di Marco, S.; Bacceli, M.; Di Serio, M.G.; Di Loreto, G.; Cellini, M.; Simone, N. *Lactiplantibacillus plantarum* used as single, multiple, and mixed starter combined with *Candida boidinii* for table olive fermentations: Chemical, textural, and sensorial characterization of final products. *Fermentation* **2021**, *7*, 239. [CrossRef]
13. Arroyo-López, F.N.; Romero-Gil, V.; Bautista-Gallego, J.; Rodríguez-Gómez, F.; Jiménez-Díaz, R.; García-García, P.; Querol, A.; Garrido-Fernández, A. Yeasts in table olive processing: Desirable or spoilage microorganisms? *Int. J. Food Microbiol.* **2012**, *160*, 42–49. [CrossRef] [PubMed]
14. Porru, C.; Rodríguez-Gómez, F.; Benítez-Cabello, A.; Jiménez-Díaz, R.; Zara, G.; Budroni, M.; Mannazzu, I.; Arroyo-López, F.N. Genotyping, identification and multifunctional features of yeasts associated to *Bosana* naturally black table olive fermentations. *Food Microbiol.* **2018**, *69*, 33–42. [CrossRef]
15. Camiolo, S.; Porru, C.; Benítez-Cabello, A.; Rodríguez-Gómez, F.; Calero-Delgado, B.; Porceddu, A.; Budroni, M.; Mannazzu, I.; Jiménez-Díaz, R.; Arroyo-López, F.N. Genome overview of eight *Candida boidinii* strains isolated from human activities and wild environments. *Stand. Genomic Sci.* **2017**, *12*, 70. [CrossRef]
16. Hernández, A.; Martín, A.; Aranda, E.; Pérez-Nevado, F.; Córdoba, M.G. Identification and characterization of yeast isolated from the elaboration of seasoned green table olives. *Food Microbiol.* **2007**, *24*, 346–351. [CrossRef]
17. Garrido-Fernández, A.; Fernandez Diez, M.J.; Adams, M.R. *Table Olives: Production and Processing*, 1st ed.; Chapman & Hall: London, UK, 1997.
18. Chytiri, A.; Tasioula-Margari, M.; Bleve, G.; Kontogianni, V.G.; Kallimanis, A.; Kontominas, M.G. Effect of different inoculation strategies of selected yeast and LAB cultures on *Conservolea* and *Kalamàta* table olives considering phenol content, texture, and sensory attributes. *J. Sci. Food Agric.* **2019**, *100*, 926–935. [CrossRef]
19. Blana, V.A.; Grounta, A.; Tassou, C.C.; Nychas, G.J.E.; Panagou, E.Z. Inoculated fermentation of green olives with potential probiotic *Lactobacillus pentosus* and *Lactobacillus plantarum* starter cultures isolated from industrially fermented olives. *Food Microbiol.* **2014**, *38*, 208–218. [CrossRef]
20. De Castro, A.; Montaño, A.; Casado, F.J.; Sánchez, A.H.; Rejano, L. Utilization of *Enterococcus casseliflavus* and *Lactobacillus pentosus* as starter cultures for Spanish-style green olive fermentation. *Food Microbiol.* **2002**, *19*, 637–644. [CrossRef]
21. Liu, G.L.; Chi, Z.; Wang, G.-Y.; Wang, Z.P.; Li, Y.; Chi, Z.-M. Yeast killer toxins, molecular mechanisms of their action and their applications. *Crit. Rev. Biotechnol.* **2015**, *35*, 222–234. [CrossRef]
22. El-Banna, A.A.; El-Sahn, M.A.; Shehata, M.G. Yeasts Producing Killer Toxins: An Overview. *Alexandria J. Food Sci. Technol.* **2011**, *8*, 41–53. [CrossRef]

23. Mannazzu, I.; Domizio, P.; Carboni, G.; Zara, S.; Zara, G.; Comitini, F.; Budroni, M.; Ciani, M. Yeast killer toxins: From ecological significance to application. *Crit. Rev. Biotechnol.* **2019**, *39*, 603–617. [CrossRef] [PubMed]
24. Sanz-Cortés, F.; Martinez-Calvo, J.; Badenes, M.L.; Bleiholder, H.; Hack, H.; Llacer, G.; Meier, U. Phenological growth stages of olive trees (*Olea europaea*). *Ann. Appl. Biol.* **2002**, *140*, 151–157. [CrossRef]
25. Fibiani, M.; Paolo, D.; Leteo, F.; Campanelli, G.; Picchi, V.; Bianchi, G.; Lo Scalzo, R. Influence of year, genotype and cultivation system on nutritional values and bioactive compounds in tomato (*Solanum lycopersicum* L.). *Food Chem.* **2022**, *389*, 133090. [CrossRef]
26. Bleve, G.; Tufariello, M.; Durante, M.; Grieco, F.; Ramires, F.A.; Mita, G.; Tasioula-Margari, M.; Logrieco, A.F. Physico-chemical characterization of natural fermentation process of *Conservolea* and *Kalamàta* table olives and developement of a protocol for the pre-selection of fermentation starters. *Food Microbiol.* **2015**, *46*, 368–382. [CrossRef]
27. Bleve, G.; Tufariello, M.; Durante, M.; Perbellini, E.; Ramires, F.A.; Grieco, F.; Cappello, M.S.; de Domenico, S.; Mita, G.; Tasioula-Margari, M.; et al. Physico-chemical and microbiological characterization of spontaneous fermentation of *Cellina di Nardò* and *Leccino* table olives. *Front. Microbiol.* **2014**, *5*, 570. [CrossRef]
28. Wang, J.; Kim, Y.M.; Rhee, H.S.; Lee, M.W.; Park, J.M. Bioethanol production from mannitol by a newly isolated bacterium, *Enterobacter* sp. JMP3. *Bioresour. Technol.* **2013**, *135*, 199–206. [CrossRef]
29. Russo, A.; Migliorini, M.; Cecchi, L.; Cherubini, C.; Giusti, M.; Zanoni, B.; Perri, E.; Nardi, M.; Sindona, G. Identification and assay of carbohydrates in olive drupes by cesium attachment electrospray tandem mass spectrometry (ESI-MS/MS). *J. Mass Spectrom.* **2012**, *47*, 1242–1246. [CrossRef]
30. Brooks, W.F., Jr. The Enterobacteriaceae: A Preliminary Study of New Methods of Identification: Rapid Biochemical Testing and Salt Tolerance. Master Thesis, Medical University of South Carolina, Charleston, SC, USA, 1970. Available online: https://medica-musc.researchcommons.org/theses/167 (accessed on 20 January 2023).
31. Zarour, K.; Vieco, N.; Pérez-Ramos, A.; Nácher-Vázquez, M.; Mohedano, M.L.; Lopez, P. Food Ingredients Synthesized by Lactic Acid Bacteria. Microbial Production of Food Ingredients and Additives. In *Handbook of Food Bioengineering, Microbial Production of Food Ingredients and Additives*; Holban, A.M., Grumezescu, A.M., Eds.; Elsevier Inc.: Amsterdam, The Netherlands; Academic Press: Cambridge, MA, USA, 2017; pp. 89–124.
32. Bevilacqua, A.; Campaniello, D.; Speranza, B.; Sinigaglia, M.; Corbo, M.R. Survival of *Listeria monocytogenes* and *Staphylococcus aureus* in synthetic brines. Studying the effects of salt, temperature and sugar through the approach of the design of experiments. *Front. Microbiol.* **2018**, *9*, 240. [CrossRef]
33. Massenti, R.; Scalisi, A.; Marra, F.P.; Caruso, T.; Marino, G.; Lo Bianco, R. Physiological and Structural Responses to Prolonged Water Deficit in Young Trees of Two Olive Cultivars. *Plants* **2022**, *11*, 1695. [CrossRef]
34. Benítez-Cabello, A.; Ramiro-García, J.; Romero-Gil, V.; Medina, E.; Arroyo-López, F.N. Fungal biodiversity in commercial table olive packages. *Food Microbiol.* **2022**, *107*, 104082. [CrossRef] [PubMed]
35. Emmanouilidou, M.G.; Koukourikou-Petridou, M.; Gerasopoulos, D.; Kyriacou, M.C. Evolution of physicochemical constitution and cultivar-differential maturity configuration in olive (*Olea europaea* L.) fruit. *Sci. Hortic.* **2020**, *272*, 109516. [CrossRef]
36. Behera, B.C. Citric acid from *Aspergillus niger*: A comprehensive overview. *Crit. Rev. Microbiol.* **2020**, *46*, 727–749. [CrossRef] [PubMed]
37. Bavaro, S.; Susca, A.; Frisvad, J.C.; Tufariello, M.; Chytiri, A.; Perrone, G.; Mita, G.; Logrieco, A.F.; Bleve, G. Isolation, characterization, and selection of molds associated to fermented black table olives. *Front. Microbiol.* **2017**, *8*, 1356. [CrossRef] [PubMed]
38. Bonatsou, S.L.; Iliopoulos, V.; Mallouchos, A.; Gogou, E.; Oikonomopoulou, V.; Krokida, M.; Taoukis, P.; Panagou, E.Z. Effect of osmotic dehydration of olives as pre-fermentation treatment and partial substitution of sodium chloride by monosodium glutamate in the fermentation profile of *Kalamata* natural black olives. *Food Microbiol.* **2017**, *63*, 72–83. [CrossRef]
39. Lanza, B. Abnormal fermentations in table-olive processing: Microbial origin and sensory evaluation. *Front. Microbiol.* **2013**, *4*, 91. [CrossRef]
40. Servili, M.; Settanni, L.; Veneziani, G.; Esposto, S.; Massitti, O.; Taticchi, A.; Urbani, S.; Montedoro, G.F.; Corsetti, A. The use of *Lactobacillus pentosus* 1MO to shorten the debittering process time of black table olives (Cv. *Itrana* and *Leccino*): A pilot-scale application. *J. Agric. Food Chem.* **2006**, *54*, 3869–3875. [CrossRef] [PubMed]
41. Ranalli, A.; Marchegiani, D.; Contento, S.; Girardi, F.; Nicolosi, M.P.; Brullo, M.D. Variations of iridoid oleuropein in Italian olive varieties during growth and maturation. *Eur. J. Lipid Sci. Technol.* **2009**, *111*, 678–687. [CrossRef]
42. Perpetuini, G.; Caruso, G.; Urbani, S.; Schirone, M.; Esposto, S.; Ciarrocchi, A.; Prete, R.; Garcia-Gonzalez, N.; Battistelli, N.; Gucci, R.; et al. Changes in polyphenolic concentrations of table olives (cv. *Itrana*) produced under different irrigation regimes during spontaneous or inoculated fermentation. *Front. Microbiol.* **2018**, *9*, 1287. [CrossRef]
43. Anagnostopoulos, D.A.; Kamilari, E.; Tsaltas, D. Evolution of Bacterial Communities, Physicochemical Changes and Sensorial Attributes of Natural Whole and Cracked Picual Table Olives during Spontaneous and Inoculated Fermentation. *Front. Microbiol.* **2020**, *11*, 1128. [CrossRef]
44. Ramírez, E.; Brenes, M.; García, P.; Medina, E.; Romero, C. Oleuropein hydrolysis in natural green olives: Importance of the endogenous enzymes. *Food Chem.* **2016**, *206*, 204–209. [CrossRef] [PubMed]

45. D'Antuono, I.; Bruno, A.; Linsalata, V.; Minervini, F.; Garbetta, A.; Tufariello, M.; Mita, G.; Logrieco, A.F.; Bleve, G.; Cardinali, A. Fermented Apulian table olives: Effect of selected microbial starters on polyphenols composition, antioxidant activities and bioaccessibility. *Food Chem.* **2018**, *248*, 137–145. [CrossRef] [PubMed]
46. Hu, T.; He, X.W.; Jiang, J.G.; Xu, X.L. Hydroxytyrosol and its potential therapeutic effects. *J. Agric. Food Chem.* **2014**, *62*, 1449–1455. [CrossRef] [PubMed]
47. Gallazzi, M.; Festa, M.; Corradino, P.; Sansone, C.; Albini, A.; Noonan, D.M. An extract of olive mill wastewater downregulates growth, adhesion and invasion pathways in lung cancer cells: Involvement of CXCR4. *Nutrients* **2020**, *12*, 903. [CrossRef] [PubMed]

Disclaimer/Publisher's Note: The statements, opinions and data contained in all publications are solely those of the individual author(s) and contributor(s) and not of MDPI and/or the editor(s). MDPI and/or the editor(s) disclaim responsibility for any injury to people or property resulting from any ideas, methods, instructions or products referred to in the content.

fermentation

Review

Biocontainment Techniques and Applications for Yeast Biotechnology

Guilherme Pavão [1,2], Isabela Sfalcin [1,2] and Diego Bonatto [1,2,*]

1 Laboratório de Biologia Molecular e Computacional, Centro de Biotecnologia da UFRGS, Departamento de Biologia Molecular e Biotecnologia, Universidade Federal do Rio Grande do Sul, Porto Alegre 91509-900, RS, Brazil; eps.guilherme@gmail.com (G.P.); isabelasfalcin@gmail.com (I.S.)
2 Bioprocess and Biotechnology for Food Research Center (Biofood), Food Science and Technology Institute (ICTA), Federal University of Rio Grande do Sul, Av. Bento Gonçalves 9500, Porto Alegre 91501-970, RS, Brazil
* Correspondence: diego.bonatto@ufrgs.br; Tel.: +55-51-3308-9410

Abstract: Biocontainment techniques for genetically modified yeasts (GMYs) are pivotal due to the importance of these organisms for biotechnological processes and also due to the design of new yeast strains by using synthetic biology tools and technologies. Due to the large genetic modifications that many yeast strains display, it is highly desirable to avoid the leakage of GMY cells into natural environments and, consequently, the spread of synthetic genes and circuits by horizontal or vertical gene transfer mechanisms within the microorganisms. Moreover, it is also desirable to avoid patented yeast gene technologies spreading outside the production facility. In this review, the different biocontainment technologies currently available for GMYs were evaluated. Interestingly, uniplex-type biocontainment approaches (UTBAs), which rely on nutrient auxotrophies induced by gene mutation or deletion or the expression of the simple kill switches apparatus, are still the major biocontainment approaches in use with GMY. While bacteria such as *Escherichia coli* account for advanced biocontainment technologies based on synthetic biology and multiplex-type biocontainment approaches (MTBAs), GMYs are distant from this scenario due to many reasons. Thus, a comparison of different UTBAs and MTBAs applied for GMY and genetically engineered microorganisms (GEMs) was made, indicating the major advances of biocontainment techniques for GMYs.

Keywords: biocontainment; genetically engineered microorganisms; genetically modified yeasts; synthetic biology; gene circuits; auxotrophies; kill switches

Citation: Pavão, G.; Sfalcin, I.; Bonatto, D. Biocontainment Techniques and Applications for Yeast Biotechnology. *Fermentation* **2023**, *9*, 341. https://doi.org/10.3390/fermentation9040341

Academic Editor: Ronnie Willaert

Received: 6 March 2023
Revised: 26 March 2023
Accepted: 27 March 2023
Published: 29 March 2023

Copyright: © 2023 by the authors. Licensee MDPI, Basel, Switzerland. This article is an open access article distributed under the terms and conditions of the Creative Commons Attribution (CC BY) license (https://creativecommons.org/licenses/by/4.0/).

1. Introduction

Genetically engineered or modified microorganisms (GEMs/GMMs) are the workhorse for basic and applied research as well as in industry. GEMs are employed in the food industry to improve protein synthesis or to generate small molecules that impact the nutritional value of a food, such as flavor enhancers, oligosaccharides, vitamins, and amino acids [1]. GEMs are also employed for food enzyme production, including enzymes such as lactase, amylase, proteases, and phospholipases [1,2]. Another biotechnological field that benefits from the application of GEMS is bioremediation, which allows for the removal of pollutants (e.g., heavy metals) from water and soil [3]; finally, different GEMs species are employed for the production of clinically important peptides/proteins (e.g., insulin) [4], vaccines [5], biofertilizers [6,7], biocontrol [8], and biofuels [9].

Considering eukaryotic GEMs, the yeasts constitute an important group of microorganisms with many industrial applications. Yeasts are used as a cell factory for the production of chemicals and biologicals due to their ability to grow on inexpensive culture media and them having well-established fermentative technology [10]. One of the most popular yeasts employed as a GEM is *Saccharomyces cerevisiae*. *S. cerevisiae* has many different biotechnology applications, from the production of traditional fermented foods [11,12] to cell factories for the synthesis of chemicals and pharmaceuticals, such as bioethanol, propanol, butanol, artemisinic acid, and insulin precursor [10,13,14]. In addition to its biotechnological importance, *S. cerevisiae* is considered an eukaryotic model organism for basic and applied research [15,16]. Other yeast species that have biotechnological importance include the halophilic/halotolerant yeast species such as *Debaryomyces hansenii*, which can metabolize different types of carbon sources and synthesize high levels of lipids, xylitol, and flavonoids, making it an attractive model for metabolic engineering [17,18]. Methylotrophic yeasts (e.g., *Pichia pastoris* and *Ogataea polymorpha*) are very efficient in producing heterologous protein on an industrial scale, with many strains and molecular tools available for genetic/synthetic biology engineering [19,20]. Other examples of non-*Saccharomyces* species with industrial importance include *Arxula adeninivorans* and *Yarrowia lipolytica* for the production of therapeutic heterologous proteins [21]. In addition, yeast hybrid strains/chassis from the *Saccharomyces sensu stricto* complex for industrial purposes are gaining industrial importance [22]. Finally, advances in synthetic biology are leading to the generation of "synthetic microorganisms" by using the so-called "bottom-up approach", where isolated and well characterized biochemical components are modularly assembled in order to design artificial cells with specific phenotypes [23]. Another synthetic biology approach is the "top-down method", where a microbial genome is reduced to its essential genes, allowing the factory of any desired DNA sequence to "tailor-make" a specific phenotype [24]. The top-down approach is currently being applied by the international Synthetic Yeast Genome project (known as the "Yeast 2.0" or "Sc2.0" project), whose major objectives are to redesign all chromosomes of *S. cerevisiae* and generate the first synthetic eukaryote [24,25].

Despite the importance of genetically modified yeasts (GMYs), there are concerns that the widespread use of GMYs could lead to a potential exchange of modified DNA molecules with other microorganisms in an ecosystem [26,27]. These concerns can be potentialized by considering that yeasts can exchange parts of their genomes with other microorganisms by interkingdom horizontal gene transfer [28], resulting in the widespread use of a transgene and/or a synthetic DNA molecule (Figure 1). Thus, highly effective biocontainment/safeguard strategies are needed to keep GMYs restricted to a laboratory and/or industrial environments (defined here as "production facilities") [27,29] and restrict the "escape frequency" of GMYs. The escape frequency is the consequence of a combination of different molecular mechanisms (e.g., mutagenesis, gene loss, recombination) as well as environmental/evolutionary processes that result in the spread of GMY/GEM into the environment (Figure 1). The National Institutes of Health (NIH) guidelines for research involving recombinant or synthetic nucleic acid molecules recommend an escape frequency of 1 cell per 10^8 cells as the standard [30].

Another major issue regarding the massive use of GMY/GEM is that the simple physico-chemical inactivation of cell biomass (e.g., by heat and/or pH treatment) does not avoid the DNA stability/persistence in the environment after its release from the cell (Figure 1). Data gathered from the PCR analysis of soil fertilized with the waste product of industrial GMY/GEM fermentations showed the presence of genes related to antibiotic resistance commonly found in synthetic plasmids [31]. Additionally, engineered DNA molecules can be detected in soil several days or months after its introduction due to the intrinsic high stability of DNA molecules [32,33]. Interestingly, none of these studies observed the horizontal transfer of released engineered DNA molecules to the native soil microbiomes; however, the chance of horizontal gene transfer from released synthetic DNA molecules in the environment cannot be excluded, especially if those molecules confer a phenotype advantage to the native microbiome.

Different molecular approaches can be employed for GMY biocontainments, including auxotrophic mutations, synthetic molecular mechanisms (xenobiology), and kill switches based on the expression of toxin-antitoxin [27,34] or exo-/endonucleases [29,35] (Figure 1). However, all of these approaches are not failproof, since molecular and evolutive mechanisms lead to the inactivation of biocontainment devices [36] (Figure 1). This is especially true when a biocontainment approach makes use of a single component/device (defined in this work as a "uniplex-type biocontainment approach" or UTBA; Table 1 and Figure 1), such as natural auxotrophic markers. On the other hand, the combination of different biocontainment approaches ("multiplex-type biocontainment approach" or MTBA; Table 1) reduces its potential escape frequency at the expense of low cell fitness (Table 1) [27,34]. Paradoxically, the reduction in cell fitness increases the frequency of inactivated biocontainment devices [37]. Thus, an effective biocontainment strategy should consider different escape mechanisms, including mutagenic drift, the environmental supplementation of nutrients and/or essential molecules, and horizontal gene transfer (HGT) [38].

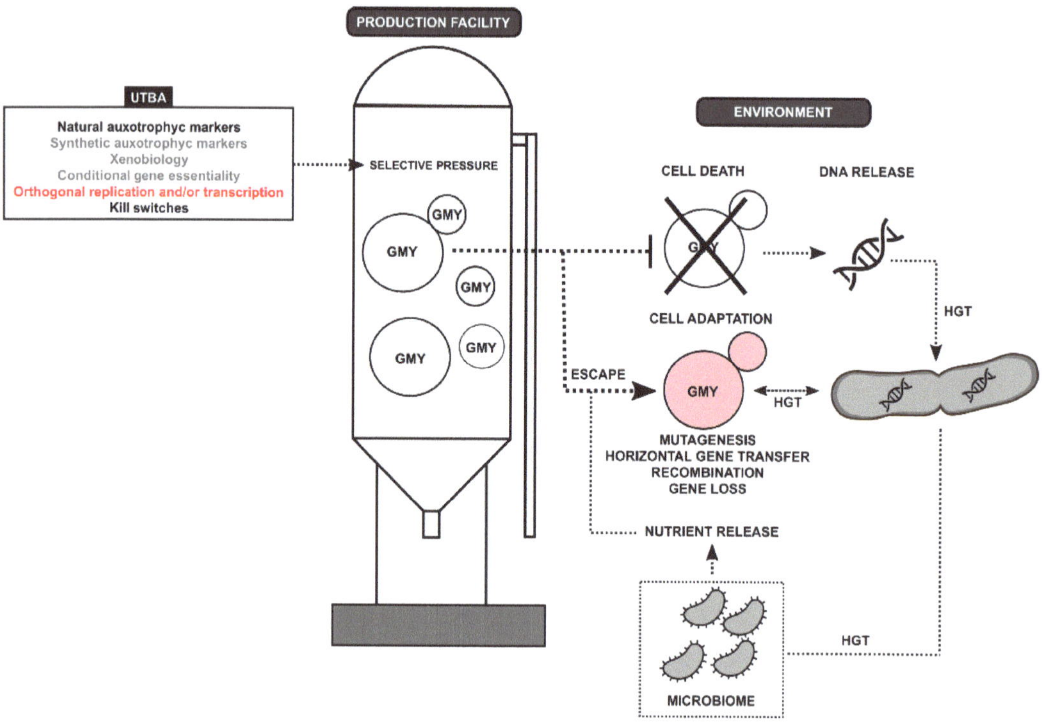

Figure 1. Diagram of major uniplex-type biocontainment approaches (UTBAs) applied for genetically modified yeast (GMY). Under different selective pressures induced by UTBAs, GMY cells are kept active/live in a production facility, where the absence of these selective pressures leads to cell death in an environment. However, a small fraction of GMY cells have the potential to escape the UTBA mechanisms, where different molecular mechanisms, such as mutagenesis, horizontal gene transfer (HGT), DNA recombination, and gene loss, result in GMY adaptability to the environment. In some cases, the environment-associated microbiome could promote GMY adaptability by releasing nutrients that bypass UTBA (e.g., natural auxotrophies). On the other hand, GMY cell death can release synthetic/transgenic DNA molecules that remain stable in the environment and can be incorporated by HGT into different environment-associated microorganisms, thus changing the microbiome. The colors of UTBAs (inset) indicate the experimental status of the technology for GMYs: black, technology extensively tested inside and outside the production facility; gray, experimental technology not tested outside the production facility; red, technology not applied for GMYs and other microorganisms.

Table 1. Advantages and disadvantages of different biocontainment approaches for genetically modified yeasts.

Disadvantages	Advantages	Biocontainment
1. Can be inactivated by molecular and evolutive mechanisms. 2. Presence of metabolites from other organisms can surpass biocontainment. 3. Potential horizontal gene transfer events can surpass biocontainment mechanisms.	1. Can be easy to implement for different yeast species. 2. Some approaches were extensively tested outside production facilities. 3. Orthogonal components diminish the impact of the biocontainment apparatus on cellular physiology. 4. Can employ synthetic small molecules or DNA parts to prevent biocontainment escape. 5. Some UTBAs can be combined into an MTBA.	Uniplex-type approach (UTBA)
1. Complex design and implementation. 2. Lower cell fitness and promotes the selection of inactive biocontainment devices. 3. Not extensively tested outside production facilities.	1. Designed with redundant or safeguard mechanisms for mutation tolerance. 2. Orthogonal circuits that respond to specific environmental inputs to induce cell death outside the production facility. 3. Can combine molecular mechanisms targeting different cellular components. 4. Can apply synthetic molecules or biological components.	Multiplex-type approach (MTBA)

The purpose of this review manuscript is to focus on different UTBA and MTBA strategies for GMY biocontainment and on what approaches have been or have not been applied for industrially important GMYs.

2. Uniplex-Type Biocontainment Approaches

2.1. Natural Auxotrophic Markers

Natural auxotrophy (Figure 1) is a simple and widespread UTBA technique commonly applied to prevent the release and proliferation of GMYs outside the production facility. In general terms, natural auxotrophy can be defined as a nutritional deficiency induced by mutated genes resulting in a GMY strain that depends on the addition of the nutrient on its growth media [39]. The yeast *S. cerevisiae* has been used as a model for auxotrophic markers due to the facility of inducing deletion and/or point mutations in different genes associated with nutrient metabolism [39]. In this sense, genes linked to the metabolism of amino acids such as L-histidine, L-leucine, L-tryptophan, and L-methionine [39] and nitrogen bases (e.g., adenine and uracil) [40] have been used as auxotrophic markers for decades.

Considering nitrogen base auxotrophies, the allele *ade2-1*, which contains a nonsense mutation (Glu64STOP) [41], is widely found in different laboratorial *S. cerevisiae* strains and confers adenine dependence. A major phenotypic characteristic of *ade2-1* is the development of a red ochre color due to the intracellular accumulation of oxidized adenine-associated metabolic intermediates [40,42], a phenotype that can be useful for red–white colony screening [43], redox biology [44], and drug discovery [40].

The depletion of adenine reserves in yeast cells increases trehalose synthesis and leads to cell cycle arrest, recapitulating the protective effects observed for desiccation stress tolerance. As a consequence, the adenine-deficient cells become viable for a longer period of time and evade biocontainment [42]. Complementally, it was observed that adenine auxotrophy increases the mutagenesis rate in yeast cells [45], a condition that is observed for other auxotrophic markers such as *leu2* and/or *lys2* alleles. Yeast cells carrying *leu2* and/or *lys2* alleles, when subjected to L-lysine or L-leucine starvation, display an increase in the number of respiratory deficient cells (*rho*$^-$ cells) due to the accumulation of mutations in the mitochondrial genome, a condition termed "adaptive mutation" [46]. Additionally, adaptive mutations are linked to auxotrophy marker reversion to the wild-type state, as observed for the *his4* marker in yeast under selective pressure [47].

In order to avoid the reversion of auxotrophic markers, a series of deletion gene markers have been generated for *S. cerevisiae* [48], as well as for other non-*Saccharomyces* species, such as *Kluyveromyces lactis* [49]. However, it has been observed in some non-*Saccharomyces* species that auxotrophic gene deletion leads to a bradytroph/leak auxotroph phenotype, such as those observed for *PHA2* in *Pichia pastoris*, which is linked to L-phenylalanine biosynthesis [50]. In this case, the auxotrophic *phe*$^-$ cells are able to survive under L-phenylalanine starvation due to alternative L-phenylalanine biosynthesis mechanisms [50].

Considering the impact of auxotrophy in yeast metabolism, the effect of adaptive mutation in the selection of prototrophic cells, and bradytrophy due to the presence of poorly described biochemical pathways in non-*Saccharomyces* species, natural auxotrophy *per se* has strong limitations regarding its use as a biocontainment strategy. In addition, the presence of other microorganisms in the environment and/or production facility (Figure 1) can provide the auxotrophy-dependent nutrients for GMY mutants, bypassing the auxotrophy requirement [51].

A solution for overcoming the natural auxotrophies limitations is the use of synthetic amino acids and nitrogen bases that are not available outside the production facility and cannot be biologically supplied. This approach, termed "xenobiology" or "synthetic auxotrophies" (Figure 1), makes use of top-down synthetic biology techniques for engineering GMYs and creating new biocontainment strategies [35]. Another approach to circumventing the limitations of natural auxotrophies is the use of a so-called "conditional gene essentiality" (Figure 1), which employs promoter engineering to modify genes linked to nutrient metabolism and strictly regulate their expression by using synthetic molecules and/or orthogonal RNA polymerases [52].

2.2. Xenobiology and Synthetic Auxotrophies

Xenobiology focuses on the development of synthetic biological devices and systems that utilize non-canonical amino acids (ncAAs), nucleic acids with a non-standard sugar backbone (xeno-nucleic acids or XNA), and non-natural nitrogen base pairs for different purposes, including the design of synthetic metabolic processes (neo-metabolism) [53] and biocontainment [52]. Many of the xenobiology devices are designed by principle using orthogonality, where synthetic components (e.g., proteins, RNAs, DNAs, and small molecules) are engineered for a purposed function and will not interfere with the natural biochemistry of a host cell [53,54]. The orthogonality also ensures that these components will not be used by natural biological systems, making it useful as a biocontainment strategy [55,56].

Biocontainment-based xenobiology is mostly centered on the use of different ncAAs for protein synthesis for both GEM and GMY. For example, the application of genetic code expansion (GCE) or orthogonal translation systems (OTSs) techniques [57,58] has been used with success in *Escherichia coli* biocontainment [56].

In bacteria, GCE/OTS relies on an orthogonal pyrrolysyl-tRNA synthetase/tRNA$^{Pyl}{}_{CUA}$ (PylRS/tRNA$^{Pyl}{}_{CUA}$) pair derived from *Methanosarcina barkeri*, *M. mazei*, or *Methanocaldococcus jannaschii* to incorporate ncAAs in proteins [59,60], in the reassignment of the UAG amber codon, a rare stop codon in both *E. coli* and *S. cerevisiae* [61], and in the deletion of the release factor 1 (RF1). Considering GCE for *E. coli* biocontainment, the genome of this bacterium was refactored by the introduction of a reassigned UAG codon into 22 essential genes together with a *Methanocaldococcus jannaschii* PylRS that is able to incorporate L-phenylalanine derivatives into tRNA$^{Pyl}{}_{CUA}$ [56], generating a synthetic auxotrophy. Data from this work indicated that the synthetic *E. coli* auxotroph cells were dependent on the addition of L-phenylalanine derivatives; moreover, these synthetic auxotrophs have undetectable escape frequencies in both solid and liquid culture media [56]. Additional work related to the creation of *E. coli* synthetic auxotroph strains has been carried out [38], pointing to the feasibility of this biocontainment technique for bacteria.

The generation of yeast ncAAS-dependent synthetic auxotrophs was achieved with very limited success [58,62]. Yeasts are naturally able to incorporate ncAAs into proteins [61,63] and a GCE/OTS technique for *S. cerevisiae* was developed by Chin et al. [62]. In this work, the authors engineered orthogonal codons, anticodons, and tRNA synthetases, including an *E. coli* tyrosine-tRNA synthetase (TyrRS) and an amber suppressor $tRNA^{Tyr}{}_{CUA}$, to generate a library of TyrRS mutants that pair only with ncAAs [62]. Once this TyrRS library was transformed into an *S. cerevisiae* strain, five different ncAAs were incorporated into the human superoxide dismutase 1 protein (hSOD1) [62].

Other relevant works related to the development of GCE/OTS for *S. cerevisiae* and *Pichia pastoris* have been conducted by using the *E. coli* TyrRS/$tRNA^{Tyr}{}_{CUA}$ or leucyl-(Leu)RS/$tRNA^{Leu5}{}_{CUA}$ pairs for ncAA incorporation into proteins [60,64–72]. However, the use of GCE/OTS in yeast has some major challenges, including the low expression of $tRNA^{Pyl}{}_{CUA}$ in yeast due to the absence of intragenic promoter sequences A- and B-boxes [58,60]; moreover, the eukaryotic release factor 1 (eRF1), codified in yeast by the *SUP45* essential gene [73], composes the Sup45p-Sup35p complex that is necessary to end translation by binding into all three stop codons in yeast cells [74]. Comparatively, the *E. coli* RF1, which recognizes the UAG/UAA codons, can be deleted without major physiological impacts into the cell due to the functional superimposition with the release factor 2 (RF2) [75].

In order to circumvent the limitations inherently associated with the implementation of a GCE/OST-based biocontainment in GMY, different approaches were applied, such as (i) prospecting new archeal PylRS with higher ncAA incorporation efficiency in *S. cerevisiae* [76], (ii) selecting yeast strains with specific mutations (e.g., *yil014c-a*Δ and *alo1*Δ) for increasing ncAAs incorporation into $tRNA^{Pyl}{}_{CUA}$ [77], (iii) improving *E. coli* TyrRS and LeuRS with enhanced ncAAs polyspecificity and efficiency by using random mutagenesis and directed selection [78], (iv) the use of an OTS based on the recognition of quadruplet codons by an engineered orthogonal ribosome that allows for the expansion of the genetic code to 256 codons (this quadruplet codon-base OTS has been implemented more or less successfully in *E. coli* [58,79,80] and mammalian cells [81]), and (v) the use of synthetic/unnatural nitrogen base pairs (UBPs) to expand the genetic code and increase the repertoire of ncAAs that can be used for protein synthesis in *E. coli* [82].

There are different approaches when considering the use of UPBs and XNAs for xenobiology and GMY/GEM biocontainment [83]. In fact, it is expected that XNA technology could be efficient for GMY/GEM biocontainment since the building blocks of XNAs (e.g., nucleobases, sugar moieties and phosphate-modified groups) cannot be found in natural environments and the GMY/GEM cells should be able to incorporate these building blocks into new XNA polymers by the usage of specialized polymerization enzymes and transmembrane proteins that are able to take up these precursors [84,85]. Thus, UBPs and XNAs can be like a "genetic firewall" [84], where HGT events could be avoided regarding the restrained aspects of XNA technology.

The developments of XNA technologies follow two major mainstreams: (i) the use of unnatural nucleobases, sugar moieties, and phosphate-modified groups (XNA substrates) to incorporate into XNAs or hybrid XNA/DNA/RNA polymers by canonical DNA and/or RNA polymerases [55,86]; and (ii) the design of new DNA and/or RNA enzymes (XNAzymes) that are able to metabolize XNA polymers and/or substrates with an "alien" chemistry, such as threose nucleic acid (TNA), cyclohexenyl nucleic acid (CeNA), arabino nucleic acid (ANA), 2′-fluoro-arabino nucleic acid (FANA), glycol nucleic acid (GNA), and locked nucleic acid (LNA) [87–94]. In all cases, the XNAs should display orthogonality in vivo with little or, preferentially, no interaction with the canonical components of the DNA/RNA metabolism [95]. Unfortunately, many different XNA technologies were not implemented in vivo, which renders their usage for GMY/GEM biocontainment still far away from a technical and practical viewpoint [96]. Finally, both GCE/OTS and XNA technologies were not tested in an open and uncontrolled environment (Figure 1), making their behavior unpredictable for real biocontainment applications [26].

Another approach related to the induction of synthetic auxotrophies in *E. coli* is based on the selection of essential proteins whose structure and activity are dependent on the presence of small molecule ligands, such as the so called "synthetic auxotrophs based on ligand-dependent essential genes" (SLiDE) technique [97]. In this work, the authors were able to select five essential genes in *E. coli* by applying protein engineering and saturation mutagenesis and generate proteins dependent on benzothiazole. The authors related a very low escape frequency (<3×10^{-11}) under laboratory assays [97]. Similar works on bacterial synthetic auxotrophs have been conducted, including phosphite-dependent *Synechococcus elongatus* [98] and *Pseudomonas putida* [99]. In *S. cerevisiae*, a series of mutations in the *CDC10* gene have been identified, which codify an essential septin protein that can be rescued in the presence of small molecules, such as guanidinium ion [100,101]. Although the authors of this work have not applied the *CDC10* conditional mutants for biocontainment, the data may indicate new techniques based on chemical rescue for GMY biocontainment.

2.3. Conditional Gene Essentiality

The conditional essentiality is, similar to auxotrophy or xenobiology, a plethora of molecular techniques where essential genes that codify for proteins and enzymes needed for cell growth and maintenance are modified to have its expression regulated by external agents [52]. In this regard, the conditional gene essentiality has been applied for biocontainment in *S. cerevisiae*, *E. coli*, and other bacteria with a low escape frequency (Figure 1) [27,52]. In *S. cerevisiae*, conditional gene essentiality was employed to regulate the expression of histone genes [27], fatty acid synthetase, mitochondrial and cytoplasmic histidine tRNA synthetase (HisRS), RNA polymerase II subunit B, and interorganellar chaperones and GTPases related to vesicular transport and fusion [102]. These genes have their original regulatory sequences replaced by a combination of promoters whose activities are modulated by galactose and estradiol [27]. Moreover, the orthogonality is ensured in this biocontainment system by providing a "fail-safe" recombination-induced lethality mechanism based on small molecule-dependent Cre recombinase [27,103].

2.4. Orthogonal DNA Replication and RNA Transcription

Another potential strategy for GMY biocontainment is to implement orthogonal DNA and RNA polymerases [51]. In yeast, an orthogonal DNA replication system derived from the *Kluyveromyces lactis* cytoplasmic plasmid pGKL1/2, named "OrthoRep", has been applied for studies of the in vivo continuous evolution of target genes due to the high error-prone activity of DNA polymerase 1 (DNAP1) and DNA polymerase 2 (DNAP2), both codified by pGLK1/2 [104–106]. Similar in vitro works were also described by using the orthogonal DNA polymerase derived from bacteriophage phi29 to generate a self-contained synthetic transcription and translation-coupled DNA replication system [107–109]. Additionally, orthogonal DNA replication systems, including XNA polymerases and/or engineered DNA polymerases [110], are being developed for xenobiology applications. However, the use of different orthogonal DNA replication mechanisms as a biocontainment system for GMY/GEM was not reported until now, pointing to an unexplored research field (Figure 1).

Like orthogonal DNA replication systems, orthogonal RNA transcription systems are also underexplored in the context of GMY/GEM biocontainment. Some examples include orthogonal genic circuits based on the use of bacteriophage T7 RNA polymerase in bacteria and yeast for different biotechnological purposes [111,112]. However, no biocontainment applications using T7 RNA polymerase-based techniques for GMY/GEM were described until now (Figure 1).

2.5. Nuclease-Based Kill Switches

Kill switches are defined as gene circuits activated by specific environmental inputs, resulting in the expression of lethal genes that lead to cell death [113]. Different kill switch designs have been used for GEM biocontainment (Figure 1), including gene circuits that respond to environmental changes by activating/deactivating a type II toxin–antitoxin system CcdB/CcdA [113]. Besides this, specific and unspecific nucleases, auxotrophies, as well as type I toxin–antitoxin pairs have been employed to engineer kill switch circuits for biocontainment purposes, many of them in an MTBA format (Table 1) [34,56,114–117] in order to reduce the GEM escape frequency.

Considering nucleases for kill switch design, its usefulness for bacteria biocontainment has been shown. For example, the expression of the *Eco*RI endonuclease–*Eco*RI methylase pair combined with conditional gene essentiality in *E. coli* considerably lowers the escape frequency [34]. In this sense, combining *Eco*RI and mf-Lon protease in a strictly regulated gene circuit allowed for the development of the "Deadman" kill switch, which considerably lowered the escape frequency associated with a high genetic stability [114]. Other non-specific nucleases, such as *nucA* from *Serratia marcescens* [118] or nuclease A from *Staphylococcus aureus*, have been employed in different bacterial biocontainment projects [117].

In *S. cerevisiae*, a kill switch based on the conditional expression of *S. marcescens nucA* under the control of a glucose-repressed *ADH2/GAPDH* hybrid promoter has been proposed [119]. This work showed that, under non-repressible conditions, the *nucA* was efficiently expressed, leading to yeast cell death in both laboratorial and soil microcosm assays [119]. Interestingly, type II restriction enzymes have been expressed in yeast to study DNA damage and repair mechanisms [120–122]. Data from the expression of type II restriction enzymes in yeast cells point to a low survival and DNA damage tolerance, especially when blunt ends are induced by type II restriction enzymes (e.g., *Pvu*II) [122]; however, the use of type II restriction enzymes as part of a kill switch mechanism in yeast cells for biocontainment purposes was not described until now.

In addition to conventional exo- and endonucleases, the CRISPR-associated nucleases (Cas) have been used in *E. coli* to design kill switches. These kill switch gene circuits make use of Cas3 [123] or Cas9 [124] and were successfully used to biocontain *E. coli* cells, both displaying an escape frequency $\leq 10^{-8}$ cells. A major advantage of the Cas-based kill switches is the use of guide RNAS (gRNAs) to select specific DNA sequence targets or microorganism strains, allowing for selectively eliminating the target strain from the microbiome [124,125]. Until now, no Cas-based kill switches were reported for GMY.

2.6. Kill Switches Based on Type I and II Toxin–Antitoxin Systems

Another kill switch extensively applied in GEM biocontainment is based on type I and II toxin–antitoxin (TA) systems. These kill switch-based TA systems efficiently lead to controlled cell death outside the production facility and can be easily engineered in GEM/GYM cells (Figure 1). What makes the TA systems so attractive for biocontainment applications is their pleiotropy, which targets different molecular mechanisms, such as DNA and mRNA synthesis, the cell cycle, nucleotide synthesis, and protein translation [126,127].

The prokaryotic TA system is classified into seven types numbered from I to VII [128,129], which act in different cell mechanisms. For example, the type I toxin–antitoxin system prevents toxin RNA translation through the binding of sRNA [128], while the type II toxin–antitoxin system is based on the interaction of an endoribonuclease and an inhibitor, forming a stable complex [130]. Type III toxin–antitoxin systems consist of the inhibition of the toxin protein by an antitoxin RNA [131]. Type IV toxin–antitoxin systems are composed of two proteins that do not form a complex; instead, the antitoxin acts as an antagonist of the toxin at its cellular targets [132]. The type V toxin–antitoxin system differs from the others by having its antitoxin cleave the toxin mRNA [133]. Type VI is composed of an antitoxin-based proteolytic adapter that degrades the toxin-based protein [134], while type VII is composed of an antitoxin that enzymatically neutralizes the toxin by post-translational modification [129].

Type I and II are historically employed for biocontainment gene circuit design, especially for *E. coli* [135]. For example, the type I *hok-sok* pair has been applied for the design of the conditional suicide of plasmid-containing *E. coli* cells in phosphate- [136] or tryptophan-limited [135] environments. Additionally, type I TA systems have been used in synthetic biology projects related to GEM biocontainment [117]. Similarly, type II TA systems have been applied into kill switch circuits based on the use of the *ccdB–ccdA* pair regulated by a bi-stable cI/Cro memory switch ("essentializer" circuit) and by a cold-inducible promoter ("cryodeath" circuit) [113]. Finally, type II TA systems have been applied for the design of "plasmid addiction" systems to prevent the loss of plasmids under the production facility [137]

In addition to GEM, the use of type I and II TA systems was proven to be effective for GMY biocontainment, such as the expression of the *relE–relB* system in *S. cerevisiae* [138]. When not being neutralized by the RelB antitoxin, the RelE toxin inhibits protein synthesis by cleaving mRNA that are being translated on the ribosome [139]. When RelB is neutralizing RelE, it displaces the toxin's C-terminal and forms the RelBE complex [140]. To express this TA system in *S. cerevisiae*, two recombinant plasmids were constructed, one containing the *relE* toxin gene under the control of the galactose-induced *GAL1* promoter (pKP727) and another containing *relE*, regulated by the *GAL1* promoter, and the *relB* antitoxin gene, under the control of the methionine-repressible *MET25* promoter (pKP1006). Yeast cells transformed with the plasmid pKP727 showed visible growth inhibitory effects induced by the *relE* toxin; on the other hand, yeast transformed with the plasmid pKP1006 showed a higher growth rate. These data indicate that the toxic effects of *relE* can be minimized by the *relB* antitoxin in yeast [138]. Similar to the *relE–relB* system, the ribonuclease-associated toxin Kid (killing determinant) and its antitoxin Kis (killing suppressor) [141] have been used in *S. cerevisiae* as a potential biocontainment system. For this purpose, an expression system containing the antitoxin Kis controlled by the *MET25* promoter and the Kid toxin controlled by the Cu^{2+}-induced *CUP1* promoter was cloned into the recombinant integrative plasmid pRS303. In the presence of both methionine and Cu^{2+}, the expression of the Kis antitoxin is inhibited, while the expression of the Kid toxin is induced, leading to cell death [142].

Another TA system already tested in yeast is based on the ε-ζ (*epsilon-zeta*) genes from the gram-positive bacteria *Streptococcus pyogenes* [143]. The ε-ζ genes are organized in an operon together with the ω (*omega*) gene, which codify a repressor that modulates the transcription of the ζ toxin, the ε antitoxin, and its own [143]. This TA system was expressed in a yeast two-hybrid system without including the ω sequence, through which the toxicity of the ζ protein was shown as well as the efficiency of the ε protein in antagonizing its toxin [144]. Despite the efficiency of the ε-ζ pair in inducing yeast cell death, its use as a biocontainment system for GMY was not attempted until now.

Finally, it was observed that the expression of the *chpK–chpI* TA pair from *Lepstospira interrogans* in *E. coli* and *S. cerevisiae* modulates their cell growth [145,146]. However, the application of the *chpK–chpI* TA pair as a biocontainment system was not reported until now. In addition to *S. cerevisiae*, the expression of the endoribonuclease-associated toxin *mazF* from the *E. coli mazEF* module [147] in *Pichia pastoris* results in cell death and can be useful as a counter-selectable marker for genetic modification [148].

3. Multiplex-Type Biocontainment Approaches

Multiplex-type biocontainment approaches (MTBA) (Table 1) have several advantages over UTBAs. First, the combined use of different molecular mechanisms can reduce the selective evolutionary pressure observed for the majority of UTBAs that results in mutagenic drift, HGT events, and small molecule supplementation by an ecologic niche that inactivates the biocontainment-associated gene or gene circuits [38]. Interestingly, an early work on *E. coli* showed that duplicating the biocontainment-associated gene could lower the mutational events that lead to biocontainment inactivation [149]. The same authors suggest that the use of multiple copies of biocontainment-associated genes, with different regulatory mechanisms and independent molecular targets, could also diminish the populational evolutionary pressure to negatively select mutated, non-functional biocontainment circuits [149]. In this sense, the use of synthetic biology approaches, such as redesigning the organism genome by incorporating expanded genetic code alphabets, the application of orthogonal DNA/RNA devices and/or systems, or XNA technologies, could be a solution for reducing the evolutionary pressure to inactivate the biocontainment in GMY/GEM populations [27,52–54,83,84]. Unfortunately, such approaches are far away for direct biocontainment applications in eukaryotic cells, especially yeasts. Even for prokaryotes, these technologies are mainly restricted to *E. coli*, and many of them were never tested in non-laboratorial conditions. Even considering the production facility environment, it has been reported that synthetic transcriptional circuits based on the Boolean gates "AND" and "NOR" have altered responses in laboratory-scale fermentations, negatively impacting both biomass production and gene circuit expression [150]. However, some successful applications of multiplex technologies in biocontainment were already described, which include the "Deadman" and "Passcode" kill switches in *E. coli* [114] and the transcriptional and recombinational control of essential genes in *S. cerevisiae* [27]. Another approach is the use of overlapping "safeguards" composed of engineered riboregulators that control the expression of essential genes and engineered nucleases that cleave the *E. coli* genome in the absence of exogenously supplied synthetic small molecules [34]. These examples reinforce the idea that redundancy combined with the use of different molecular approaches are crucial factors for the design of a resilient biocontainment technology.

4. Conclusions and Perspectives

More than ever, the development of new biocontainment technologies for GMY is imperative, facing the availability of synthetic biology tools and/or fermentation technologies. The data gathered so far showed that the technologies available for GMY biocontainment are far away from what is available for GEMs (e.g., *E. coli*). While technologies such as xenobiology and MTBAs are available for *E. coli*, the same is not true for yeasts that mainly rely on natural auxotrophies or other non-redundant UTBAs for biocontainment. The major factors that contribute to the lack of new GMY-linked biocontainment technologies are the complexity of the eukaryotic genome, the intracellular compartmentalization and biochemical specialization of organelles, as well as natural mechanisms that increase the genetic diversity of the yeast population, such as meiosis or even non-homologous recombination mechanisms, which lead to the inactivation of biocontainment circuits. However, with the advances in top-down synthetic biology approaches for eukaryote genomes (e.g., "Yeast 2.0"), it is expected that technologies such as xenobiology and genetic code refactoring will become available for the development of different UTBAs and MTBAs for GMY.

Funding: This work was supported by the Conselho Nacional de Desenvolvimento Científico e Tecnológico—CNPq (grant number 314558/2020-9), by the "Programa Pesquisador Gaúcho—PqG" from Fundação de Amparo à Pesquisa do Estado do Rio Grande do Sul—FAPERGS (Edital FAPERGS 07/2021 and grant number 21/2551-0001958-1), and by the "Programa de Redes Inovadoras de Tecnologias Estratégicas do Rio Grande Do Sul—RITEs-RS" from Fundação de Amparo à Pesquisa do Estado do Rio Grande do Sul—FAPERGS (Edital FAPERGS 06/2021 and grant number 22/2551-0000397-4).

Conflicts of Interest: The authors declare no conflict of interest. The sponsors had no role in the study design; the collection, analysis, and interpretation of the data; the writing of the report; or the decision to submit the article for publication.

References

1. Hanlon, P.; Sewalt, V. GEMs: Genetically Engineered Microorganisms and the Regulatory Oversight of Their Uses in Modern Food Production. *Crit. Rev. Food Sci. Nutr.* **2021**, *61*, 959–970. [CrossRef] [PubMed]
2. Olempska-Beer, Z.S.; Merker, R.I.; Ditto, M.D.; DiNovi, M.J. Food Processing Enzymes from Recombinant Microorganisms—A Review. *Regul. Toxicol. Pharmacol.* **2006**, *45*, 144–158. [CrossRef] [PubMed]
3. Chen, R.; Tu, H.; Chen, T. Potential Application of Living Microorganisms in the Detoxification of Heavy Metals. *Foods* **2022**, *11*, 1905. [CrossRef] [PubMed]
4. Singh, V.K.; Rai, S.; Singh, D.; Upadhyay, R.S. Application of Soil Microorganisms for Agricultural and Environmental Sustainability: A Review. In *Plant, Soil and Microbes in Tropical Ecosystems*; Dubey, S.K., Verma, S.K., Eds.; Rhizosphere Biology; Springer: Singapore, 2021; pp. 151–175; ISBN 9789811633645.
5. Shin, M.-K.; Yoo, H.S. Animal Vaccines Based on Orally Presented Yeast Recombinants. *Vaccine* **2013**, *31*, 4287–4292. [CrossRef] [PubMed]

6. Adetunji, C.O.; Anani, O.A.; Olaniyan, O.T.; Bodunrinde, R.E.; Osemwegie, O.O.; Ubi, B.E. Chapter 14—Sustainability of Biofertilizers and Other Allied Products from Genetically Modified Microorganisms. In *Biomass, Biofuels, Biochemicals*; Varjani, S., Pandey, A., Bhaskar, T., Mohan, S.V., Tsang, D.C.W., Eds.; Elsevier: Amsterdam, The Netherlands, 2022; pp. 363–393; ISBN 978-0-323-89855-3.
7. Ali, R.; Zulaykha, K.D.; Sajjad, N. Genetically Modified Microbes as Biofertilizers. In *Bioremediation and Biotechnology, Vol 4: Techniques for Noxious Substances Remediation*; Bhat, R.A., Hakeem, K.R., Eds.; Springer Nature: Cham, Switzerland, 2020; pp. 275–293; ISBN 978-3-030-48690-7.
8. Kebede, G.; Abera, S.; Haregu, S.; Yeshitila, A.; Palanivel, H. Genetically Modified Bacteria for Alleviating Agrochemical Impact on the Environment. In *Agrochemicals in Soil and Environment: Impacts and Remediation*; Naeem, M., Bremont, J.F.J., Ansari, A.A., Gill, S.S., Eds.; Springer Nature: Singapore, 2022; pp. 565–583; ISBN 9789811693106.
9. Ruchala, J.; Kurylenko, O.O.; Dmytruk, K.V.; Sibirny, A.A. Construction of Advanced Producers of First- and Second-Generation Ethanol in *Saccharomyces cerevisiae* and Selected Species of Non-Conventional Yeasts (*Scheffersomyces stipitis, Ogataea polymorpha*). *J. Ind. Microbiol. Biotechnol.* **2020**, *47*, 109–132. [CrossRef]
10. Wang, G.; Huang, M.; Nielsen, J. Exploring the Potential of *Saccharomyces cerevisiae* for Biopharmaceutical Protein Production. *Curr. Opin. Biotechnol.* **2017**, *48*, 77–84. [CrossRef] [PubMed]
11. Kavšček, M.; Stražar, M.; Curk, T.; Natter, K.; Petrovič, U. Yeast as a Cell Factory: Current State and Perspectives. *Microb. Cell Factories.* **2015**, *14*, 94. [CrossRef]
12. Budroni, M.; Zara, G.; Ciani, M.; Comitini, F. Saccharomyces and Non-*Saccharomyces* Starter Yeasts. In *Brewing Technology*; IntechOpen: London, UK, 2017; ISBN 978-953-51-3342-1. [CrossRef]
13. Parapouli, M.; Vasileiadis, A.; Afendra, A.-S.; Hatziloukas, E. *Saccharomyces cerevisiae* and Its Industrial Applications. *AIMS Microbiol.* **2020**, *6*, 1–31. [CrossRef]
14. Nielsen, J. Production of Biopharmaceutical Proteins by Yeast. *Bioengineered* **2013**, *4*, 207–211. [CrossRef]
15. Petranovic, D.; Nielsen, J. Can Yeast Systems Biology Contribute to the Understanding of Human Disease? *Trends Biotechnol.* **2008**, *26*, 584–590. [CrossRef]
16. Mustacchi, R.; Hohmann, S.; Nielsen, J. Yeast Systems Biology to Unravel the Network of Life. *Yeast* **2006**, *23*, 227–238. [CrossRef] [PubMed]
17. Andreu, C.; Zarnowski, R.; Del Olmo, M. Recent Developments in the Biology and Biotechnological Applications of Halotolerant Yeasts. *World J. Microbiol. Biotechnol.* **2022**, *38*, 27. [CrossRef]
18. Strucko, T.; Andersen, N.L.; Mahler, M.R.; Martínez, J.L.; Mortensen, U.H. A CRISPR/Cas9 Method Facilitates Efficient Oligo-Mediated Gene Editing in *Debaryomyces hansenii*. *Synth. Biol.* **2021**, *6*, ysab031. [CrossRef] [PubMed]
19. Potvin, G.; Ahmad, A.; Zhang, Z. Bioprocess Engineering Aspects of Heterologous Protein Production in *Pichia pastoris*: A Review. *Biochem. Eng. J.* **2012**, *64*, 91–105. [CrossRef]
20. Yan, C.; Yu, W.; Zhai, X.; Yao, L.; Guo, X.; Gao, J.; Zhou, Y.J. Characterizing and Engineering Promoters for Metabolic Engineering of *Ogataea polymorpha*. *Synth. Syst. Biotechnol.* **2022**, *7*, 498–505. [CrossRef] [PubMed]
21. Gellissen, G.; Kunze, G.; Gaillardin, C.; Cregg, J.; Berardi, E.; Veenhuis, M.; Vanderklei, I. New Yeast Expression Platforms Based on Methylotrophic *Hansenula polymorpha* and *Pichia pastoris* and on Dimorphic *Arxula adeninivorans* and *Yarrowia lipolytica*—A Comparison. *FEMS Yeast Res.* **2005**, *5*, 1079–1096. [CrossRef]
22. Borneman, A.R.; Pretorius, I.S. Genomic Insights into the Saccharomyces Sensu Stricto Complex. *Genetics* **2015**, *199*, 281–291. [CrossRef]
23. Hirschi, S.; Ward, T.R.; Meier, W.P.; Müller, D.J.; Fotiadis, D. Synthetic Biology: Bottom-Up Assembly of Molecular Systems. *Chem. Rev.* **2022**, *122*, 16294–16328. [CrossRef]
24. Venter, J.C.; Glass, J.I.; Hutchison, C.A.; Vashee, S. Synthetic Chromosomes, Genomes, Viruses, and Cells. *Cell* **2022**, *185*, 2708–2724. [CrossRef]
25. Pretorius, I.S.; Boeke, J.D. Yeast 2.0—Connecting the Dots in the Construction of the World's First Functional Synthetic Eukaryotic Genome. *FEMS Yeast Res.* **2018**, *18*, foy032. [CrossRef]
26. Arnolds, K.L.; Dahlin, L.R.; Ding, L.; Wu, C.; Yu, J.; Xiong, W.; Zuniga, C.; Suzuki, Y.; Zengler, K.; Linger, J.G.; et al. Biotechnology for Secure Biocontainment Designs in an Emerging Bioeconomy. *Curr. Opin. Biotechnol.* **2021**, *71*, 25–31. [CrossRef] [PubMed]
27. Cai, Y.; Agmon, N.; Choi, W.J.; Ubide, A.; Stracquadanio, G.; Caravelli, K.; Hao, H.; Bader, J.S.; Boeke, J.D. Intrinsic Biocontainment: Multiplex Genome Safeguards Combine Transcriptional and Recombinational Control of Essential Yeast Genes. *Proc. Natl. Acad. Sci. USA* **2015**, *112*, 1803–1808. [CrossRef] [PubMed]
28. Gonçalves, P.; Gonçalves, C. Horizontal Gene Transfer in Yeasts. *Curr. Opin. Genet. Dev.* **2022**, *76*, 101950. [CrossRef]
29. Schmidt, M.; de Lorenzo, V. Synthetic Bugs on the Loose: Containment Options for Deeply Engineered (Micro)organisms. *Curr. Opin. Biotechnol.* **2016**, *38*, 90–96. [CrossRef]
30. NIH Guidelines for Research Involving Recombinant or Synthetic Nucleic Acid Molecules (NIH Guidelines)—April 2019. 2019. Available online: https://osp.od.nih.gov/wp-content/uploads/NIH_Guidelines.pdf (accessed on 28 February 2023).
31. Andersen, J.T.; Schäfer, T.; Jørgensen, P.L.; Møller, S. Using Inactivated Microbial Biomass as Fertilizer: The Fate of Antibiotic Resistance Genes in the Environment. *Res. Microbiol.* **2001**, *152*, 823–833. [CrossRef]
32. Nielsen, K.M.; Johnsen, P.J.; Bensasson, D.; Daffonchio, D. Release and Persistence of Extracellular DNA in the Environment. *Environ. Biosaf. Res.* **2007**, *6*, 37–53. [CrossRef] [PubMed]
33. Recorbet, G.; Picard, C.; Normand, P.; Simonet, P. Kinetics of the Persistence of Chromosomal DNA from Genetically Engineered *Escherichia coli* Introduced into Soil. *Appl. Environ. Microbiol.* **1993**, *59*, 4289–4294. [CrossRef]

34. Gallagher, R.R.; Patel, J.R.; Interiano, A.L.; Rovner, A.J.; Isaacs, F.J. Multilayered Genetic Safeguards Limit Growth of Microorganisms to Defined Environments. *Nucleic Acids Res.* **2015**, *43*, 1945–1954. [CrossRef]
35. Lee, J.W.; Chan, C.T.Y.; Slomovic, S.; Collins, J.J. Next-Generation Biocontainment Systems for Engineered Organisms. *Nat. Chem. Biol.* **2018**, *14*, 530–537. [CrossRef]
36. Knudsen, S.M.; Karlström, O.H. Development of Efficient Suicide Mechanisms for Biological Containment of Bacteria. *Appl. Environ. Microbiol.* **1991**, *57*, 85–92. [CrossRef]
37. Kong, W.; Wanda, S.-Y.; Zhang, X.; Bollen, W.; Tinge, S.A.; Roland, K.L.; Curtiss, R. Regulated Programmed Lysis of Recombinant Salmonella in Host Tissues to Release Protective Antigens and Confer Biological Containment. *Proc. Natl. Acad. Sci. USA* **2008**, *105*, 9361–9366. [CrossRef] [PubMed]
38. Mandell, D.J.; Lajoie, M.J.; Mee, M.T.; Takeuchi, R.; Kuznetsov, G.; Norville, J.E.; Gregg, C.J.; Stoddard, B.L.; Church, G.M. Biocontainment of Genetically Modified Organisms by Synthetic Protein Design. *Nature* **2015**, *518*, 55–60. [CrossRef] [PubMed]
39. Pronk, J.T. Auxotrophic Yeast Strains in Fundamental and Applied Research. *Appl. Environ. Microbiol.* **2002**, *68*, 2095. [CrossRef] [PubMed]
40. Kokina, A.; Ozolina, Z.; Liepins, J. Purine Auxotrophy: Possible Applications beyond Genetic Marker. *Yeast* **2019**, *36*, 649–656. [CrossRef]
41. Ralser, M.; Kuhl, H.; Ralser, M.; Werber, M.; Lehrach, H.; Breitenbach, M.; Timmermann, B. The *Saccharomyces cerevisiae* W303-K6001 Cross-Platform Genome Sequence: Insights into Ancestry and Physiology of a Laboratory Mutt. *Open Biol.* **2012**, *2*, 120093. [CrossRef]
42. Kokina, A.; Kibilds, J.; Liepins, J. Adenine Auxotrophy—Be Aware: Some Effects of Adenine Auxotrophy in *Saccharomyces cerevisiae* Strain W303-1A. *FEMS Yeast Res.* **2014**, *14*, 697–707. [CrossRef]
43. Weng, Y.; Nickoloff, J.A. Nonselective URA3 Colony-Color Assay in Yeast Ade1 or Ade2 Mutants. *BioTechniques* **1997**, *23*, 237–242. [CrossRef]
44. Bharathi, V.; Girdhar, A.; Prasad, A.; Verma, M.; Taneja, V.; Patel, B.K. Use of Ade1 and Ade2 Mutations for Development of a Versatile Red/White Colour Assay of Amyloid-Induced Oxidative Stress in *Saccharomyces cerevisiae*: Use of Ade1 and Ade2 Mutations for Development of Colour Assay. *Yeast* **2016**, *33*, 607–620. [CrossRef]
45. Achilli, A.; Matmati, N.; Casalone, E.; Morpurgo, G.; Lucaccioni, A.; Pavlov, Y.I.; Babudri, N. The Exceptionally High Rate of Spontaneous Mutations in the Polymerase Delta Proofreading Exonuclease-Deficient *Saccharomyces cerevisiae* Strain Starved for Adenine. *BMC Genet.* **2004**, *5*, 34. [CrossRef]
46. Heidenreich, E.; Wintersberger, U. Starvation for a Specific Amino Acid Induces High Frequencies of Rho− Mutants in *Saccharomyces cerevisiae*. *Curr. Genet.* **1997**, *31*, 408–413. [CrossRef]
47. Hall, B.G. Selection-Induced Mutations Occur in Yeast. *Proc. Natl. Acad. Sci. USA* **1992**, *89*, 4300–4303. [CrossRef] [PubMed]
48. Brachmann, C.B.; Davies, A.; Cost, G.J.; Caputo, E.; Li, J.; Hieter, P.; Boeke, J.D. Designer Deletion Strains Derived from *Saccharomyces cerevisiae* S288C: A Useful Set of Strains and Plasmids for PCR-Mediated Gene Disruption and Other Applications. *Yeast* **1998**, *14*, 115–132. [CrossRef]
49. Heinisch, J.J.; Buchwald, U.; Gottschlich, A.; Heppeler, N.; Rodicio, R. A Tool Kit for Molecular Genetics of *Kluyveromyces lactis* Comprising a Congenic Strain Series and a Set of Versatile Vectors. *FEMS Yeast Res.* **2010**, *10*, 333–342. [CrossRef] [PubMed]
50. Ahmad, M.; Winkler, C.M.; Kolmbauer, M.; Pichler, H.; Schwab, H.; Emmerstorfer-Augustin, A. *Pichia pastoris* Protease-Deficient and Auxotrophic Strains Generated by a Novel, User-Friendly Vector Toolbox for Gene Deletion. *Yeast* **2019**, *36*, 557–570. [CrossRef]
51. Zhu, X.; Zhang, Z.; Jia, B.; Yuan, Y. Current Advances of Biocontainment Strategy in Synthetic Biology. *Chin. J. Chem. Eng.* **2023**, *56*, 141–151. [CrossRef]
52. Kim, D.; Lee, J.W. Genetic Biocontainment Systems for the Safe Use of Engineered Microorganisms. *Biotechnol. Bioprocess Eng.* **2020**, *25*, 974–984. [CrossRef]
53. Nieto-Domínguez, M.; Nikel, P.I. Intersecting Xenobiology and Neometabolism To Bring Novel Chemistries to Life. *ChemBioChem* **2020**, *21*, 2551–2571. [CrossRef]
54. Odar, C.; Winkler, M.; Wiltschi, B. Fluoro Amino Acids: A Rarity in Nature, yet a Prospect for Protein Engineering. *Biotechnol. J.* **2015**, *10*, 427–446. [CrossRef]
55. Handal-Marquez, P.; Anupama, A.; Pezo, V.; Marlière, P.; Herdewijn, P.; Pinheiro, V.B. Beneath the XNA World: Tools and Targets to Build Novel Biology. *Curr. Opin. Syst. Biol.* **2020**, *24*, 142–152. [CrossRef]
56. Rovner, A.J.; Haimovich, A.D.; Katz, S.R.; Li, Z.; Grome, M.W.; Gassaway, B.M.; Amiram, M.; Patel, J.R.; Gallagher, R.R.; Rinehart, J.; et al. Recoded Organisms Engineered to Depend on Synthetic Amino Acids. *Nature* **2015**, *518*, 89–93. [CrossRef]
57. de la Torre, D.; Chin, J.W. Reprogramming the Genetic Code. *Nat. Rev. Genet.* **2021**, *22*, 169–184. [CrossRef] [PubMed]
58. Sanders, J.; Hoffmann, S.A.; Green, A.P.; Cai, Y. New Opportunities for Genetic Code Expansion in Synthetic Yeast. *Curr. Opin. Biotechnol.* **2022**, *75*, 102691. [CrossRef] [PubMed]
59. Wan, W.; Tharp, J.M.; Liu, W.R. Pyrrolysyl-TRNA Synthetase: An Ordinary Enzyme but an Outstanding Genetic Code Expansion Tool. *Biochim. Biophys. Acta* **2014**, *1844*, 1059–1070. [CrossRef] [PubMed]
60. Hancock, S.M.; Uprety, R.; Deiters, A.; Chin, J.W. Expanding the Genetic Code of Yeast for Incorporation of Diverse Unnatural Amino Acids via a Pyrrolysyl-TRNA Synthetase/TRNA Pair. *J. Am. Chem. Soc.* **2010**, *132*, 14819–14824. [CrossRef] [PubMed]
61. Wiltschi, B. Incorporation of Non-Canonical Amino Acids into Proteins in Yeast. *Fungal Genet. Biol.* **2016**, *89*, 137–156. [CrossRef]
62. Chin, J.W.; Cropp, T.A.; Anderson, J.C.; Mukherji, M.; Zhang, Z.; Schultz, P.G. An Expanded Eukaryotic Genetic Code. *Science* **2003**, *301*, 964–967. [CrossRef]
63. Budisa, N.; Wenger, W.; Wiltschi, B. Residue-Specific Global Fluorination of *Candida antarctica* Lipase B in *Pichia pastoris*. *Mol. Biosyst.* **2010**, *6*, 1630–1639. [CrossRef]
64. Ai, H.-W.; Shen, W.; Brustad, E.; Schultz, P.G. Genetically Encoded Alkenes in Yeast. *Angew. Chem. Int. Ed.* **2010**, *49*, 935–937. [CrossRef]

65. Wang, Q.; Wang, L. New Methods Enabling Efficient Incorporation of Unnatural Amino Acids in Yeast. *J. Am. Chem. Soc.* **2008**, *130*, 6066–6067. [CrossRef]
66. Chen, S.; Schultz, P.G.; Brock, A. An Improved System for the Generation and Analysis of Mutant Proteins Containing Unnatural Amino Acids in *Saccharomyces cerevisiae*. *J. Mol. Biol.* **2007**, *371*, 112–122. [CrossRef]
67. Brustad, E.; Bushey, M.L.; Brock, A.; Chittuluru, J.; Schultz, P.G. A Promiscuous Aminoacyl-TRNA Synthetase That Incorporates Cysteine, Methionine, and Alanine Homologs into Proteins. *Bioorg. Med. Chem. Lett.* **2008**, *18*, 6004–6006. [CrossRef] [PubMed]
68. Tippmann, E.M.; Schultz, P.G. A Genetically Encoded Metallocene Containing Amino Acid. *Tetrahedron* **2007**, *63*, 6182–6184. [CrossRef]
69. Lemke, E.A.; Summerer, D.; Geierstanger, B.H.; Brittain, S.M.; Schultz, P.G. Control of Protein Phosphorylation with a Genetically Encoded Photocaged Amino Acid. *Nat. Chem. Biol.* **2007**, *3*, 769–772. [CrossRef] [PubMed]
70. Summerer, D.; Chen, S.; Wu, N.; Deiters, A.; Chin, J.W.; Schultz, P.G. A Genetically Encoded Fluorescent Amino Acid. *Proc. Natl. Acad. Sci. USA* **2006**, *103*, 9785–9789. [CrossRef]
71. Lee, H.S.; Guo, J.; Lemke, E.A.; Dimla, R.D.; Schultz, P.G. Genetic Incorporation of a Small, Environmentally Sensitive, Fluorescent Probe into Proteins in *Saccharomyces cerevisiae*. *J. Am. Chem. Soc.* **2009**, *131*, 12921–12923. [CrossRef]
72. Young, T.S.; Ahmad, I.; Brock, A.; Schultz, P.G. Expanding the Genetic Repertoire of the Methylotrophic Yeast *Pichia pastoris*. *Biochemistry* **2009**, *48*, 2643–2653. [CrossRef]
73. Himmelfarb, H.J.; Maicas, E.; Friesen, J.D. Isolation of the SUP45 Omnipotent Suppressor Gene of *Saccharomyces cerevisiae* and Characterization of Its Gene Product. *Mol. Cell. Biol.* **1985**, *5*, 816–822.
74. Stansfield, I.; Jones, K.M.; Kushnirov, V.V.; Dagkesamanskaya, A.R.; Poznyakovski, A.I.; Paushkin, S.V.; Nierras, C.R.; Cox, B.S.; Ter-Avanesyan, M.D.; Tuite, M.F. The Products of the SUP45 (ERF1) and SUP35 Genes Interact to Mediate Translation Termination in *Saccharomyces cerevisiae*. *EMBO J.* **1995**, *14*, 4365–4373. [CrossRef]
75. Lajoie, M.J.; Rovner, A.J.; Goodman, D.B.; Aerni, H.-R.; Haimovich, A.D.; Kuznetsov, G.; Mercer, J.A.; Wang, H.H.; Carr, P.A.; Mosberg, J.A.; et al. Genomically Recoded Organisms Expand Biological Functions. *Science* **2013**, *342*, 357–360. [CrossRef]
76. Stieglitz, J.T.; Lahiri, P.; Stout, M.I.; Van Deventer, J.A. Exploration of Methanomethylophilus Alvus Pyrrolysyl-TRNA Synthetase Activity in Yeast. *ACS Synth. Biol.* **2022**, *11*, 1824–1834. [CrossRef]
77. Zackin, M.T.; Stieglitz, J.T.; Van Deventer, J.A. Genome-Wide Screen for Enhanced Noncanonical Amino Acid Incorporation in Yeast. *ACS Synth. Biol.* **2022**, *11*, 3669–3680. [CrossRef]
78. Stieglitz, J.T.; Van Deventer, J.A. High-Throughput Aminoacyl-TRNA Synthetase Engineering for Genetic Code Expansion in Yeast. *ACS Synth. Biol.* **2022**, *11*, 2284–2299. [CrossRef]
79. Dunkelmann, D.L.; Oehm, S.B.; Beattie, A.T.; Chin, J.W. A 68-Codon Genetic Code to Incorporate Four Distinct Non-Canonical Amino Acids Enabled by Automated Orthogonal MRNA Design. *Nat. Chem.* **2021**, *13*, 1110–1117. [CrossRef] [PubMed]
80. Hankore, E.D.; Zhang, L.; Chen, Y.; Liu, K.; Niu, W.; Guo, J. Genetic Incorporation of Noncanonical Amino Acids Using Two Mutually Orthogonal Quadruplet Codons. *ACS Synth. Biol.* **2019**, *8*, 1168–1174. [CrossRef] [PubMed]
81. Niu, W.; Schultz, P.G.; Guo, J. An Expanded Genetic Code in Mammalian Cells with a Functional Quadruplet Codon. *ACS Chem. Biol.* **2013**, *8*, 1640–1645. [CrossRef] [PubMed]
82. Fischer, E.C.; Hashimoto, K.; Zhang, Y.; Feldman, A.W.; Dien, V.T.; Karadeema, R.J.; Adhikary, R.; Ledbetter, M.P.; Krishnamurthy, R.; Romesberg, F.E. New Codons for Efficient Production of Unnatural Proteins in a Semi-Synthetic Organism. *Nat. Chem. Biol.* **2020**, *16*, 570–576. [CrossRef]
83. Nie, P.; Bai, Y.; Mei, H. Synthetic Life with Alternative Nucleic Acids as Genetic Materials. *Molecules* **2020**, *25*, 3483. [CrossRef]
84. Torres, L.; Krüger, A.; Csibra, E.; Gianni, E.; Pinheiro, V.B. Synthetic Biology Approaches to Biological Containment: Pre-Emptively Tackling Potential Risks. *Essays Biochem.* **2016**, *60*, 393–410. [CrossRef]
85. Herdewijn, P.; Marlière, P. Toward Safe Genetically Modified Organisms through the Chemical Diversification of Nucleic Acids. *Chem. Biodivers.* **2009**, *6*, 791–808. [CrossRef]
86. Dhami, K.; Malyshev, D.A.; Ordoukhanian, P.; Kubelka, T.; Hocek, M.; Romesberg, F.E. Systematic Exploration of a Class of Hydrophobic Unnatural Base Pairs Yields Multiple New Candidates for the Expansion of the Genetic Alphabet. *Nucleic Acids Res.* **2014**, *42*, 10235–10244. [CrossRef]
87. Pinheiro, V.B.; Holliger, P. The XNA World: Progress towards Replication and Evolution of Synthetic Genetic Polymers. *Curr. Opin. Chem. Biol.* **2012**, *16*, 245–252. [CrossRef] [PubMed]
88. Hendrix, C.; Rosemeyer, H.; Verheggen, I.; Van Aerschot, A.; Seela, F.; Herdewijn, P. 1′, 5′-Anhydrohexitol Oligonucleotides: Synthesis, Base Pairing and Recognition by Regular Oligodeoxyribonucleotides and Oligoribonucleotides. *Chem.–A Eur. J.* **1997**, *3*, 110–120. [CrossRef]
89. Schöning, K.U.; Scholz, P.; Guntha, S.; Wu, X.; Krishnamurthy, R.; Eschenmoser, A. Chemical Etiology of Nucleic Acid Structure: The α-Threofuranosyl-(3′→2′) Oligonucleotide System. *Science* **2000**, *290*, 1347–1351. [CrossRef] [PubMed]
90. Wang, J.; Verbeure, B.; Luyten, I.; Lescrinier, E.; Froeyen, M.; Hendrix, C.; Rosemeyer, H.; Seela, F.; Van Aerschot, A.; Herdewijn, P. Cyclohexene Nucleic Acids (CeNA): Serum Stable Oligonucleotides That Activate RNase H and Increase Duplex Stability with Complementary RNA. *J. Am. Chem. Soc.* **2000**, *122*, 8595–8602. [CrossRef]
91. Noronha, A.M.; Wilds, C.J.; Lok, C.N.; Viazovkina, K.; Arion, D.; Parniak, M.A.; Damha, M.J. Synthesis and Biophysical Properties of Arabinonucleic Acids (ANA): Circular Dichroic Spectra, Melting Temperatures, and Ribonuclease H Susceptibility of ANA·RNA Hybrid Duplexes. *Biochemistry* **2000**, *39*, 7050–7062. [CrossRef]

92. Kalota, A.; Karabon, L.; Swider, C.R.; Viazovkina, E.; Elzagheid, M.; Damha, M.J.; Gewirtz, A.M. 2′-Deoxy-2′-Fluoro-β-d-Arabinonucleic Acid (2′F-ANA) Modified Oligonucleotides (ON) Effect Highly Efficient, and Persistent, Gene Silencing. *Nucleic Acids Res.* **2006**, *34*, 451–461. [CrossRef]
93. Zhang, L.; Peritz, A.; Meggers, E. A Simple Glycol Nucleic Acid. *J. Am. Chem. Soc.* **2005**, *127*, 4174–4175. [CrossRef]
94. Singh, S.K.; Koshkin, A.A.; Wengel, J.; Nielsen, P. LNA (Locked Nucleic Acids): Synthesis and High-Affinity Nucleic Acid Recognition. *Chem. Commun.* **1998**, *4*, 455–456. [CrossRef]
95. Liu, C.; Cozens, C.; Jaziri, F.; Rozenski, J.; Maréchal, A.; Dumbre, S.; Pezo, V.; Marlière, P.; Pinheiro, V.B.; Groaz, E.; et al. Phosphonomethyl Oligonucleotides as Backbone-Modified Artificial Genetic Polymers. *J. Am. Chem. Soc.* **2018**, *140*, 6690–6699. [CrossRef]
96. Diwo, C.; Budisa, N. Alternative Biochemistries for Alien Life: Basic Concepts and Requirements for the Design of a Robust Biocontainment System in Genetic Isolation. *Genes* **2019**, *10*, 17. [CrossRef]
97. Lopez, G.; Anderson, J.C. Synthetic Auxotrophs with Ligand-Dependent Essential Genes for a BL21(DE3) Biosafety Strain. *ACS Synth. Biol.* **2015**, *4*, 1279–1286. [CrossRef] [PubMed]
98. Motomura, K.; Sano, K.; Watanabe, S.; Kanbara, A.; Gamal Nasser, A.-H.; Ikeda, T.; Ishida, T.; Funabashi, H.; Kuroda, A.; Hirota, R. Synthetic Phosphorus Metabolic Pathway for Biosafety and Contamination Management of Cyanobacterial Cultivation. *ACS Synth. Biol.* **2018**, *7*, 2189–2198. [CrossRef] [PubMed]
99. Asin-Garcia, E.; Batianis, C.; Li, Y.; Fawcett, J.D.; de Jong, I.; dos Santos, V.A.P.M. Phosphite Synthetic Auxotrophy as an Effective Biocontainment Strategy for the Industrial Chassis *Pseudomonas putida*. *Microb. Cell Factories* **2022**, *21*, 156. [CrossRef] [PubMed]
100. Johnson, C.R.; Steingesser, M.G.; Weems, A.D.; Khan, A.; Gladfelter, A.; Bertin, A.; McMurray, M.A. Guanidine Hydrochloride Reactivates an Ancient Septin Hetero-Oligomer Assembly Pathway in Budding Yeast. *eLife* **2020**, *9*, e54355. [CrossRef]
101. Hassell, D.S.; Steingesser, M.G.; Denney, A.S.; Johnson, C.R.; McMurray, M.A. Chemical Rescue of Mutant Proteins in Living *Saccharomyces cerevisiae* Cells by Naturally Occurring Small Molecules. *G3 GenesGenomesGenetics* **2021**, *11*, jkab252. [CrossRef]
102. Agmon, N.; Tang, Z.; Yang, K.; Sutter, B.; Ikushima, S.; Cai, Y.; Caravelli, K.; Martin, J.A.; Sun, X.; Choi, W.J.; et al. Low Escape-Rate Genome Safeguards with Minimal Molecular Perturbation of *Saccharomyces cerevisiae*. *Proc. Natl. Acad. Sci. USA* **2017**, *114*, E1470–E1479. [CrossRef]
103. Lindstrom, D.L.; Gottschling, D.E. The Mother Enrichment Program: A Genetic System for Facile Replicative Life Span Analysis in *Saccharomyces cerevisiae*. *Genetics* **2009**, *183*, 413–422. [CrossRef]
104. Arzumanyan, G.A.; Gabriel, K.N.; Ravikumar, A.; Javanpour, A.A.; Liu, C.C. Mutually Orthogonal DNA Replication Systems In Vivo. *ACS Synth. Biol.* **2018**, *7*, 1722–1729. [CrossRef]
105. Ravikumar, A.; Arzumanyan, G.A.; Obadi, M.K.A.; Javanpour, A.A.; Liu, C.C. Scalable, Continuous Evolution of Genes at Mutation Rates above Genomic Error Thresholds. *Cell* **2018**, *175*, 1946–1957.e13. [CrossRef]
106. Strak, M.J.R.; Boyd, A.; Mileham, A.J.; Ramonos, M.A. The Plasmid-Encoded Killer System of *Kluyveromyces lactis*: A Review. *Yeast* **1990**, *6*, 1–29. [CrossRef]
107. van Nies, P.; Westerlaken, I.; Blanken, D.; Salas, M.; Mencía, M.; Danelon, C. Self-Replication of DNA by Its Encoded Proteins in Liposome-Based Synthetic Cells. *Nat. Commun.* **2018**, *9*, 1583. [CrossRef] [PubMed]
108. Sakatani, Y.; Ichihashi, N.; Kazuta, Y.; Yomo, T. A Transcription and Translation-Coupled DNA Replication System Using Rolling-Circle Replication. *Sci. Rep.* **2015**, *5*, 10404. [CrossRef] [PubMed]
109. Libicher, K.; Hornberger, R.; Heymann, M.; Mutschler, H. In Vitro Self-Replication and Multicistronic Expression of Large Synthetic Genomes. *Nat. Commun.* **2020**, *11*, 904. [CrossRef]
110. Medina, E.; Yik, E.J.; Herdewijn, P.; Chaput, J.C. Functional Comparison of Laboratory-Evolved XNA Polymerases for Synthetic Biology. *ACS Synth. Biol.* **2021**, *10*, 1429–1437. [CrossRef] [PubMed]
111. Kar, S.; Ellington, A.D. Construction of Synthetic T7 RNA Polymerase Expression Systems. *Methods* **2018**, *143*, 110–120. [CrossRef] [PubMed]
112. Morse, N.J.; Wagner, J.M.; Reed, K.B.; Gopal, M.R.; Lauffer, L.H.; Alper, H.S. T7 Polymerase Expression of Guide RNAs In Vivo Allows Exportable CRISPR-Cas9 Editing in Multiple Yeast Hosts. *ACS Synth. Biol.* **2018**, *7*, 1075–1084. [CrossRef]
113. Stirling, F.; Bitzan, L.; O'Keefe, S.; Redfield, E.; Oliver, J.W.K.; Way, J.; Silver, P.A. Rational Design of Evolutionarily Stable Microbial Kill Switches. *Mol. Cell* **2017**, *68*, 686–697.e3. [CrossRef] [PubMed]
114. Chan, C.T.Y.; Lee, J.W.; Cameron, D.E.; Bashor, C.J.; Collins, J.J. "Deadman" and "Passcode" Microbial Kill Switches for Bacterial Containment. *Nat. Chem. Biol.* **2016**, *12*, 82–86. [CrossRef]
115. Halvorsen, T.M.; Ricci, D.P.; Park, D.M.; Jiao, Y.; Yung, M.C. Comparison of Kill Switch Toxins in Plant-Beneficial *Pseudomonas fluorescens* Reveals Drivers of Lethality, Stability, and Escape. *ACS Synth. Biol.* **2022**, *11*, 3785–3796. [CrossRef]
116. Wright, O.; Delmans, M.; Stan, G.-B.; Ellis, T. GeneGuard: A Modular Plasmid System Designed for Biosafety. *ACS Synth. Biol.* **2015**, *4*, 307–316. [CrossRef]
117. Whitford, C.M.; Dymek, S.; Kerkhoff, D.; März, C.; Schmidt, O.; Edich, M.; Droste, J.; Pucker, B.; Rückert, C.; Kalinowski, J. Auxotrophy to Xeno-DNA: An Exploration of Combinatorial Mechanisms for a High-Fidelity Biosafety System for Synthetic Biology Applications. *J. Biol. Eng.* **2018**, *12*, 13. [CrossRef] [PubMed]
118. Ahrenholtz, I.; Lorenz, M.G.; Wackernagel, W. A Conditional Suicide System in *Escherichia coli* Based on the Intracellular Degradation of DNA. *Appl. Environ. Microbiol.* **1994**, *60*, 3746–3751. [CrossRef] [PubMed]
119. Balan, A.; Schenberg, A.C.G. A Conditional Suicide System for *Saccharomyces cerevisiae* Relying on the Intracellular Production of the *Serratia marcescens* Nuclease. *Yeast* **2005**, *22*, 203–212. [CrossRef]

120. Lewis, L.K.; Kirchner, J.M.; Resnick, M.A. Requirement for End-Joining and Checkpoint Functions, but Not RAD52-Mediated Recombination, after EcoRI Endonuclease Cleavage of *Saccharomyces cerevisiae* DNA. *Mol. Cell. Biol.* **1998**, *18*, 1891–1902. [CrossRef]
121. Barnes, G.; Rine, J. Regulated Expression of Endonuclease EcoRI in *Saccharomyces cerevisiae*: Nuclear Entry and Biological Consequences. *Proc. Natl. Acad. Sci. USA* **1985**, *82*, 1354–1358. [CrossRef] [PubMed]
122. Westmoreland, J.W.; Summers, J.A.; Holland, C.L.; Resnick, M.A.; Lewis, L.K. Blunt-Ended DNA Double-Strand Breaks Induced by Endonucleases PvuII and EcoRV Are Poor Substrates for Repair in *Saccharomyces cerevisiae*. *DNA Repair* **2010**, *9*, 617–626. [CrossRef] [PubMed]
123. Caliando, B.J.; Voigt, C.A. Targeted DNA Degradation Using a CRISPR Device Stably Carried in the Host Genome. *Nat. Commun.* **2015**, *6*, 6989. [CrossRef]
124. Rottinghaus, A.G.; Ferreiro, A.; Fishbein, S.R.S.; Dantas, G.; Moon, T.S. Genetically Stable CRISPR-Based Kill Switches for Engineered Microbes. *Nat. Commun.* **2022**, *13*, 672. [CrossRef]
125. Rottinghaus, A.G.; Vo, S.; Moon, T.S. Computational Design of CRISPR Guide RNAs to Enable Strain-Specific Control of Microbial Consortia. *Proc. Natl. Acad. Sci. USA* **2023**, *120*, e2213154120. [CrossRef]
126. Pantoja Angles, A.; Valle-Pérez, A.U.; Hauser, C.; Mahfouz, M.M. Microbial Biocontainment Systems for Clinical, Agricultural, and Industrial Applications. *Front. Bioeng. Biotechnol.* **2022**, *10*, 830200. [CrossRef]
127. Yeo, C.C.; Abu Bakar, F.; Chan, W.T.; Espinosa, M.; Harikrishna, J.A. Heterologous Expression of Toxins from Bacterial Toxin-Antitoxin Systems in Eukaryotic Cells: Strategies and Applications. *Toxins* **2016**, *8*, 49. [CrossRef] [PubMed]
128. Yamaguchi, Y.; Park, J.-H.; Inouye, M. Toxin-Antitoxin Systems in Bacteria and Archaea. *Annu. Rev. Genet.* **2011**, *45*, 61–79. [CrossRef] [PubMed]
129. Wang, X.; Yao, J.; Sun, Y.-C.; Wood, T.K. Type VII Toxin/Antitoxin Classification System for Antitoxins That Enzymatically Neutralize Toxins. *Trends Microbiol.* **2021**, *29*, 388–393. [CrossRef] [PubMed]
130. Kamruzzaman, M.; Wu, A.Y.; Iredell, J.R. Biological Functions of Type II Toxin-Antitoxin Systems in Bacteria. *Microorganisms* **2021**, *9*, 1276. [CrossRef]
131. Fineran, P.C.; Blower, T.R.; Foulds, I.J.; Humphreys, D.P.; Lilley, K.S.; Salmond, G.P.C. The Phage Abortive Infection System, ToxIN, Functions as a Protein-RNA Toxin-Antitoxin Pair. *Proc. Natl. Acad. Sci. USA* **2009**, *106*, 894–899. [CrossRef]
132. Masuda, H.; Tan, Q.; Awano, N.; Wu, K.-P.; Inouye, M. YeeU Enhances the Bundling of Cytoskeletal Polymers of MreB and FtsZ, Antagonizing the CbtA (YeeV) Toxicity in *Escherichia coli*. *Mol. Microbiol.* **2012**, *84*, 979–989. [CrossRef]
133. Wang, X.; Lord, D.M.; Cheng, H.-Y.; Osbourne, D.O.; Hong, S.H.; Sanchez-Torres, V.; Quiroga, C.; Zheng, K.; Herrmann, T.; Peti, W.; et al. A New Type V Toxin-Antitoxin System Where MRNA for Toxin GhoT Is Cleaved by Antitoxin GhoS. *Nat. Chem. Biol.* **2012**, *8*, 855–861. [CrossRef]
134. Aakre, C.D.; Phung, T.N.; Huang, D.; Laub, M.T. A Bacterial Toxin Inhibits DNA Replication Elongation through a Direct Interaction with the β Sliding Clamp. *Mol. Cell* **2013**, *52*, 617–628. [CrossRef]
135. Bej, A.K.; Perlin, M.H.; Atlas, R.M. Model Suicide Vector for Containment of Genetically Engineered Microorganisms. *Appl. Environ. Microbiol.* **1988**, *54*, 2472–2477. [CrossRef]
136. Schweder, T.; Schmidt, I.; Herrmann, H.; Neubauer, P.; Hecker, M.; Hofmann, K. An Expression Vector System Providing Plasmid Stability and Conditional Suicide of Plasmid-Containing Cells. *Appl. Microbiol. Biotechnol.* **1992**, *38*, 91–93. [CrossRef]
137. Kroll, J.; Klinter, S.; Schneider, C.; Voss, I.; Steinbüchel, A. Plasmid Addiction Systems: Perspectives and Applications in Biotechnology. *Microb. Biotechnol.* **2010**, *3*, 634–657. [CrossRef]
138. Kristoffersen, P.; Jensen, G.B.; Gerdes, K.; Piškur, J. Bacterial Toxin-Antitoxin Gene System as Containment Control in Yeast Cells. *Appl. Environ. Microbiol.* **2000**, *66*, 5524–5526. [CrossRef] [PubMed]
139. Pedersen, K.; Zavialov, A.V.; Pavlov, M.Y.; Elf, J.; Gerdes, K.; Ehrenberg, M. The Bacterial Toxin RelE Displays Codon-Specific Cleavage of MRNAs in the Ribosomal A Site. *Cell* **2003**, *112*, 131–140. [CrossRef]
140. Li, G.-Y.; Zhang, Y.; Inouye, M.; Ikura, M. Inhibitory Mechanism of *Escherichia coli* RelE-RelB Toxin-Antitoxin Module Involves a Helix Displacement Near an MRNA Interferase Active Site. *J. Biol. Chem.* **2009**, *284*, 14628–14636. [CrossRef] [PubMed]
141. Monti, M.C.; Hernández-Arriaga, A.M.; Kamphuis, M.B.; López-Villarejo, J.; Heck, A.J.R.; Boelens, R.; Díaz-Orejas, R.; van den Heuvel, R.H.H. Interactions of Kid–Kis Toxin–Antitoxin Complexes with the ParD Operator-Promoter Region of Plasmid R1 Are Piloted by the Kis Antitoxin and Tuned by the Stoichiometry of Kid–Kis Oligomers. *Nucleic Acids Res.* **2007**, *35*, 1737–1749. [CrossRef]
142. de la Cueva-Méndez, G.; Mills, A.D.; Clay-Farrace, L.; Díaz-Orejas, R.; Laskey, R.A. Regulatable Killing of Eukaryotic Cells by the Prokaryotic Proteins Kid and Kis. *EMBO J.* **2003**, *22*, 246–251. [CrossRef]
143. Zielenkiewicz, U.; Cegłowski, P. The Toxin-Antitoxin System of the Streptococcal Plasmid PSM19035. *J. Bacteriol.* **2005**, *187*, 6094–6105. [CrossRef] [PubMed]
144. Zielenkiewicz, U.; Kowalewska, M.; Kaczor, C.; Cegłowski, P. In Vivo Interactions between Toxin-Antitoxin Proteins Epsilon and Zeta of Streptococcal Plasmid PSM19035 in *Saccharomyces cerevisiae*. *J. Bacteriol.* **2009**, *191*, 3677–3684. [CrossRef] [PubMed]
145. Picardeau, M.; Le Dantec, C.; Richard, G.-F.; Saint Girons, I. The Spirochetal ChpK-Chromosomal Toxin–Antitoxin Locus Induces Growth Inhibition of Yeast and Mycobacteria. *FEMS Microbiol. Lett.* **2003**, *229*, 277–281. [CrossRef]
146. Picardeau, M.; Ren, S.; Saint Girons, I. Killing Effect and Antitoxic Activity of The *Leptospira Interrogans* Toxin-Antitoxin System In *Escherichia coli*. *J. Bacteriol.* **2001**, *183*, 6494–6497. [CrossRef]

147. Engelberg-Kulka, H.; Hazan, R.; Amitai, S. MazEF: A Chromosomal Toxin-Antitoxin Module That Triggers Programmed Cell Death in Bacteria. *J. Cell Sci.* **2005**, *118*, 4327–4332. [CrossRef]
148. Yang, J.; Jiang, W.; Yang, S. MazF as a Counter-Selectable Marker for Unmarked Genetic Modification of *Pichia pastoris*. *FEMS Yeast Res.* **2009**, *9*, 600–609. [CrossRef]
149. Knudsen, S.; Saadbye, P.; Hansen, L.H.; Collier, A.; Jacobsen, B.L.; Schlundt, J.; Karlström, O.H. Development and Testing of Improved Suicide Functions for Biological Containment of Bacteria. *Appl. Environ. Microbiol.* **1995**, *61*, 985–991. [CrossRef]
150. Moser, F.; Broers, N.J.; Hartmans, S.; Tamsir, A.; Kerkman, R.; Roubos, J.A.; Bovenberg, R.; Voigt, C.A. Genetic Circuit Performance under Conditions Relevant for Industrial Bioreactors. *ACS Synth. Biol.* **2012**, *1*, 555–564. [CrossRef]

Disclaimer/Publisher's Note: The statements, opinions and data contained in all publications are solely those of the individual author(s) and contributor(s) and not of MDPI and/or the editor(s). MDPI and/or the editor(s) disclaim responsibility for any injury to people or property resulting from any ideas, methods, instructions or products referred to in the content.

fermentation

Article

Single-Cell Optical Nanomotion of *Candida albicans* in Microwells for Rapid Antifungal Susceptibility Testing

Vjera Radonicic [1,2,3], Charlotte Yvanoff [1,2,3], Maria Ines Villalba [3,4], Bart Devreese [2,5], Sandor Kasas [3,4,6,†] and Ronnie G. Willaert [1,2,3,*,†]

1 Research Group Structural Biology Brussels, Vrije Universiteit Brussel, 1050 Brussels, Belgium
2 Alliance Research Group VUB-UGent NanoMicrobiology (NAMI), Vrije Universiteit Brussel, 1050 Brussels, Belgium
3 International Joint Research Group VUB-EPFL NanoBiotechnology & NanoMedicine (NANO), Vrije Universiteit Brussel, 1050 Brussels, Belgium
4 Laboratory of Biological Electron Microscopy, Ecole Polytechnique Fédérale de Lausanne (EPFL), University of Lausanne, 1015 Lausanne, Switzerland
5 Laboratory of Microbiology, Department of Biochemistry and Microbiology, Ghent University, 9000 Ghent, Belgium
6 Centre Universitaire Romand de Médecine Légale, UFAM, Université de Lausanne, 1015 Lausanne, Switzerland
* Correspondence: ronnie.willaert@vub.be; Tel.: +32-2-6291846
† These authors contributed equally to this work.

Abstract: *Candida albicans* is an emerging multidrug-resistant opportunistic pathogen representing an important source of invasive disease in humans and generating high healthcare costs worldwide. The development of a rapid and simple antifungal susceptibility test (AFST) could limit the spread of this disease, increase the efficiency of treatment, and lower the risk of developing resistant strains. We developed a microfluidic chip containing an array of microwells that were designed to trap the cells and perform rapid antifungal susceptibility tests using optical nanomotion detection (ONMD). Yeast cell entrapment in a microwell allows for a very rapid exchange of growth medium with the antifungal, which enables performing single-cell ONMD measurements on the same cell before and after antifungal treatment. The exposure to a low concentration of the antifungal caspofungin or flucanozole induced a significant decrease in the nanomotion signal, demonstrating the high sensitivity of this technique. We used this chip to quantify the real-time response of individual *C. albicans* cells to the antifungal treatment in as fast as 10 min. This simple and label-free technique could be further developed into a simple-to-use device that allows the performance of fast AFST as part of a routine hospital procedure in developed and also eventually developing world countries.

Keywords: cellular nanomotion; single cell; optical nanomotion detection; microfluidic chip; microwells; yeast; *Candida albicans*; antifungal; antifungal susceptibility testing

Citation: Radonicic, V.; Yvanoff, C.; Villalba, M.I.; Devreese, B.; Kasas, S.; Willaert, R.G. Single-Cell Optical Nanomotion of *Candida albicans* in Microwells for Rapid Antifungal Susceptibility Testing. *Fermentation* **2023**, *9*, 365. https://doi.org/10.3390/fermentation9040365

Academic Editor: Alexander Rapoport

Received: 13 February 2023
Revised: 1 April 2023
Accepted: 2 April 2023
Published: 7 April 2023

Copyright: © 2023 by the authors. Licensee MDPI, Basel, Switzerland. This article is an open access article distributed under the terms and conditions of the Creative Commons Attribution (CC BY) license (https://creativecommons.org/licenses/by/4.0/).

1. Introduction

Most human fungal infections are caused by the species that belong to the genus *Candida*. The most frequent species that cause opportunistic infections include *C. albicans*, *C. glabrata*, *C. auris*, *C. tropicalis*, *C. parapsilosis*, and *C. krusei* [1]. *C. albicans* are frequently found as an important part of the microbial flora of healthy persons and can induce systemic and superficial infections under certain environmental conditions [2]. This species causes a variety of diseases that range from vaginal infections, which affect up to 75% of women at least once in their lifetime [3], to oral infections and may also, under favourable conditions, enter the bloodstream leading to deep-tissue infections [2]. *C. albicans* may also play a role in the persistence or worsening of some chronic inflammatory bowel diseases [3]. Furthermore, invasive candidiasis (IC) is an important nosocomial infection with high morbidity and mortality rates in hospitalised children (oncology patients) [4]. Finally,

Candida species present a major source of extended infections in intensive care patients (ICU), especially in patients who are immunocompromised or elderly [2]

The fast emergence of multi-resistant pathogenic *Candida* species is caused by the extensive and sometimes unnecessary use of broad-spectrum antifungals [2]. Hence, the development of rapid antifungal sensitivity tests could allow for the identification and use of more selective antifungals, thereby reducing the spread of pathogenic fungi and their evolution toward a multi-resistant phenotype [5–7]. A "rapid" AST is usually defined as a test that can be performed within a single working shift, i.e., within 8 h or less [8].

Standard methods to perform AFST rely on measuring fungal growth in the presence of antifungals over a long period (usually a few days) [9]. Traditional antimicrobial susceptibility testing (AST) methods include disc diffusion and broth dilution assays, which rely on the observation of the growth of microorganisms in the presence of specific antimicrobials under defined test conditions. Commercially available tests, such as Sensititre YeastOne, Etest, and the fully automated VITEK2 (bioMérieux), MicroScan WalkAway (Siemens Healthcare Diagnostics), BD Phoenix (BD Diagnostics) and Sensititre ARIS 2X (Trek Diagnostic Systems), are also widely used to test antifungal susceptibility [8,10]. Some of these tests require a long period of analysis time. VITEK2 (bioMérieux), MicroScan WalkAway (Siemens Healthcare Diagnostics) and BD Phoenix (BD Diagnostics) can generate rapid results (3.5–16 h). However, it must be noted that they require a standardised microbial inoculum (24–48 h or more) prior to their inoculation into the AST system [8].

More rapidly emerging AST methods with different approaches have been reported, including the use of laser scatter technology [11], the sensing of bacterial vibrations using phase-noise measurements on resonant piezo-electric substrates [12], protein-adsorbed magnetic nanoparticle-mediated protocols [13], field-effect enzymatic detection [14], glucose metabolisation [15], optical screening and diffusometry [16], nanometre-scale bacterial deformation measurement [17], mass spectrometry [18], nanowire sensors [19], electrochemical sensing [20], real-time laser scattering [21], fluorescence-activated cell sorting (FACS) [22], Raman and infrared spectroscopy at the single-cell level [23–25] nanotube-assisted microwave electroporation [24], hydrodynamic trapping [26] and phenotypic AST [27]. New and more advanced diagnostic methods use matrix-assisted laser desorption ionisation time-of-flight mass spectroscopy (MALDI-TOF-MS). Very recently, a microdroplet-based MALDI-TOF-MS approach for the performance of rapid AST was presented [28], which required an incubation period of 3 or 4 h for the data acquisition. However, clinical trials have still not demonstrated the direct detection of resistance factors [29].

Over the last few years, atomic force microscope (AFM)-based sensitivity tests have been developed to assess the sensitivity of microorganisms to a given drug in a timeframe of minutes [30,31]. This method consists of attaching the cells to an AFM cantilever, treating them with an antimicrobial drug, and monitoring their oscillations (referred to as "cellular nanomotion" since the displacements of the cantilever are of the nanometer order) as a function of time. Numerous studies demonstrated that living organisms attached to AFM cantilevers induce oscillations of the lever that immediately stop when the organisms die [32,33]. Cellular nanomotion detection (NMD) proved the detection of living organisms in a chemistry-independent manner and in a very time-efficient way [34,35]. Nevertheless, this method has some drawbacks. AFMs are expensive and relatively complex devices that require expertise to operate. Additionally, such nanomotion detectors are relatively complex to scale up and can only monitor one bacterial/fungal species at a time. Finally, it is not a single-cell method.

Other fast AST methods have also been developed to detect the nanomotion of living microorganisms that are attached to a surface. These include the plasmonic imaging of the z motion of attached bacteria [36], the sensing of attached bacterial vibrations with the phase noise of a resonant crystal [12], tracking the x-y motion of attached uropathogenic *Escherichia coli* [37], and subcellular fluctuation imaging, which is based on total internal reflection microscopy (TIRM) [38], as well as optically tracking bacterial responses on micropillar architectures using intrinsic phase-shift spectroscopy [39].

Recently, we found that a simple optical microscope equipped with a video camera can also detect living cells' nanometric scale oscillations at a subpixel resolution using dedicated software [40]. We applied this "optical nanomotion detection" (ONMD) method to assess the susceptibility of various yeasts to antifungals. Previously, we developed a microfluidic chip with a "no-flow" chamber that allowed the exchange of a growth medium with antimicrobial agents without disturbing the nanomotion of the cells [41]. The no-flow chamber allowed us to observe the nanomotion of a single cell before, during, and after treatment with an antifungal. This measurement was relatively fast, simple, and inexpensive.

Here, we further improved this technique by designing a microwell-based chip and testing its efficiency. The new device traps single cells in microwells and permits the observation of their nanomotion before, during, and after treatment with a chemical compound. The opportunistic yeast *C. albicans* was used to evaluate the device and the effect of caspofungin on individual yeast cells, which were successfully assessed by monitoring their nanomotion pattern as a function of time. We successfully explored the effects of caspofungin on wild-type CAF2-1, fluconazole, and caspofungin on azole-hypersensitive DSY1024 yeast strains and showed that the susceptibility of the cells to caspofungin and fluconazole could be assessed within 10 min. We are convinced that the simplicity and cost efficiency of this new technique will open novel avenues in the field of rapid AFST.

2. Materials and Methods

2.1. Yeast Cultivation

For the selection of the microwell size and shape, we used *Saccharomyces cerevisiae* BY4742 (genotype: *MATα his3Δ1 leu2Δ0 lys2Δ0 ura3Δ0*). Yeast cells were cultured by the overnight inoculation of 20 mL YPD medium (10 g/L yeast extract, 20 g/L peptone, 20 g/L dextrose) with a colony from a YPD agar (YPD containing 20 g/L agar) plate in 100 mL Erlenmeyer flasks with ridges for better aeration at 30 °C and 180 revolutions per minute (rpm) (Innova 4400, New Brunswick, Edison, NJ, USA). The overnight cultures were diluted in a YPD medium to obtain an optical density at 600 nm ($OD_{600\ nm}$) of 0.5 and were then grown in Erlenmeyer flasks for 1 h at 30 °C and 180 rpm. The cultures were further diluted, depending on the cell concentration ($OD_{600\ nm}$ value), to obtain an optimal number of cells and allow for single-cell visualisation in each imaging microwell.

For the antifungal susceptibility testing experiments, we used *C. albicans* strains CAF2-1 (genotype: *Δura3::imm434/URA3*) [42,43] and DSY1024 (mutant for efflux systems; genotype: *Δcdr1:hisG/Δcdr1::hisG; Δcdr2::hisG; Δflu1::hisG/Δflu1::hisG; Δmdr1::hisG-URA3-hisG/Δmdr1::hisG*) [42]. The yeast cells were cultured by inoculating 20 mL of a YPD medium with a colony from a YPD agar plate. The cultures were grown overnight in 100 mL Erlenmeyer flasks with ridges for better aeration at 30 °C and 180 revolutions per minute (rpm). The overnight cultures were diluted in a YPD medium to obtain an optical density at 600 nm ($OD_{600\ nm}$) of 0.1 and were then grown in Erlenmeyer flasks for 1 h at 30 °C and 180 rpm. The cultures were further diluted to obtain an optical density at 600 nm ($OD_{600\ nm}$) of 0.03. This allowed us to obtain single cells per well.

2.2. Microfluidic Chip Construction

The microwell layer of the microfluidic chip was fabricated using soft lithography. The design of the microfluidic chip (20 × 87 microwells) was made in DesignSpark Mechanical (www.rs-online.com, 1 May 2022) and then transferred onto a film photomask (0.18 mm polyester with photo emulsion layer) using a laser-photoplotter (Selba S.A., Versoix, Switzerland). The mould for the microwell layer was created using photolithography onto a silicon wafer (4-inch Test CZ-Si wafer, Microchemicals GmbH, Ulm, Germany). The wafer was cleaned in acetone (Carl Roth, Karlsruhe, Germany) and 2-propanol (Carl Roth,

Karlsruhe, Germany) for 5 min each before being rinsed with ultrapure water and, finally, air blow-dried. The wafer was then heated on a hot plate (Torrey Pines Scientific, Carlsbad, CA, USA) at 100 °C to ensure that all the residual solvents were evaporated. The wafer was then exposed to air plasma at 100 W, 50 kHz, for 1 min (Plasma System Cute, Femto Science, Dongtangiheung-Ro, Korea) to ensure more uniformed spin-coating. The mould was fabricated by spin-coating the negative photoresist layer SU-8 3050 (Kayaku Advanced Materials, Westborough, MA, USA) onto the wafer at 3500 rpm for 30 s, which was then soft-baked for 15 min at 95 °C on a hot plate, to reach an approximative thickness of 56 µm. The wafer and the wells' photomask were brought into hard contact in the mask aligner UV-KUB3 (Kloé, Saint-Mathieu-de-Tréviers, France) and was illuminated at 365 nm ultraviolet (UV) light (intensity of 35 mW/cm^2) for 12 s. Next, the wafer was post-baked for 5 min at 95 °C on a hot plate. Finally, the wafer was immersed in a SU-8 developer (Kayaku Advanced Materials, Westborough, MA, USA) for 6 min and then washed with 2-propanol and blow-dried. The dimensions of the wells were measured with a 3-D profilometer (Profilm 3 D, Filmetrics, San Diego, CA, USA).

The SU-8 mould was cleaned from (in-)organic residue by rinsing it with acetone, 2-propanol, and ultrapure water and blow-dried with air. The mould was then exposed to air plasma at 100 W, 50 kHz, for 5 min (Plasma System Cute, Femto Science, Dongtangiheung-Ro, Korea). Next, a sacrificial photoresist layer AZ 1514 H (MicroChemicals GmbH, Ulm, Germany) was spun-coated onto the mould at 4000 rpm for 2 min to obtain a thickness of approximately 3 µm before being soft-baked for 2 min at 100 °C. The polydimethylsiloxane (PDMS) (Sylgard 184 Silicone Elastomer Kit, Dow Inc., Midland, MI, USA) was prepared by mixing the base and the curing agent at a 10:1 ratio which was then degassed in a glass bell for 20 min. Afterward, the PDMS was poured onto the mould and covered with a coverslip glass (25 × 75 mm, 170 µm thickness) and pressed to obtain a thin layer of PDMS (membrane) (Figure 1). By pressing the coverslip glass, air bubbles were introduced, and the chip was degassed again in the glass bell. Afterward, the PDMS was cured for 10 min at 100 °C on a hot plate. The excess PDMS around the coverslip glass was cut off and the mould was submerged in acetone. The acetone dissolved the sacrificial photoresist layer, and the PDMS membrane with wells attached to the coverslip glass, which was easily removed from the mould. Finally, the microfluidic chip was assembled by aligning and sticking the coverslip glass with the PDMS microwells a sticky µ-Slide VI 0.4 channel slides with a double-sided sticky tape of 130 µm thickness (Ibidi GmbH, Gräfelfing, Germany) making six different channels with microwells (Figure 1).

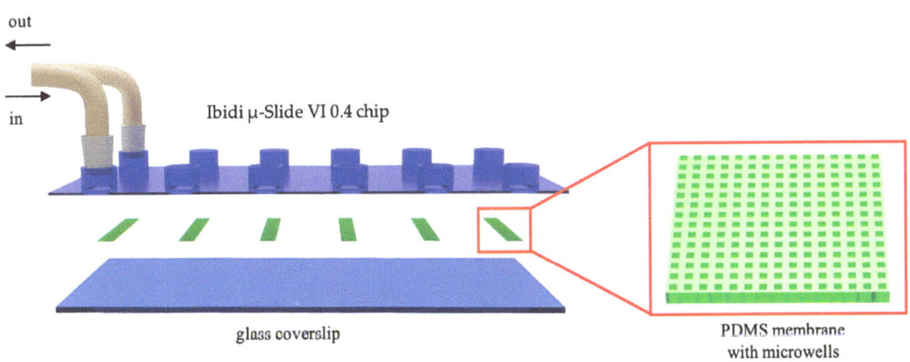

Figure 1. Assembly of the microfluidic chip for cellular nanomotion measurement. The bottom of the chip is composed of 6 PDMS membranes attached to a glass coverslip in such a way that they are aligned with the 6 channels of a sticky Ibidi µ-Slide VI 0.4 chip, used as the top plate. Each PDMS membrane consists of a 20 × 87 array of microwells.

2.3. Computational Simulation of Fluid Flow in the Microfluidic Chip

To simulate the fluid flow in the designed microwells in the channel, fluid flow simulations were performed in 3D using COMSOL Multiphysics 6.1 (COMSOL, Stockholm, Sweden) and by selecting "fluid flow" (single-phase flow, laminar flow) physics. Water at 20 °C was selected as the working liquid. A laminar fluid flow through the channels was simulated with laminar inflow as a boundary condition at the inlet. The inlet flow rate was set at 120 μL/min, which corresponds to the inlet flow rate that was used during the actual experiments. Pressure (atmospheric pressure) was selected as the boundary condition at the outlet, and backflow was suppressed. The physics-controlled mesh was selected as the mesh settings with a normal element size. Particle tracing for the fluid flow physics module of COMSOL was used to calculate the motion of the particles in a background laminar fluid flow in the channel and the microwell. A hundred particles were released at the inlet, with a diameter of 4 μm which corresponded to the size of a yeast cell, and 1 cell was positioned in the microwell. The density of the particles was set at 1117 kg/m^3, which is the average value for yeast cells found in the literature [44–47].

2.4. Microfluidic Chip Setup

The inlet and outlet of the imaging chambers were connected to a pressure-driven pump (LineUp™ Push-Pull, Fluigent, Le Kremlin-Bicêtre, France) via fluorinated ethylene propylene (FEP) tubing with an internal diameter of 0.51 mm. The microfluidic chip setup was then mounted onto an inverted Nikon Eclipse Ti2 epifluorescence microscope (Nikon, Tokyo, Japan) to perform ONMD with bright-field microscopy. The microfluidic chip and tubing were first flushed with ethanol 70% (*v*/*v*) to sterilize them and remove any air bubbles. Then, the tubing and the chip were flushed with a YPD medium to remove any residual ethanol. Next, the yeast liquid culture in YPD was filled into the channels at a flow rate of 120 μL/min and used the pressure-driven pump for 5 min. Then, the pressure pump was turned off to allow the cells to sediment in the wells. Afterward, the YPD medium was flushed through again to clear out any surplus cells. The first ONMD movie was acquired and labeled as "0 min". The solution containing the antifungal caspofungin (Sigma-Aldrich, Overijse, Belgium) or fluconazole (Sigma-Aldrich, Overijse, Belgium) was then filled in from the inlet at a flow rate of 120 μL/min, using the pressure-driven pump for 10 min (until all the liquid in the channel was exchanged with the treatment). Approximately 5 min after turning off the pumps, the second movie was acquired, which corresponded to 10 min after treatment. Each subsequent movie was taken at 1 h time intervals for 5 h. This estimate was based on the time required to pump the compound through the tubing with a flow rate of 120 μL/min and the time it took for the compound to reach a maximal concentration at the bottom of the well.

To observe how the channels and wells were filled, we used the fluorescent dye propidium iodide (PI, Sigma-Aldrich, Overijse, Belgium). The chip was filled with the dye at a flow rate of 120 μL/min (the flow rate used for filling in the antifungal) and 10 μL/min. The movie was acquired at a frame rate of 1 fps (frame/second), at the magnification of 60×, and was later analysed in Fiji ImageJ. We analysed the brightness intensity of the dye in two planes: at the bottom of the well and in the channel and 200 μm above the well (area 100 × 100 px).

2.5. Nanomotion Measurement and Analysis

Cellular nanomotion was monitored by recording bright-field movies of 300 frames at a framerate of 21 fps with an EMCCD camera (Andor) using a 40× objective (1 pixel corresponded to 0.2 μm). The movies were analysed using the cross-correlation analysis method [48] described previously [40]. The ONMD algorithm (MATLAB, MathWorks) works by calculating the x-y displacement of individual cells for each frame. The trajectories of the tracked cells and the root mean square of the displacement are then saved to an MS Excel file. The distribution of the displacements per frame was represented as violin and box-and-whisker (10th to 90th percentile) plots, whereas the total displacements for all cells were represented as box-and-whisker plots (Prism8, GraphPad). The changes in pixels from frame to frame (differential image) could also be calculated and presented utilizing the software and using a colour scale: the pixels that changed the most were indicated in red, whereas the ones that changed the least were shown in blue (in house MATLAB program).

2.6. Cell Nanomotion and Viability Evaluation

To perform a viability test wild type, strain *C. albicans* CAF2-1 was used. The microfluidic chip with microwells was first sterilized with 70% ethanol and then rinsed with a YPD growth medium, and the microwells in the chip were filled with cells (as described in 2.4). The first set of videos was made and designated as time zero. Next, the chip was filled with propidium iodide (PI) solution (final concentration 5 µg/mL dissolved in water), which was left to incubate for 15 min to stain dead cells [49]. After a set of pictures were made, the liquid in the microfluidic chip was exchanged with 70% (v/v) ethanol, and the cells were treated for 1 h. A new set of videos was acquired to determine the nanomotion signal of dead cells. Next, the cells were once again stained with PI by exchanging the liquid in the chip, and the viability of the cells was determined.

3. Results

3.1. Construction of the Microfluidic Chip

Using soft lithography, we constructed the mould used for the fabrication of the PDMS membrane containing microwells. Using 3-D profilometer imaging, it was shown that the dimensions of the pillars on the mould (Figure 2A) (corresponding to the microwells in the PDMS membrane (Figure 2C, zoom)) matched the design specifications. The PDMS membrane was deposited onto a glass slide, and a sticky Ibidi µ-Slide VI 0.4 was used as a top plate (Figure 2B,C). This resulted in a sealed microfluidic chip with 6 channels with wells (Figure 2C).

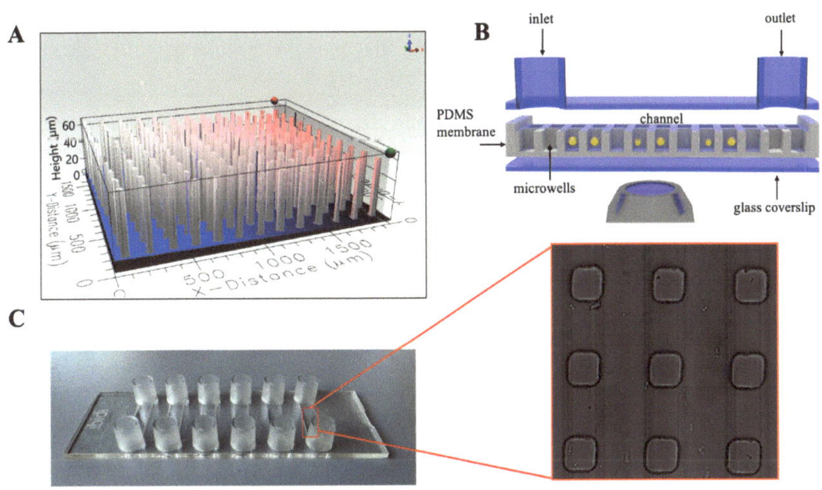

Figure 2. (**A**) Dimensions of a 3D model of the SU8 mould with pillars (3-D profilometer). (**B**) Schematic cross-section of the assembled microfluidic chip filled with yeast cells (yellow dots) (**C**) Picture of the complete microfluidic chip and bright-field image of the microwells filled with cells.

3.1.1. Selection of the Microwell Shape and Dimensions

To assess the optimal size of the microwells, we observed the nanomotion of single *S. cerevisiae* cells in wells of different sizes and shapes. Firstly, we evaluated the influence of the diameter of cylindrical microwells on yeast nanomotion patterns. We tested commercially available cylindrical microwells (MWA-015-08-02, Microsurfaces, Flemington, Australia) with a diameter of 15 µm and a depth of 10 µm and in-house constructed cylindrical microwells with a diameter of 30 µm and depth of 30 µm as well as a diameter of 50 µm and depth of 50 µm. Figure 3A shows that the nanomotion of cells significantly dropped with smaller diameters. These smaller volumes most likely constrained the x-y displacement of the yeasts. Hence, we selected a 50 µm size design to continue the experiments.

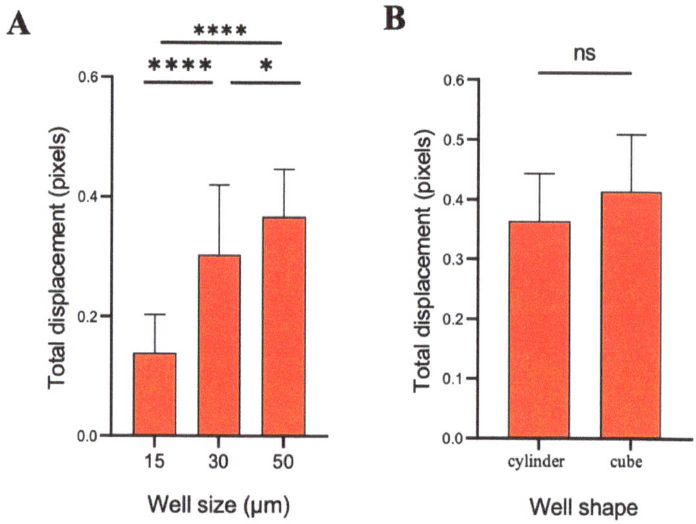

Figure 3. (**A**) Comparison of nanomotion signal between 3 different diameters of cylindrical mi crowells (15 µm, 30 µm and 50 µm) (n = 26). (**B**) Comparison of nanomotion signal between cylindrical (50 × 50 µm) and rectangular (50 × 50 × 50 µm) microwells (n = 26). Wilcoxon test: **** $p < 0.0001$; * $p < 0.1$; ns: not significant.

Besides the size, we evaluated the optimum shape of the well too. We considered two shapes: a cylindrical and cubic shape, where the diameter/sides of the wells were equal to the depth of the wells. There was not a significant difference in nanomotion between the two shapes (Figure 3B), but the volume of the cubic wells was larger than cylindrical wells, which provided the cells with more room for nanomovements as well as undisturbed growth.

3.1.2. Microfluidic Chip Design and Modelling of Fluid Flow and Mass Transport

The microfluidic chip was designed to enable the easy exchange of liquids without flushing out the cells that were trapped in the microwells. We simulated the fluid flow in a microwell (cubic with a width of 50 µm and a depth of 50 µm) in the Ibidi channel (Figure 4), using the flow rate to exchange the growth medium with the medium containing the antifungal. The calculated velocity field is shown in Figure 4A. The laminar flow in the channel induced a vortex in the microwell, as shown by drawing the streamlines (Figure 4B,C). Additionally, this was confirmed by simulating the circular movement of a yeast cell in the microwell upon fluid flow in the channel (Figure 4E–H and Video S1). It could also be observed (Figure 4G) that a cell that flowed over the microwell was initially attracted to the microwell before it was next repelled.

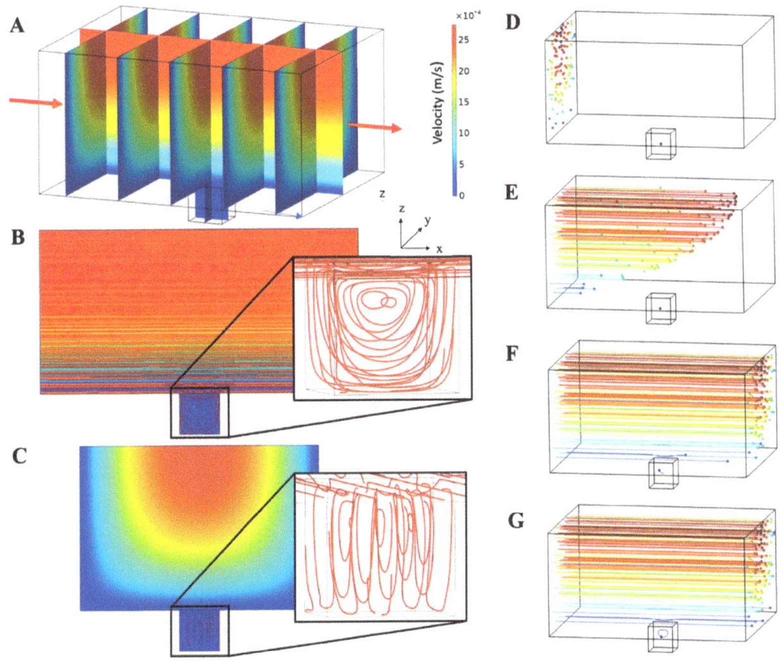

Figure 4. (**A**) Simulation of the velocity field in the channel and the microwell for an inlet flow velocity of 1.32 10^{-3} m/s. The top surface is the symmetry plane of the channel (middle plane of the channel at 200 μm). (**B**) Zx cross-section showing the velocity field and the streamlines (red colour). The insert shows a zoom of the microwell in 3D with the streamlines represented as tubes. (**C**) Zy cross-section showing the velocity field and the streamlines (red colour). The insert shows a zoom of the microwell in 3D with the streamlines represented as tubes. (**D–G**) Simulation of the flow of yeast cells with a flowing medium in the channel and the circular movement of a yeast cell in the microwell at (**D**) 0.00 s, (**E**) 0.13 s, (**F**) 0.8 s, and (**G**) 3.7 s (the colours of the streamlines correspond to the magnitude of the velocity, as shown in the scale bar of (**A**)) (see also Video S1).

The trapping of the cell in the microwells was experimentally confirmed by observing the movement of the cell during an exchange of the medium with an antifungal solution (Video S2). We observed that a yeast cell did not leave the well, even if there was a continuous flow (flow rate of 120 μL/min) in the channel above (Figure 5). The cell circled inside the well and went in and out of focus depending on its position in the well.

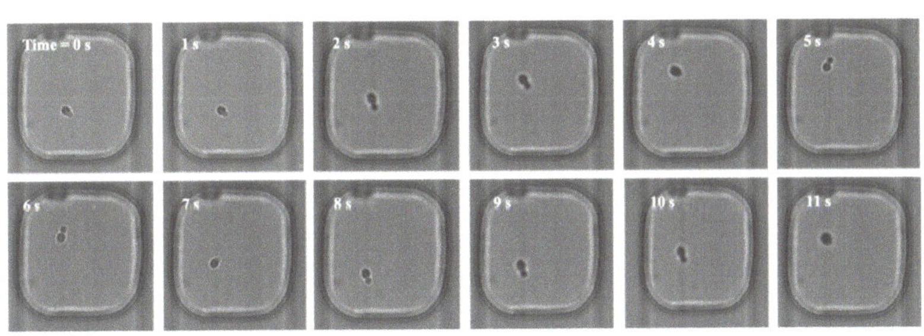

Figure 5. Progression of the circular movement of the yeast cell in the microwell during the continuous flow of liquid in the channel at a flow rate of 120 μL/min (see also Video S2).

To determine how fast the mixing in a microwell occurred when a solution flowed over the well, we observed the increase in the fluorescence of the microwell when a fluorescent solution flowed over the microwell. The evolution of the fluorescence was recorded in the middle of the well and then in the channel 200 μm above the well (Figure S1). The two data sets were normalized and plotted. We observed no difference in the increase in fluorescence between the two spots when the fluorescent solution flowed at a flow rate of 120 μL/min (Figure 6A–C). The same experiment was repeated with a lower flow rate of 10 μL/min. Here, we could observe a delay in the increase in the fluorescence of the well compared to its position on the channel (Figure 6B–D).

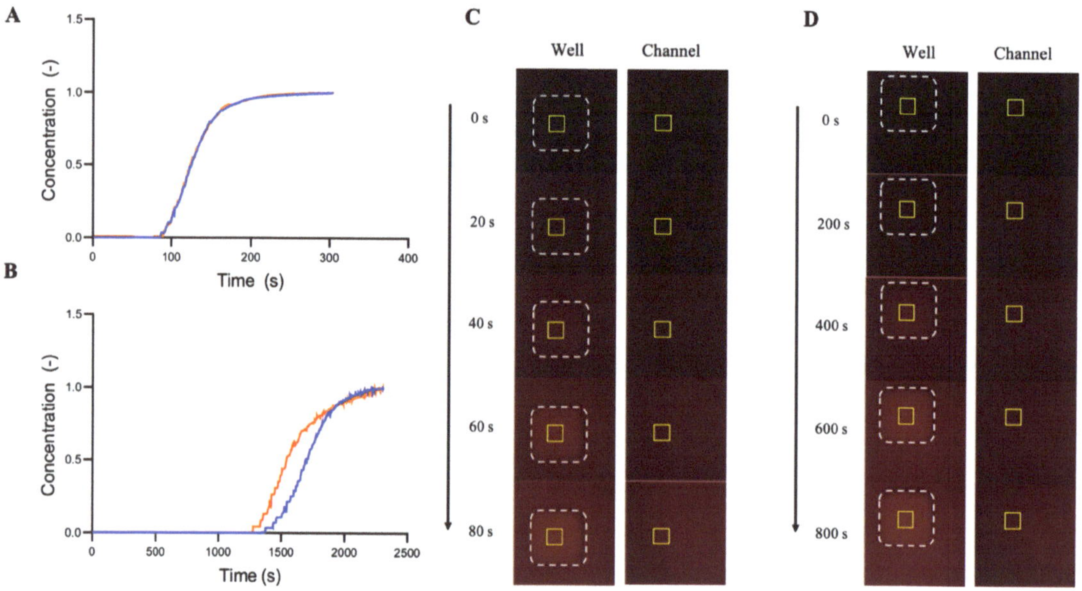

Figure 6. (**A**) Evolution of experimentally determined PI concentration in the well (blue) and middle of the channel (orange) at a flow rate of 120 μL/min (**B**) Evolution of the experimentally determined PI concentration in the well (blue) and middle of the channel (orange) at a flow rate of 10 μL/min. (**C**) Mixing of PI in the chip at a flow rate of 120 μL/min: fluorescent images at different time points; yellow squares represent the analysed area and white dashed squares indicate the edges of the well. (**D**) Mixing of PI in the chip at a flow rate of 10 μL/min: fluorescent images at different time points; yellow squares represent the analysed area. The time point of 0 s in C and D corresponds to the timepoint at which the intensity curves started to increase in A and B.

To analyse the nanomotion of the same cell over a certain period, the cells must be at a distance apart to avoid any physical contact that could interfere with nanomotion measurements. Practically, it means that the optimal distribution is one or two cells per well. The chip was filled with cell suspension at a flow rate of 120 μL/min. Analysing 275 wells, we observed that 21% of the wells were empty, 31% were filled with a single cell, 24% were filled with two cells, 12% were filled with three cells, and 12% were filled with more than three cells. Finally, 55% of the wells were optimally filled, i.e., one or two cells were present in a well (Figure 7A,B).

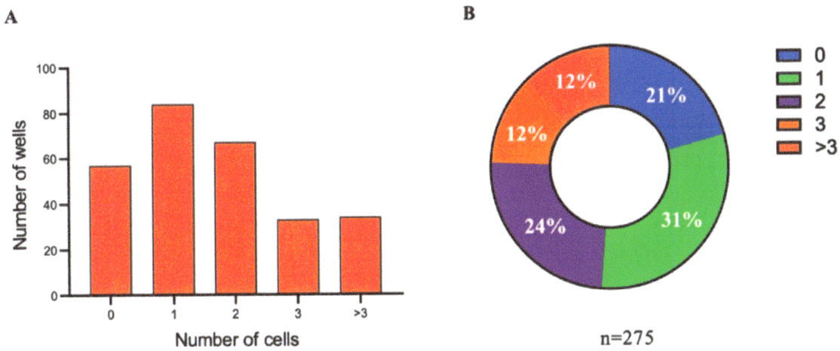

Figure 7. (**A**) The number of cells/well for n = 275 wells. (**B**) Percentage (%) of empty microwells, 1 cell/well, 2 cells/well, 3 cells/well, and more than 3 cells/well (Figure S2).

3.2. *Cell Nanomotion and Susceptibility for the Antifungals Caspofungin and Fluconazole*

The microfluidic chip permits not only monitoring single cells in different wells but also analysing more cells per experiment. Caspofungin experiments were conducted on two different *C. albicans* strains: a wild-type strain (CAF2-1) that was sensitive to caspofungin and fluconazole, and an azole hypersusceptible strain (DSY1024) that was documented to be inactivated with lower concentrations of azole antifungals since, in this strain, efflux genes were deleted [42]. We also evaluated the susceptibility of the DSY1024 strain against caspofungin.

3.2.1. *C. albicans* CAF2-1 Susceptibility for Caspofungin

As a control experiment, we first measured the nanomotion of 20 individual yeast cells of *C. albicans* CAF2-1 during their growth in a YPD medium over a period of 120 min. As shown in Figure 8A, Figure 9A, and Figure S3, there was no significant difference in the total displacement of all cells between 0 min and 120 min of growth.

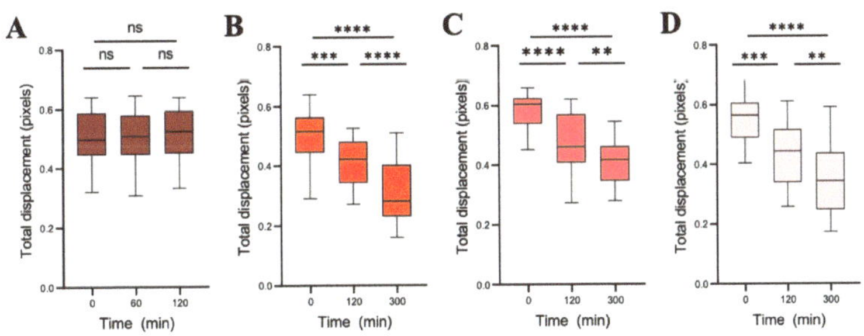

Figure 8. The average cellular nanomotion of 20 cells of *C. albicans* CAF2-1 wild-type strain. The total displacement of 20 cells was analysed during (**A**) 2 h of growth in a YPD medium (control) and during 5 h of caspofungin treatment at (**B**) 0.5 μg/mL, (**C**) 1 μg/mL, and (**D**) 100 μg/mL. Wilcoxon test: **** $p < 0.0001$; *** $p < 0.001$; ** $p < 0.01$; ns: not significant.

Next, we evaluated the effect of caspofungin on *C. albicans* CAF2-1 cells (i.e., the wild-type strain susceptible to caspofungin). Single-cell displacements were recorded every 60 min for 300 min. We used different concentrations of the antifungal compound, i.e., 0.5 μg/mL, 1 μg/mL, and 100 μg/mL. Reported MIC values of caspofungin for the strain CAF2-1 were 0.06 μg/mL [50] and 0.03 μg/mL [51] as well as the general range for *C. albicans* 0.06 μg/mL–2 μg/mL [52]. For all three concentrations, a significant decrease in the cellular displacements and total displacement was detected as soon as 2 h of treatment had elapsed (Figure 8B–D, Figure S4). The most significant drop in nanomotion after 2 h of treatment was observed using a concentration of 1 μg/mL (Figure 8C). Both 100 μg/mL and 0.5 μg/mL caspofungin treatments also caused a significant reduction in nanomotion after 2 h (Figure 8B,D).

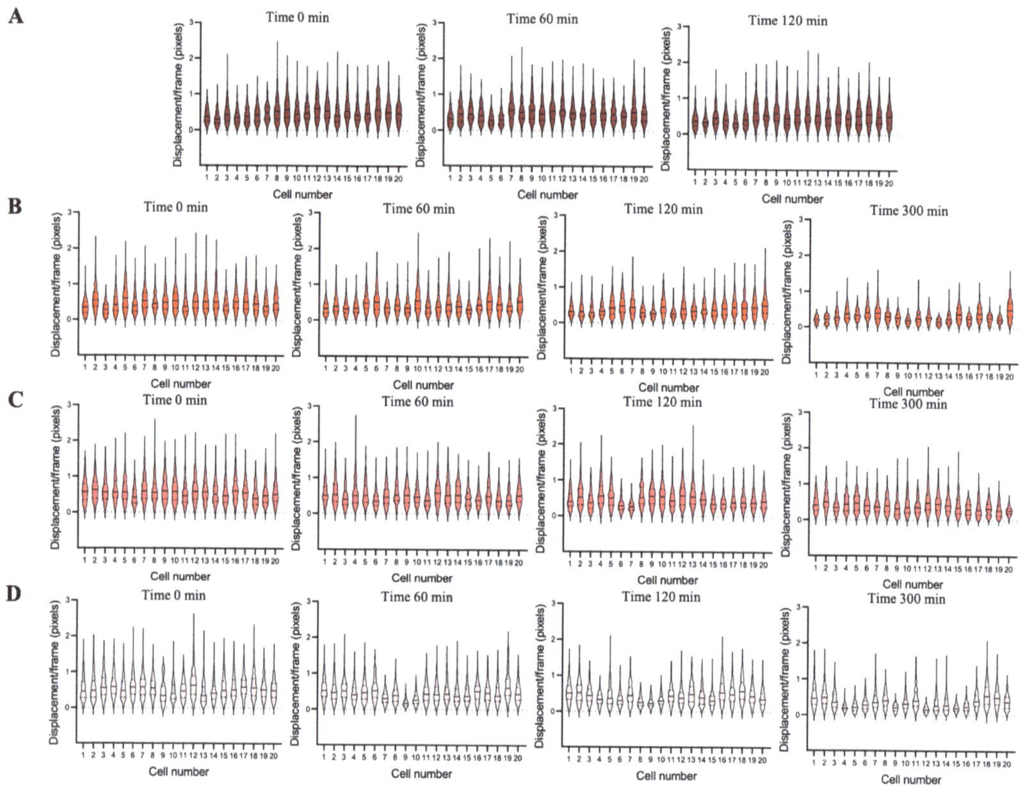

Figure 9. The displacement distribution for 20 single cells of *C. albicans* CAF2-1 wild-type strain displayed in violin plots during (**A**) 2 h of growth in a YPD medium (control) and during 5 h of treatment with the antifungal caspofungin at (**B**) 0.5 μg/mL, (**C**) 1 μg/mL, and (**D**) 100 μg/mL.

While observing the effect of 0.5 μg/mL antifungal treatment on a population of the CAF2-1 strain at a single cell level, we noticed that cells 4, 6, 15, and 20 (Figure 9B) still showed nanomotion after 300 min where the rest of the population had a significant drop, as illustrated by cells 1, 10, and 12 (Figure S5). At the single cell level at 1 μg/mL antifungal treatment after 300 min, we observed a drastic drop in almost all the cells in the population (Figure 9C). Cells 8, 9, and 10 (Figure 10 and Figure S6) presented a similar drop in nanomotion from 10 min after treatment. Looking at the single cell level at 100 μg/mL of antifungal treatment, we observed that cells 1, 2, 18, and 19 (Figure 9D) had a less drastic drop in nanomotion after 300 min in comparison to other cells in the population. Furthermore, in the separately represented cells 5, 10, and 12 (Figure S7), we could observe a difference in the nanomotion drop over the period. Cell 10 had a nearly linear drop in contrast to the other two cells, where the nanomotion dropped down almost immediately (after 10 min) and fluctuated over time.

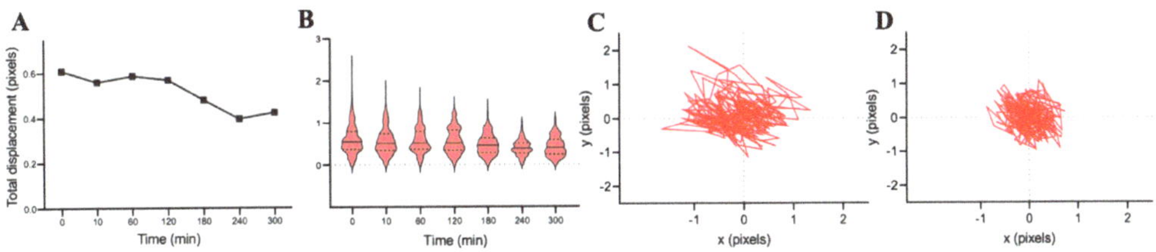

Figure 10. Effect of antifungal caspofungin (1 µg/mL) on the cellular nanomotion of cell 8 from Figure 9C (*C. albicans* CAF2−1). (**A**) Total displacement (pixels) as a function of time for cell 8 (see also a movie of cell 8 that was recorded before treatment (Video S3) and at 10 min (Video S4). (**B**) Distribution of the displacement/frame as a function of time for cell 8. (**C**) X−y displacement of cell 8 before treatment, time point 0 min). (**D**) X−y displacement of cell 8 after 300 min of treatment.

3.2.2. *C. albicans* CAF2-1 Susceptibility for Fluconazole

We evaluated the effect of fluconazole on *C. albicans'* CAF2-1 cells. Single-cell displacements were recorded every 60 min for 300 min. We used antifungal concentrations of 0.5 µg/mL, 1 µg/mL, and 100 µg/mL. A reported MIC value of fluconazole for the strain CAF2-1 was 0.25 µg/mL [53]. For all three concentrations, a significant decrease in cellular displacements and the total displacement was detected as soon as after 2 h of treatment (Figure 11B–D, Figure S8). The most significant drop in nanomotion after 2 h of treatment was observed using a concentration of 100 µg/mL (Figure 11D). Both 0.5 µg/mL and 1.0 µg/mL fluconazole treatments also caused a significant reduction in nanomotion after 2 h (Figure 11B,C).

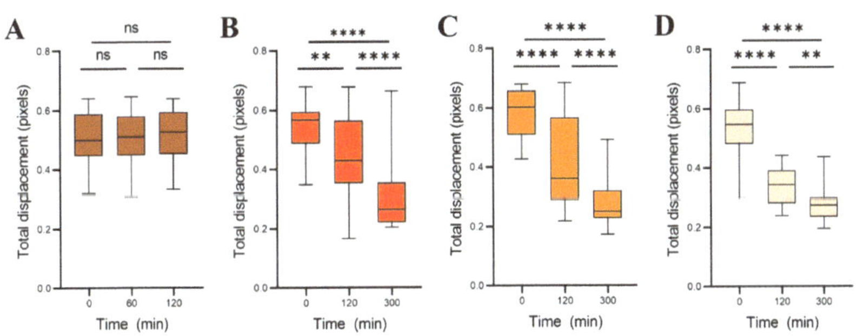

Figure 11. The average cellular nanomotion of 20 cells of *C. albicans* CAF2-1 wild-type strain. The total displacement of 20 cells was analysed during (**A**) 2 h of growth in a YPD medium (control) and during 5 h of fluconazole treatment at (**B**) 0.5 µg/mL, (**C**) 1 µg/mL, and (**D**) 100 µg/mL. Wilcoxon test: **** $p < 0.0001$; ** $p < 0.01$; ns: not significant.

While observing the effect of 0.5 µg/mL antifungal treatment on a population of the CAF2-1 strain at a single cell level, we noticed that cells 4, 9, 13, and 16 (Figure 12B) still showed nanomotion after 300 min where the rest of the population experienced a significant drop. In cells 5, 6 and 7, we observed that the nanomotion signal starts dropping after 10 min of treatment (Figure S9). At the single cell level at 1 µg/mL antifungal treatment after 300 min, we observed a drop in nanomotion except for cells 8, 12, and 13 (Figure 12C). Furthermore, in the separately represented cells 17 (Figure 13), 18, and 19 (Figure S10), we could observe a difference in nanomotion drop over the period. Cell 17 and 19 had a nearly linear drop in contrast to cell 18, where the nanomotion dropped down after 2 h. For treatment with 100 µg/mL antifungal treatment, we observed a significant drop in all the cells (Figure 11D, Figure S11).

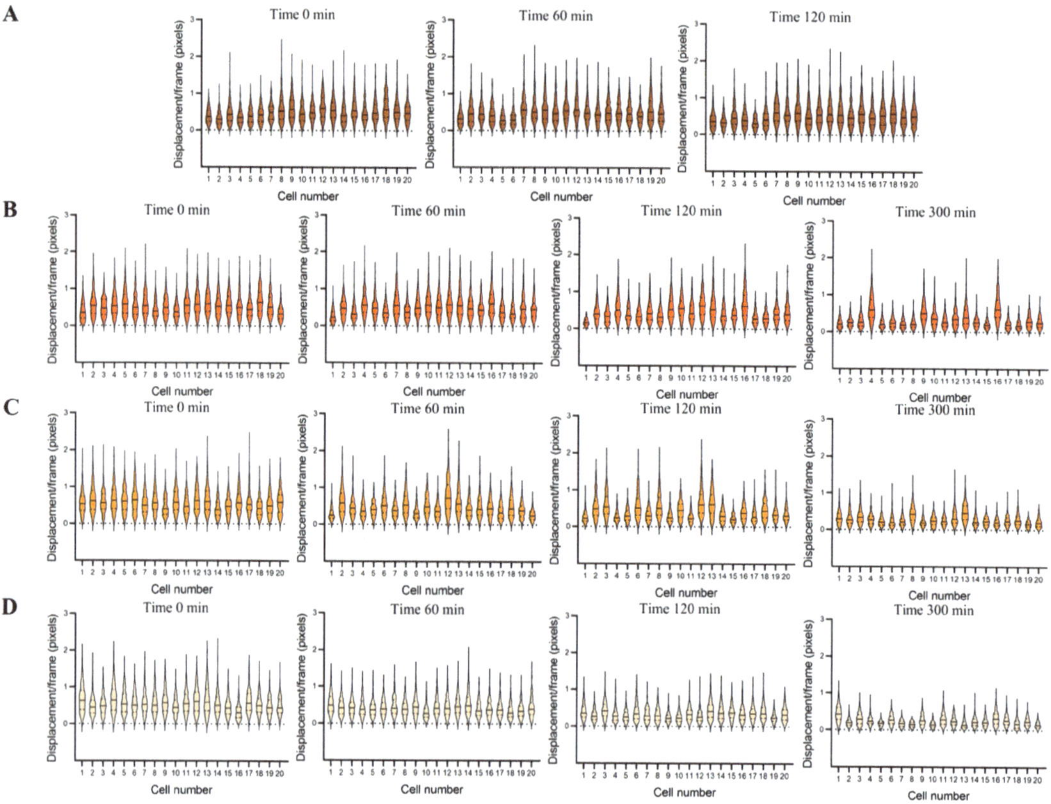

Figure 12. The displacement distribution for 20 single cells of *C. albicans* CAF2-1 wild-type strain displayed in violin plots during (**A**) 2 h of growth in a YPD medium (control) and during 5 h of treatment with the antifungal fluconazole at (**B**) 0.5 µg/mL, (**C**) 1 µg/mL, and (**D**) 100 µg/mL.

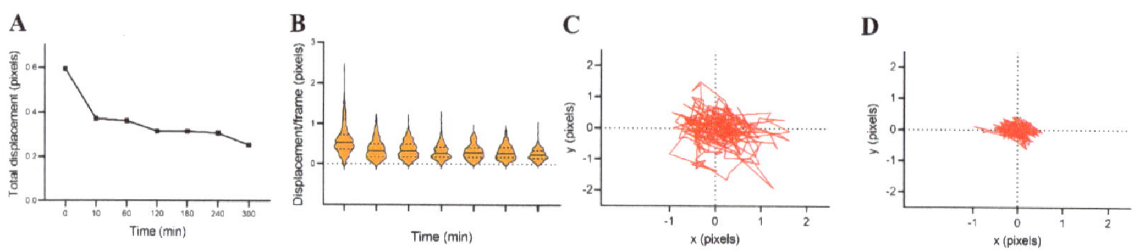

Figure 13. Effect of the antifungal fluconazole (1 µg/mL) on the cellular nanomotion of cell 17 with *C. albicans* CAF2−1. (**A**) Total displacements (pixels) as a function of time for cell 17. (**B**) Distribution of the displacement/frame as a function of time for cell 17. (**C**) X−y displacement at the time point 0 min for cell 17. (**D**) X−y displacement at the time point 300 min for cell 17.

3.2.3. *C. albicans* DSY1024 Susceptibility for Caspofungin and Fluconazole

As a control experiment, the cellular nanomotion of 20 individual yeast cells of the hypersusceptible *C. albicans* DSY1024 strain was monitored during 2 h of growth in a YPD medium. There was no significant difference in the total displacement of all the cells between 0 min and 120 min of growth (Figure 14A, Figure S12).

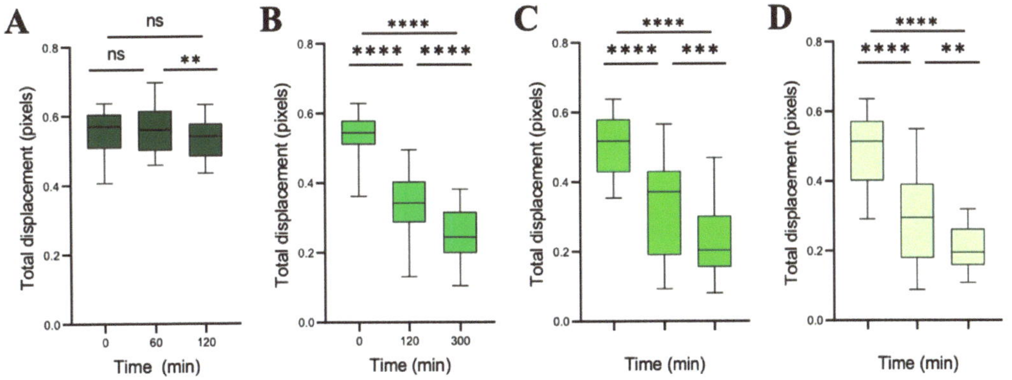

Figure 14. The average cellular nanomotion of 20 cells of *C. albicans* azole hypersusceptible DSY1024 strain treated with fluconazole. The total displacement of 20 cells was analysed during (**A**) 2 h of growth in a YPD medium (control) and during 5 h of fluconazole treatment at (**B**) 0.5 µg/mL, (**C**) 1 µg/mL, and (**D**) 100 µg/mL. Wilcoxon test: **** $p < 0.0001$; *** $p < 0.001$; ** $p < 0.01$; ns: not significant.

ONMD of *C. albicans* DSY1014 Treated with Fluconazole

The susceptibility of the azole hypersusceptible DSY1024 strain for fluconazole treatment was assessed. A reported MIC value of fluconazole for the strain DSY1024 was 0.03 µg/mL [53]. During the treatment with different concentrations of fluconazole, i.e., 0.5 µg/mL, 1 µg/mL, and 100 µg/mL, we observed a significant decrease in the cellular displacements and total displacement (Figure 14B–D, Figure 15B–D, Figure S13) compared to the control (Figure 14A). All 20 cells that were analysed showed a drastic drop in nanomotion after 300 min of treatment with 0.5 µg/mL fluconazole, which indicated that this strain was susceptible to this concentration (Figure 15B, Figure S14). While treating the strain DSY1024 with 1 µg/mL of fluconazole, we observed that after both 120 min and 300 min, all the cells in the population reduced their nanomovements (Figure 15C, Figure S15). In cell 1, we observed as well a drop in the nanomotion (Figure 16). During the treatment with 100 µg/mL fluconazole, we observed that all cells after 300 min had a significant drop in their nanomotion (Figure 15D, Figure S16).

ONMD of *C. albicans* DSY1014 Treated with Caspofungin

The susceptibility of the azole hypersusceptible DSY1024 strain for caspofungin was assessed. During the treatment with different concentrations of caspofungin, i.e., 0.5 µg/mL, 1 µg/mL, and 100 µg/mL, we observed a significant decrease in the cellular displacements and total displacement (Figure 17B–D, Figure 18B–D, Figure S17) as well as in the x-y displacement (Figure 19). Using a caspofungin concentration of 1 µg/mL, we observed the most significant drop over a period of 5 h (Figure 17C). Both responses to 100 µg/mL and 0.5 µg/mL treatments were significant even after 2 h (Figure 17B,D) but were less in comparison to 1 µg/mL.

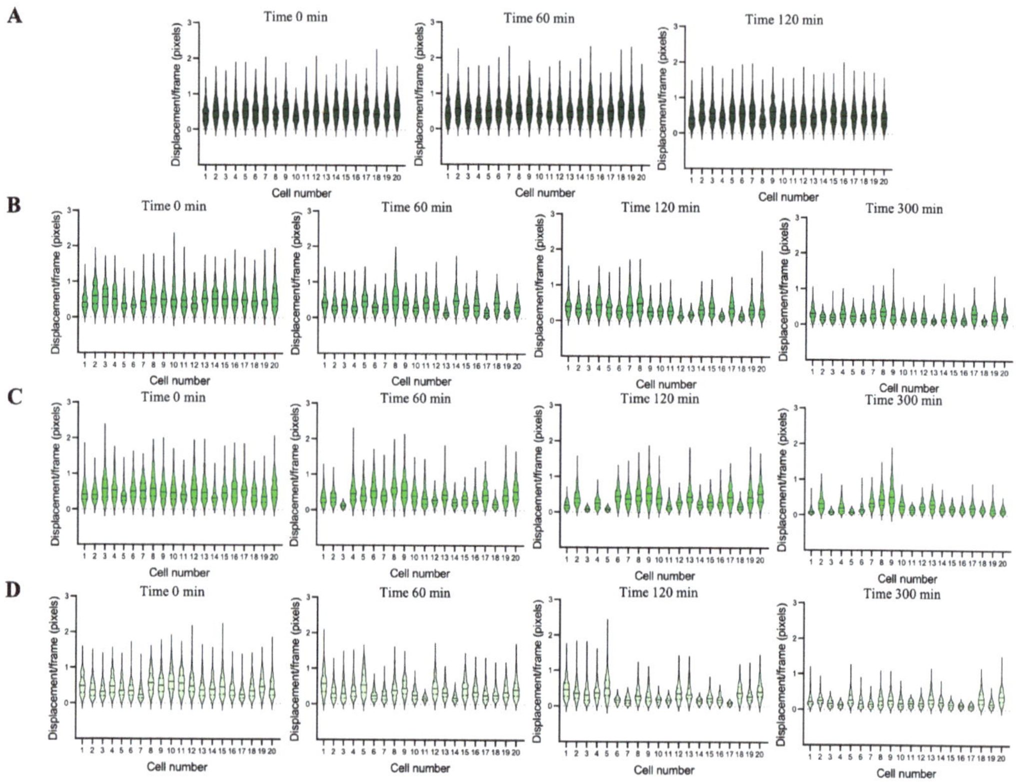

Figure 15. The displacement distribution for 20 single cells of *C. albicans* DSY1024 strain displayed in violin plots during (**A**) 2 h of growth in a YPD medium (control) and during 5 h of treatment with the antifungal fluconazole at (**B**) 0.5 μg/mL, (**C**) 1 μg/mL, and (**D**) 100 μg/mL.

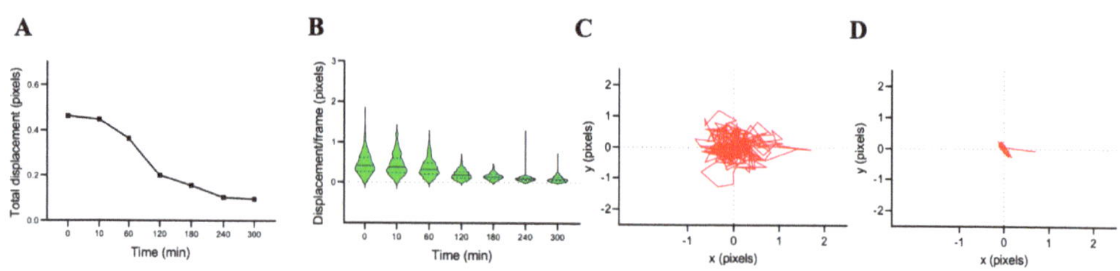

Figure 16. Effect of the antifungal fluconazole (1 μg/mL) on the cellular nanomotion of cell 1 from *C. albicans* DSY1024. (**A**) Total displacement (pixels) as a function of time for cell 1 (**B**) Distribution of the displacement/frame as a function of time for cell 1. (**C**) X−y displacement of cell 1 before treatment time point 0 min. (**D**) X−y displacement of cell 1 after 300 min of treatment.

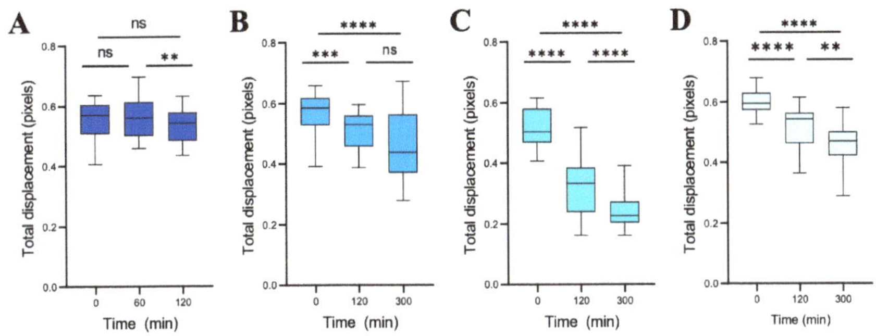

Figure 17. Average cellular nanomotion for 20 cells of *C. albicans* DSY1024 strain treated with caspofungin. The total displacement of 20 cells was analysed during (**A**) 2 h of growth in a YPD medium (control) and during 5 h of caspofungin treatment with (**B**) 0.5 µg/mL, (**C**) 1 µg/mL, and (**D**) 100 µg/mL. Wilcoxon test: **** $p < 0.0001$; *** $p < 0.001$; ** $p < 0.01$; ns: not significant.

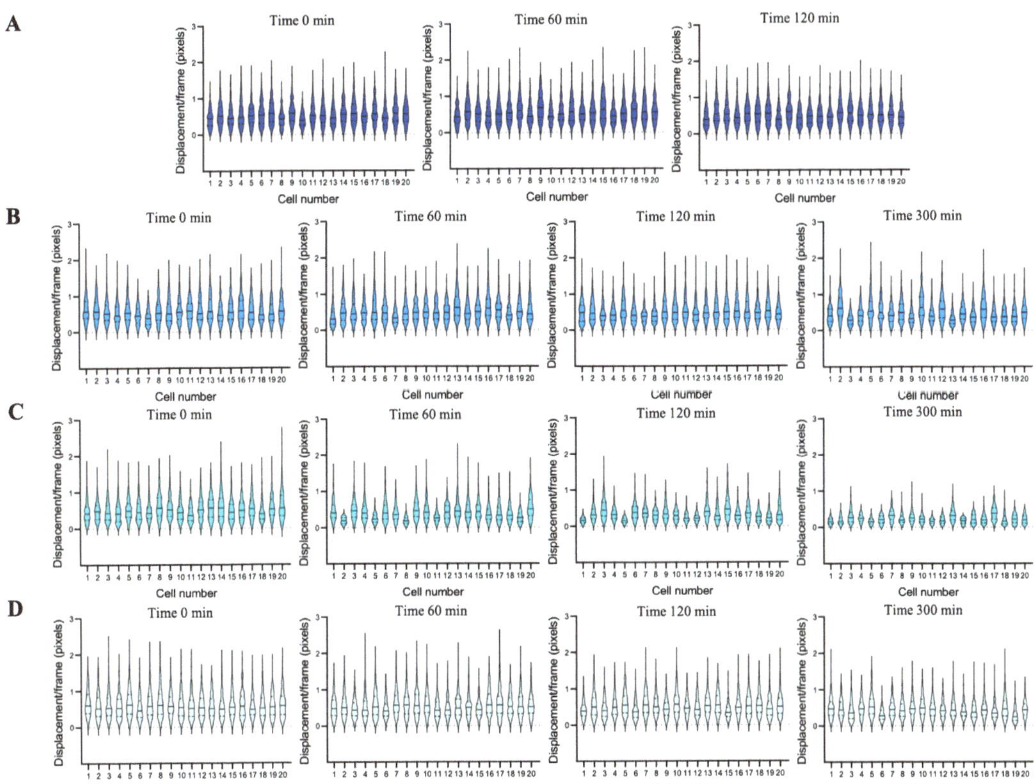

Figure 18. The displacement distribution for 20 single cells of *C. albicans* DSY1024 strain displayed in violin plots during (**A**) 2 h of growth in a YPD medium (control) and during 5 h of caspofungin treatment with (**B**) 0.5 µg/mL, (**C**) 1 µg/mL, and (**D**) 100 µg/mL.

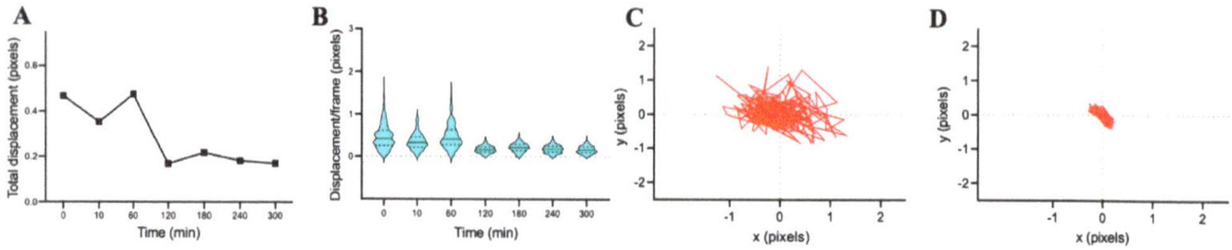

Figure 19. Effect of antifungal caspofungin (1 µg/mL) on the cellular nanomotion of cell 1 in Figure 12 C (*C. albicans* DSY1024). (**A**) Total displacements (pixels) as a function of time for cell 1. (**B**) Distribution of the displacement/frame as a function of time for cell 1. (**C**) X-y displacement of cell 1 before treatment. (**D**) X-y displacement at the time point of 300 min for cell 1.

In the population of strain DSY1024, half of the cells (cells 2, 5, 6, 7, 8, 10, 12, 15, and 16) (Figure 18B) still showed nanomovements after 300 min of treatment with 0.5 µg/mL of caspofungin, which indicated that these cells were more tolerant to this concentration of the antifungal, as represented by cell 12 (Figure S18), for which we observed an increase in nanomotion over the period. On the other hand, the rest of the cells had a significant drop in nanomotion, as represented by cells 3 and 14 (Figure S18).

After treating the strain DSY1024 with 1 µg/mL of caspofungin, we observed that after both 120 min and 300 min, all the cells in the population reduced their nanomovements (Figure 18C), and the drop was greater than in the population of the wild-type strain (Figure 9C). Furthermore, for the separately represented cell 1, we observed a large drop after 120 min, which did not change much for the next 2 h. Cells 9 and 13 experienced a more gradual drop (Figure S19). While observing the single-cell nanomotion of strain DSY1024 during the treatment with 100 µg/mL caspofungin, we observed that many cells after 300 min stayed with similar nanomovements as before the treatment (Figure 18D). The separately represented cells 2, 6, and 19 (Figure S20) were quite diverse in their response to a high concentration of caspofungin. Cell 2 had a slight drop in nanomotion in contrast to cells 6 and 19, where the drop was more drastic.

3.2.4. Sensitivity Analysis Based on the Evolution of the Slopes of the Total Displacements

The sensitivity of the individual *Candida* cells to caspofungin was characterized by analyzing the slope of their decrease in nanomotion activity (total displacement) as a function of time. We calculated the slopes of the total displacement as a function of time for 10, 60, 120, 180, 240, and 300 min after the treatment with 0.5 and 1 µg/mL caspofungin for the wild-type CAF2-1 strain (Figure 20A, Figure S21) and the hypersusceptible DSY1024 strain (Figure 20C, Figure S21C), as well as the treatment with 0.5 and 1 µg/mL fluconazole for the wild-type CAF2-1 strain (Figure 20B, Figure S21B) and the DSY1024 strain (Figure 20D, Figure S21D). The more sensitive the cells were to caspofungin or fluconazole, the steeper the slopes were. The decreases in the slopes of the hypersusceptible strain, when treated with 0.5 µg/mL of fluconazole, were significantly larger than the ones treated with caspofungin as well as the caspofungin treatment of the wild-type strain. The largest slopes were recorded 10 min or 60 min after the treatment and, in the case of the fluconazole treatment of the DSY1024 strain, this slope was the largest, which clearly demonstrates that this strain was hypersusceptible to fluconazole. Based on these results, we can conclude that the values of these slopes enable the effect of a killing compound to be quantified in a short time frame (10 to 60 min). The evolution of the average slopes for the 20 cells showed that the slope values decreased as a function of time (Figure 20A–D, Figure S21A–D), and the slope value at 10 min of treatment already gave an indication of the susceptibility of the strain. The calculation of the slopes for the control experiment (without antifungal treatment) showed fluctuating negative—positive values for the CAF2-1 strain (Figures 20E and 21F) and the DSY1024 strain (Figures 20E and 21F). The nanomotion analysis results the total displacements as a function of time indicate that this wild-type strain was somewhat more susceptible to caspofungin than to fluconazole (Figure 21A,C and Figure S22).

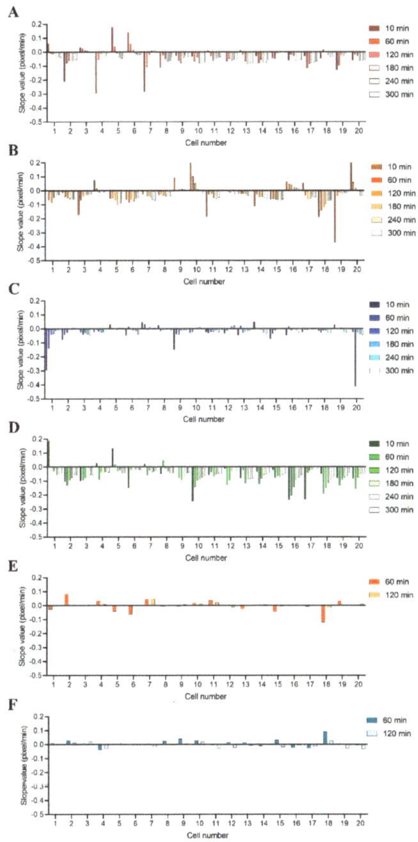

Figure 20. The slopes of the decrease in the total displacement of 20 cells as a function of time were calculated based on the time points 10, 60, 120, 180, 240, and 300 min during 0.5 µg/mL caspofungin or fluconazole treatment of the *C. albicans* strains: (**A**) Caspofungin-treated CAF2−1, (**B**) Fluconazole-treated CAF2−1, (**C**) Caspofungin-treated DSY1024, (**D**) Fluconazole-treated DSY1024 during the 2 h growth in YDP medium (**E**) CAF2−1 and (**F**) DSY1024.

3.3. Cellular Nanomotion of Living Cells Compared to Dead Cells

We performed single-cell nanomotion measurements on living cells and compared the data to the measurements on dead cells that were killed using 70% (*v*/*v*) ethanol and 1 h of incubation. Living cells were characterised by a median total displacement of 0.51 (pixels), whereas dead cells had a median total displacement of 0.06 (pixels) (Figure S23. The viability of the cells was determined using PI staining just before the nanomotion measurements as 100% and 0% after 1 h treatment with ethanol (Figure S23B).

4. Discussion

Nanomotion is a sensitive and rapid label-free technique that assesses the viability of living microbial cells [40]. Therefore, it is very well suited to conduct rapid antifungal susceptibility testing. It was demonstrated that the nanomotion of yeast cells could be linked to their metabolic state by comparing the nanomotion signal of cells that grew in the presence of sufficient nutrients to cells that lacked nutrients and by detecting a maximal cellular nanomotion at the optimal growth temperature [40]. Here, we developed a microfluidic chip equipped with microwells to trap and analyse the nanomotion of single cells. We selected cubic-shaped microwells (Figure 3B) with a width of 50 μm (Figure 3A) and a depth of 50 μm (giving a depth-to-width ratio of 1). The microwells trapped single cells homogenously and organized them over the microwell array. More than 50% of the wells could be filled with one or two cells, which is ideal for the performance of single-cell nanomotion analysis (Figure 7).

As confirmed by the simulations of the fluid flow in the microwells (Figure 4), these dimensions create a vortex when the liquid is flown over the microwell (Figure 5 and Video S1). Previously, it was demonstrated that a vortex was created in a microwell when the depth-to-width ratio equalled one or was larger than one [54]. The vortex traps the cell in the microwell and allows the quick mixing of the antifungal upon liquid exchange. In our analysis, the convective mixing time at a high flow rate (120 μL/min) was less than 2 min and did not differ between the mixing time in the channel (Figure 6). Such a fast-mixing time allowed us to record the nanomotion of the cells immediately after their exposure to an antifungal drug or any other chemical treatment.

We performed a series of experiments to validate the nanomotion method for cells in microwells as an efficient and rapid antifungal susceptibility test. As a first proof-of-concept experiment of the microwell ONMD method, we assessed the susceptibility of the *C. albicans* wild-type CAF2-1 strain to caspofungin. Caspofungin acetate belongs to a new class of antifungals referred to as echinocandins [55]. The effect of caspofungin is related to the inhibition of fungal cell wall synthesis by disrupting the cell wall glucan formation through the non-competitive inhibition of an enzyme complex 1,3-β-D-glucan synthase [55]. We selected two concentrations that were reported within or close to the minimal inhibitory concentration (MIC) range [52], i.e., 0.5 μg/mL and 1 μg/mL, and a large concentration of 100 μg/mL caspofungin (to obtain a fast response and to compare this response with other rapid nanonomotion-based methods [41]). For the 3 caspofungin concentrations, we observed a significant drop in nanomotion upon the injection of the antifungal in the chip (Figures 8 and 14). The caspofungin susceptibility for the MIC concentration could be demonstrated, indicating the method's high sensitivity.

Next, we assessed the susceptibility of the *C. albicans* CAF2-1 strain to fluconazole. Fluconazole, which belongs to the class of azole antifungals, has the useful characteristics of a wide antifungal spectrum, low toxic effects, and high bioavailability and is, therefore, widely used in the clinical treatment of *C. albicans* infections [56]. However, fluconazole is a fungistatic drug that inhibits growth but does not kill the pathogenic fungus, thereby providing an opportunity for the development of fluconazole resistance [57,58]. Fluconazole inhibits ergosterol biosynthesis by inhibiting the cytochrome P450–dependent lanosterol 14-α-demethylase (14DM) enzyme selectively [59]. Resistance can be caused by alterations in sterol biosynthesis, by mutations in the drug target enzyme 14DM, or by the overexpression of genes coding for the membrane transport proteins of the ABC (ATP-binding cassette) transporter (*CDR1/CDR2*) or the major facilitator (*MDR1*) superfamilies [60,61]. Different mechanisms are frequently combined, resulting in a stepwise development of fluconazole resistance over time. The nanomotion analysis results, including the variation in the slopes of the total displacements as a function of time indicate that this wild-type strain was somewhat more susceptible to caspofungin than to fluconazole (Figure 21A,C).

As a second proof-of-principle experiment, we assessed the susceptibility of the azole hypersusceptible DSY1024 strain for fluconazole and caspofungin. The *C. albicans* strain DSY1024 that was used in this study was hypersusceptible to fluconazole since the multidrug efflux transporter genes *CDR1*, *CDR2*, *CaMDR1*, and *FLU1* were deleted [42,62]. The disruption of *FLU1* in *C. albicans* had only a slight effect on fluconazole susceptibility; however, it resulted in hypersusceptibility to mycophenolic acid [62]. A significant drop in the nanomotion signal was observed after 2 h of the treatment of the DSY1024 strain with 0.5 μg/mL, 1 μg/mL, or 100 μg/mL fluconazole (Figures 13 and 14). The decrease in the nanomotion signal was significantly larger for the treatment with 0.5 μg/mL and 100 µg/mL fluconazole compared to the treatment with the same concentrations of caspofungin (Figure 17, Figure 18, and Figure S24). There was no significant difference at the concentration of 1 µg/mL. However, the spread of the total displacement values for 1 µg/mL fluconazole treatment was significantly larger than the spread for caspofungin treatment (Figure S24B). Previously we showed that at antifungal concentrations that were close to the MIC range, the nanomotion signal increased [40]. These results confirm that DSY1024 was hypersusceptible to fluconazole [39].

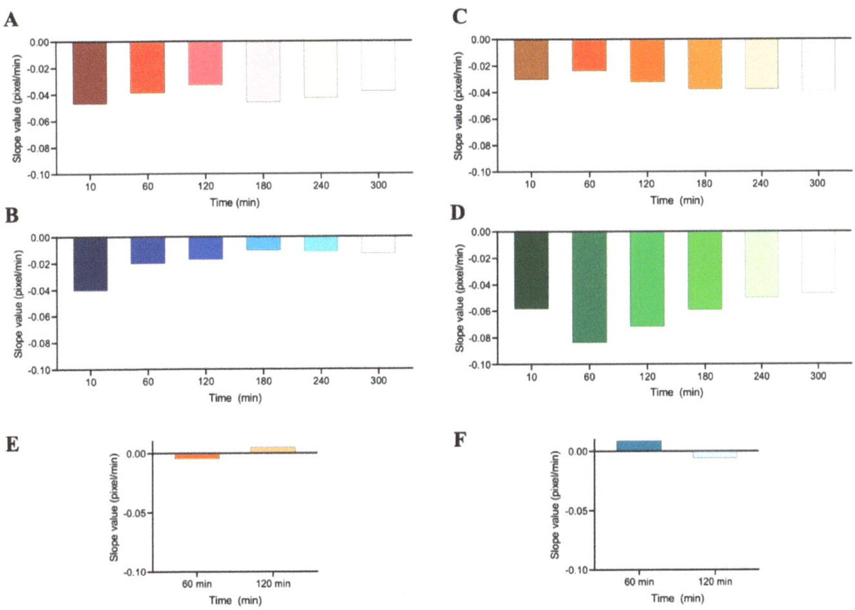

Figure 21. The slopes of the decrease in the total displacement as a function of time are calculated based on the time points 10, 60, 120, 180, 240, and 300 min during 0.5 µg/mL caspofungin or fluconazole treatment for the averages of the total displacements of all 20 cells: (**A**) Caspofungin-treated CAF2-1, (**B**) Caspofungin-treated DSY1024, (**C**) Fluconazole-treated CAF2-1, (**D**) Fluconazole-treated DSY1024 during the 2 h growth in YDP medium of (**E**) CAF2-1 and (**F**) DSY1024.

The slope of the decrease in nanomotion activity (as measured by the total displacement of the cell) as a function of time provided information on how fast the cell reacted to a specific chemical compound (Figure 20, Figure 21, Figure S21, and Figure S22) [41]. With this type of signal analysis, we highlighted the sensitivity and difference in sensitivity between the 2 *Candida* strains in as fast as 10 min. The slope for the decrease in the total displacement after 10 min or 60 min was much larger in the case of the 0.5 µg/mL fluconazole treatment of the hypersusceptible DSY1024 strain, confirming the azole hypersusceptible characteristic of this strain.

We compared the cellular nanomotion signals of living *C. albicans* CAF2-1 cells to the ones of dead cells (Figure S23). The median value of the total displacement of 20 living cells was approximately 10 times larger: 0.51 pixels for living cells versus 0.06 pixels for dead cells. The differences in nanomotion between living and dead cells were largest when living cells were in optimal conditions, i.e., when cells were in their exponential growth phase when all essential nutrients were available and when cells were grown at their optimal pH and temperature. We demonstrated previously that the nanomotion of yeast cells [40] as well as bacteria (including *Escherichia coli* and *Lactobacillus rhamnosus*) [31,63] could be linked to their metabolic state by comparing the nanomotion signal of cells that grow in the presence of sufficient nutrients (including the carbon source) to cells that lack nutrients. We previously showed for yeast cells that the cellular nanomotion was the largest at their optimal growth temperature [40]. The molecular or subcellular organelles that caused the cells to oscillate have not yet been studied in detail. Various processes could contribute to the nanomotion signal, including DNA replication, transcription, and translation, cytoskeleton rearrangements, ionic pump activity, and organelle transport [30]. The observed displacement of dead cells can be explained as the movement of the cells due to the Brownian motion.

A large variation in displacements can be observed for living cells (Figure 11, Figure 14, Figure 17, and Figure S23). Previously, we also observed this large variation using ONMD on single yeast cells (*Candida* and *Saccharomyces cerevisiae* strains) [40,41]. A large variation in the nanomotion of single living bacteria has also been reported previously using various nanomotion detection methods. We observed this recently for the non-motile bacteria *Staphylococcus aureus, L. rhamnosus,* and *Mycobacterium smegmatis* using ONMD [63]. Such a variation was observed in the nanomotion of single *E. coli* that were present in a microwell [64] or stuck to flexible graphene drums [65]. Large variations in the single-cell nanomotion of *E. coli* were loosely tethered to a glass surface and were also registered by optical microscopy [37] and plasmonic imaging [36].

Many parameters could contribute to the observed variation in the nanomotion of single bacteria. Parameters such as the spread in the metabolic status of single cells, the age of the cells, senescence, and the coexistence of cells in different moments of their cell cycle [66] can significantly affect cellular displacement. Further research is needed to evaluate the contribution of these parameters to the nanomotion behaviour of single cells.

The presence of outliners in antifungal-treated cells has also been observed previously for yeast cells [40,41] as well as for antibiotic-treated bacteria [63–65] using optical nanomotion as well as other nanomotion detection methods. A better understanding of these outliners could be obtained by studying single-cell sensitivity toward various antifungals/antibiotics with a different mode of action and in a large cell population. Future experiments also need to address the existence/occurrence of a persistent and/or a resistant sub-population.

5. Conclusions

We developed a microwell-based microfluidic chip for ONMD experiments. The microwells were integrated into a microfluidic channel, allowing us to rapidly expose the trapped yeast cells to antifungals. The method allows for the analysis of a large population of cells by conserving single-cell sensitivity. Due to the creation of a vortex in the microwell, the same cell could be observed before and after exposure to an antifungal drug or any other chemical treatment. This is a major advantage of this new method. This device permitted the performance of susceptibility testing as fast as 10 min.

In the future, a dedicated microfluidic chip could be constructed where the number of channels with microwells is increased and integrated into a chip to assess different concentrations and different antifungals. Being able to perform AST for different concentrations of antimicrobial agents can also allow us to easily determine the MICs.

This development could eventually lead to a cheap and easy operational device that can be directly implemented in hospitals or even in remote doctor's practices in developing world countries. It would allow the performance of rapid antifungal susceptibility testing in the earliest possible treatment stage in order to make the appropriate decisions for personalized effective antimicrobial therapy.

Supplementary Materials: The following supporting information can be downloaded at: https://www.mdpi.com/article/10.3390/fermentation9040365/s1, Video S1: Simulation of the movement of 100 yeast cells with the flowing medium in the channel and the circular movement of one yeast cell in the microwell due to the created fluid vortex. The scale represents the velocity (m/s); Video S2: Experimental observation of the circular movement of the yeast cell in the microwell during the continuous flow of liquid in the channel at a flow rate of 120 μL/min (movie recording at 20 fps); Video

S3: Example of the nanomotion of C. albicans CAF2-1 cell 8 in a microwell before treatment (1 μg/mL caspofungin); Video S4: Example of the nanomotion of C. albicans CAF2-1 cell 8 in a microwell after 10 min of treatment (1 μg/mL caspofungin); Figure S1. Transversal (zy) cross-section of the channel and microwell with focus plane 1 where we observed the evolution of the propidium iodide (PI) concentration in the well and focus plane 2 where we observed the evolution of the PI concentration in the channel. Figure S2. Example of a part of the microwell array showing the occupancy of the microwells by the yeast cells. Figure S3. Control experiment for the C. albicans CAF2-1 strain grown in YPD medium for 3 selected cells: cell 1, 3 and 4 (A) Total displacements (pixels) as a function of time for cell 1. (B) Distribution of the displacement/frame as a function of time for cell 1. (C) X-y displacement at the time point 0 min for cell 1. (D) X-y displacement at the time point 120 min for cell 1. (E) Total displacements (pixels) as a function of time for cell 3. (F) Distribution of the displacement/frame as a function of time for cell 3. (G) X-y displacement at the time point 0 min for cell 3. (H) X-y displacement at the time point 120 min for cell 3. (I) Total displacements (pixels) as a function of time for cell 4. (J) Distribution of the displacement/frame as a function of time for cell 4. (K) X-y displacement at the time point 0 min for cell 4. (L) X-y displacement at the time point 120 min for cell 4. Figure S3. Control experiment for the C. albicans CAF2-1 strain grown in YPD medium for 3 selected cells: cell 1, 3 and 4 (A) Total displacements (pixels) as a function of time for cell 1. (B) Distribution of the displacement/frame as a function of time for cell 1. (C) X-y displacement at the time point 0 min for cell 1. (D) X-y displacement at the time point 120 min for cell 1. (E) Total displacements (pixels) as a function of time for cell 3. (F) Distribution of the displacement/frame as a function of time for cell 3. (G) X-y displacement at the time point 0 min for cell 3. (H) X-y displacement at the time point 120 min for cell 3. (I) Total displacements (pixels) as a function of time for cell 4. (J) Distribution of the displacement/frame as a function of time for cell 4. (K) X-y displacement at the time point 0 min for cell 4. (L) X-y displacement at the time point 120 min for cell 4. Figure S5. Effect of the antifungal caspofungin (0.5 μg/mL) on the cellular nanomotion for 3 selected cells: cell 1, 10, and 12 of C. albicans CAF2-1. (A) Total displacements (pixels) as a function of time for cell 1. (B) Distribution of the displacement/frame as a function of time for cell 1. (C) X-y displacement at the time point 0 min for cell 1. (D) X-y displacement at the time point 300 min for cell 1. (E) Total displacements (pixels) as a function of time for cell 10. (F) Distribution of the displacement/frame as a function of time for cell 10. (G) X-y displacement at the time point 0 min for cell 10. (H) X-y displacement at the time point 300 min for cell 10. (I) Total displacements (pixels) as a function of time for cell 12. (J) Distribution of the displacement/frame as a function of time for cell 12. (K) X-y displacement at the time point 0 min for cell 12. (L) X-y displacement at the time point 300 min for cell 12. Figure S6. Effect of the antifungal caspofungin (1 μg/mL) on the cellular nanomotion for 2 selected cells: cell 1 and 2 of C. albicans CAF2-1. (A) Total displacements (pixels) as a function of time for cell 1. (B) Distribution of the displacement/frame as a function of time for cell 1. (C) X-y displacement at the time point 0 min for cell 1. (D) X-y displacement at the time point 300 min for cell 1. (E) Total displacements (pixels) as a function of time for cell 2. (F) Distribution of the displacement/frame as a function of time for cell 2. (G) X-y displacement at the time point 0 min for cell 2. (H) X-y displacement at the time point 300 min for cell 2. Figure S7. Effect of the antifungal caspofungin (100 μg/mL) on the cellular nanomotion for 3 selected cells: cell 5, 10 and 12 of C. albicans CAF2-1. (A) Total displacements (pixels) as a function of time for cell 5. (B) Distribution of the displacement/frame as a function of time for cell 5. (C) X-y displacement at the time point 0 min for cell 5. (D) X-y displacement at the time point 300 min for cell 5. (E) Total displacements (pixels) as a function of time for cell 10. (F) Distribution of the displacement/frame as a function of time for cell 10. (G) X-y displacement at the time point 0 min for cell 10. (H) X-y displacement at the time point 300 min for cell 10. (I) Total displacements (pixels) as a function of time for cell 12. (J) Distribution of the displacement/frame as a function of time for cell 12. (K) X-y displacement at the time point 0 min for cell 12. (L) X-y displacement at the time point 300 min for cell 12. Figure S8. Averaged cellular nanomotion for 20 cells of C. albicans CAF2-1. (A) Distribution of the displacements/frames during 2 h growth (control). (B) Distribution of the displacements/frames during 0.5 μg/mL fluconazole treatment. (C) Distribution of the displacements/frames during 1 μg/mL fluconazole treatment. (D) Distribution of the displacements/frames during 100 μg/mL fluconazole treatment. Figure S8. Averaged cellular nanomotion for 20 cells of C. albicans CAF2-1. (A) Distribution of the displacements/frames during 2 h growth (control). (B) Distribution of the displacements/frames during 0.5 μg/mL fluconazole treatment. (C) Distribution of the displacements/frames during 1 μg/mL fluconazole treatment. (D) Distribution of the displacements/frames during 100 μg/mL fluconazole treatment. Figure S10.

Effect of the antifungal fluconazole (1 µg/mL) on the cellular nanomotion for 2 selected cells: cell 18, and 19 of C. albicans CAF2-1. (A) Total displacements (pixels) as a function of time for cell 18. (B) Distribution of the displacement/frame as a function of time for cell 18. (C) X-y displacement at the time point 0 min for cell 18. (D) X-y displacement at the time point 300 min for cell 18. (E) Total displacements (pixels) as a function of time for cell 19. (F) Distribution of the displacement/frame as a function of time for cell 19. (G) X-y displacement at the time point 0 min for cell 19. (H) X-y displacement at the time point 300 min for cell 19. Figure S11. Effect of the antifungal fluconazole (100 µg/mL) on the cellular nanomotion for 3 selected cells: cell 2, 3, and 4 of C. albicans CAF2-1. (A) Total displacements (pixels) as a function of time for cell 2. (B) Distribution of the displacement/frame as a function of time for cell 2. (C) X-y displacement at the time point 0 min for cell 2. (D) X-y displacement at the time point 300 min for cell 2. (E) Total displacements (pixels) as a function of time for cell 3. (F) Distribution of the displacement/frame as a function of time for cell 3. (G) X-y displacement at the time point 0 min for cell 3. (H) X-y displacement at the time point 300 min for cell 3. (I) Total displacements (pixels) as a function of time for cell 4. (J) Distribution of the displacement/frame as a function of time for cell 4. (K) X-y displacement at the time point 0 min for cell 4. (L) X-y displacement at the time point 300 min for cell 4. Figure S12. Control experiment for the C. albicans DSY1024 strain grown in YPD medium for 3 selected cells: cell 1, 5 and 10. (A) Total displacements (pixels) as a function of time for cell 1. (B) Distribution of the displacement/frame as a function of time for cell 1. (C) X-y displacement at the time point 0 min for cell 1. (D) X-y displacement at the time point 120 min for cell 1. (E) Total displacements (pixels) as a function of time for cell 5. (F) Distribution of the displacement/frame as a function of time for cell 5. (G) X-y displacement at the time point 0 min for cell 5. (H) X-y displacement at the time point 120 min for cell 5. (I) Total displacements (pixels) as a function of time for cell 10. (J) Distribution of the displacement/frame as a function of time for cell 10. (K) X-y displacement at the time point 0 min for cell 10. (L) X-y displacement at the time point 120 min for cell 10. Figure S13. Averaged cellular nanomotion for 20 cells of C. albicans DSY1024. (A) Distribution of the displacements/frames during 2 h growth (control). (B) Distribution of the displacements/frames during 0.5 µg/mL fluconazole treatment. (C) Distribution of the displacements/frames during 1 µg/mL fluconazole treatment. (D) Distribution of the displacements/frames during 100 µg/mL fluconazole treatment. Figure S14. Effect of the antifungal fluconazole (0.5 µg/mL) on the cellular nanomotion for 3 selected cells: cell 2, 3 and 4 of C. albicans DSY1024. (A) Total displacements (pixels) as a function of time for cell 2. (B) Distribution of the displacement/frame as a function of time for cell 2. (C) X-y displacement at the time point 0 min for cell 2. (D) X-y displacement at the time point 300 min for cell 2. (E) Total displacements (pixels) as a function of time for cell 3. (F) Distribution of the displacement/frame as a function of time for cell 3. (G) X-y displacement at the time point 0 min for cell 3. (H) X-y displacement at the time point 300 min for cell 3. (I) Total displacements (pixels) as a function of time for cell 4. (J) Distribution of the displacement/frame as a function of time for cell 4. (K) X-y displacement at the time point 0 min for cell 4. (L) X-y displacement at the time point 300 min for cell 4. Figure S15. Effect of the antifungal fluconazole (1 µg/mL) on the cellular nanomotion for 2 selected cells: cell 2 and 5 of C. albicans DSY1024. (A) Total displacements (pixels) as a function of time for cell 2. (B) Distribution of the displacement/frame as a function of time for cell 2. (C) X-y displacement at the time point 0 min for cell 2. (D) X-y displacement at the time point 300 min for cell 2. (E) Total displacements (pixels) as a function of time for cell 5. (F) Distribution of the displacement/frame as a function of time for cell 5. (G) X-y displacement at the time point 0 min for cell 5. (H) X-y displacement at the time point 300 min for cell 5. Figure S16. Effect of the antifungal fluconazole (100 µg/mL) on the cellular nanomotion for 3 selected cells: cell 1, 2, and 3 of C. albicans DSY1024. (A) Total displacements (pixels) as a function of time for cell 1. (B) Distribution of the displacement/frame as a function of time for cell 1. (C) X-y displacement at the time point 0 min for cell 1. (D) X-y displacement at the time point 300 min for cell 1. (E) Total displacements (pixels) as a function of time for cell 2. (F) Distribution of the displacement/frame as a function of time for cell 2. (G) X-y displacement at the time point 0 min for cell 2. (H) X-y displacement at the time point 300 min for cell 2. (I) Total displacements (pixels) as a function of time for cell 3. (J) Distribution of the displacement/frame as a function of time for cell 3. (K) X-y displacement at the time point 0 min for cell 3. (L) X-y displacement at the time point 300 min for cell 3. Figure S17. Averaged cellular nanomotion for 20 cells of C. albicans DSY1024. (A) Distribution of the displacements/frames during 2 h growth (control). (B) Distribution of the displacements/frames during 0.5 µg/mL caspofungin treatment. (C) Distribution of the displacements/frames during 1 µg/mL caspofungin treatment. (D) Distribution of the displacements/frames during 100 µg/mL caspofungin treatment. Figure S18. Effect of the antifungal caspofungin (0.5 µg/mL) on the cellular nanomotion for 3 selected cells: cell 3, 12 and 14 of C. albicans DSY1024. (A) Total displacements (pixels) as a function of time for cell 3. (B) Distribution of the displacement/frame as a function of time for cell 3. (C) X-y displacement at the time point 0 min for cell 3. (D) X-y displacement at the time point 300 min for cell 3. (E) Total displacements (pixels) as a function of time for cell 12. (F) Distribution of the displacement/frame as a function of time for cell 12. (G) X-y displacement at the time point 0 min for cell 12. (H) X-y displacement at the time point 300 min for cell 12. (I) Total displacements (pixels) as a function of time for cell 14. (J) Distribution of the displacement/frame as a function of time for cell 14. (K) X-y displacement at the time point 0 min for cell 14. (L) X-y displacement at the time point 300 min for cell 14. Figure S19. Effect of the antifungal caspofungin (1 µg/mL) on the cellular nanomotion for 2 selected cells: cell 9 and 13 of C. albicans DSY1024. (A) Total displacements (pixels) as a function of time for cell 9. (B) Distribution of the displacement/frame as a function of time for cell 9. (C) X-y displacement at the time point 0 min for cell 9. (D) X-y displacement at the time point 300 min for cell 9. (E) Total displacements (pixels) as a function of time for cell 13. (F) Distribution of the displacement/frame as a function of time for cell 13. (G) X-y displacement at the time point 0 min for cell 13. (H) X-y displacement at the time point 300 min for cell 13. Figure S20. Effect of the antifungal caspofungin (100 µg/mL) on the cellular nanomotion for 3 selected cells: cell 2, 6, and 19 of C. albicans DSY1024. (A) Total displacements (pixels) as a function of time for cell 2. (B) Distribution of the displacement/frame as a function of time for cell 2. (C) X-y displacement at the time point 0 min for cell 2. (D) X-y displacement at the time point 300 min for cell 2. (E) Total displacements (pixels) as a function of time for cell 6. (F) Distribution of the displacement/frame as a function of time for cell 6. (G) X-y displacement at the time point 0 min for cell 6. (H) X-y displacement at the time point 300 min for cell 6. (I) Total displacements (pixels) as a function of time for cell 19. (J) Distribution of the displacement/frame as a function of time for cell 19. (K) X-y displacement at the time point 0 min for cell 19. (L) X-y displacement at the time point 300 min for cell 19. Figure S21. The slopes of the decrease in the total displacement of 20 cells as a function of time are calculated based on the time points 10, 60, 120, 180, 240, and 300 min during 1 µg/mL caspofungin or fluconazole treatment of the C. albicans strains: (A) caspofungin treated CAF2-1, (B) fluconazole treated CAF2-1, (C) caspofungin treated DSY1024, (D) fluconazole treated DSY1024; and during the 2 h growth in YDP medium (E) CAF2-1 and (F) DSY1024. Figure S22. The slopes of the decrease in the total displacement as a function of time are calculated based on the time points 10, 60, 120, 180, 240, and 300 min during 1 µg/mL caspofungin or fluconazole treatment for the averages of total displacements of all 20 cells: (A) caspofungin treated CAF2-1, (B) caspofungin treated DSY1024, (C) fluconazole treated CAF2-1, (D) fluconazole treated DSY1024; and during the 2 h growth in YDP medium of (E) CAF2-1 and (F) DSY1024. Figure S23. Comparison of the nanomotion of 20 living C. albicans CAF2-1 cells to 20 dead cell. (A) Total displacement of 20 cells of untreated living cells (time = o min) and 70 % (v/v) ethanol killed cells after 1 h of treatment; Wilcoxon test: **** $p < 0.0001$. (B) Image of PI stained dead cells in the microwells (after 1 h treatment with 70% (v/v) ethanol). (C) Total displacement of 20 single living cells (time = 0 min). (D) Total displacement of 20 single dead cells (time = 60 min). Figure S24. Comparison of the averaged cellular nanomotion for 20 cells of the C. albicans azole hypersusceptible DSY1024 strain treated with fluconazole (green) or caspofungin (blue) at a concentration of (A) 0.1 µg/mL, (B) 0.5 µg/mL, and (C) 100 µg/mL.

Author Contributions: Conceptualisation, R.G.W. and S.K.; methodology, R.G.W., V.R., C.Y., M.I.V., and S.K.; software, S.K. and M.I.V.; validation, V.R., C.Y., and R.G.W.; formal analysis, V.R.; resources, R.G.W. and S.K.; data curation, V.R.; writing—original draft preparation, R.G.W., V.R., C.Y., M.I.V., and S.K.; writing—review and editing, R.G.W., V.R., C.Y., M.I.V., B.D., and S.K.; visualisation, V.R. and R.G.W.; supervision, R.G.W., S.K., and B.D.; project administration, R.G.W. and S.K.; funding acquisition, R.G.W., S.K., and B.D. All authors have read and agreed to the published version of the manuscript.

Funding: This research was funded by the Belgian Federal Science Policy Office (Belspo) and the European Space Agency (ESA), grant number PRODEX Flumias Nanomotion; FWO, grant number I002620; FWO-SNSF, grant number 310030L_197946; FWO-SB-FDW Ph.D. grant of VR. IV and SK were supported by SNSF grants CRSII5_173863.

Institutional Review Board Statement: Not applicable.

Informed Consent Statement: Not applicable.

Data Availability Statement: All data needed to evaluate the conclusions in the paper are present in the paper and/or the Supplementary Materials. Additional data related to this paper may be requested from the authors.

Acknowledgments: The Research Council of the Vrije Universiteit Brussel (Belgium) and the University of Ghent (Belgium) are acknowledged for their support to the Alliance Research Group VUB-UGhent NanoMicrobiology (NAMI), and the International Joint Research Group (IJRG) VUBEPFL BioNanotechnology and NanoMedicine (NANO).

Conflicts of Interest: The authors declare no conflict of interest.

References

1. Lopes, J.P.; Lionakis, M.S. Pathogenesis and virulence of *Candida albicans*. *Virulence* **2021**, *13*, 89–121. [CrossRef]
2. Dadar, M.; Tiwari, R.; Karthik, K.; Chakraborty, S.; Shahali, Y.; Dhama, K. Candida albicans—Biology, molecular characterization, pathogenicity, and advances in diagnosis and control—An update. *Microb. Pathog.* **2018**, *117*, 128–138. [CrossRef]
3. Poulain, D. Candida albicans, plasticity and pathogenesis. *Crit. Rev. Microbiol.* **2015**, *41*, 208–217. [CrossRef] [PubMed]
4. Pilmis, B.; Yang, Z.; Lanternier, F.; Lortholary, O. Systemic Candidiasis. In *Infectious Diseases*; Elsevier: Amsterdam, The Netherlands, 2017; pp. 439–445.e1. [CrossRef]
5. Durand, C.; Maubon, D.; Cornet, M.; Wang, Y.; Aldebert, D.; Garnaud, C. Can We Improve Antifungal Susceptibility Testing? *Front. Cell. Infect. Microbiol.* **2021**, *11*, 720609. [CrossRef]
6. Knabl, L.; Lass-Flörl, C. Antifungal susceptibility testing in *Candida* species: Current methods and promising new tools for shortening the turnaround time. *Expert Rev. Anti-infective Ther.* **2020**, *18*, 779–787. [CrossRef]
7. CLSI. *Reference Method for Broth Dilution Antifungal Susceptibility Testing of Filamentous Fungi*; Approved Standard—2nd ed. CLSI document M38-A2; Clinical and Laboratory Standards Institute: Wayne, PA, USA, 2008.
8. Van Belkum, A.; Burnham, C.-A.D.; Rossen, J.W.A.; Mallard, F.; Rochas, O.; Dunne, W.M., Jr. Innovative and rapid antimicrobial susceptibility testing systems. *Nat. Rev. Microbiol.* **2020**, *18*, 1–13. [CrossRef] [PubMed]
9. Posteraro, B.; Torelli, R.; De Carolis, E.; Posteraro, P.; Sanguinetti, M. Antifungal susceptibility testing: Current role from the clinical laboratory perspective. *Mediterr. J. Hematol. Infect. Dis.* **2014**, *6*, e2014030. [CrossRef]
10. Pfaller, M.A. Antifungal Drug Resistance: Mechanisms, Epidemiology, and Consequences for Treatment. *Am. J. Med.* **2012**, *125*, S3–S13. [CrossRef] [PubMed]
11. Hayden, R.T.; Clinton, L.K.; Hewitt, C.; Koyamatsu, T.; Sun, Y.; Jamison, G.; Perkins, R.; Tang, L.; Pounds, S.; Bankowski, M.J. Rapid Antimicrobial Susceptibility Testing Using Forward Laser Light Scatter Technology. *J. Clin. Microbiol.* **2016**, *54*, 2701–2706. [CrossRef]
12. Johnson, W.L.; France, D.C.; Rentz, N.S.; Cordell, W.T.; Walls, F.L. Sensing bacterial vibrations and early response to antibiotics with phase noise of a resonant crystal. *Sci. Rep.* **2017**, *7*, 12138. [CrossRef]
13. Cowger, T.A.; Yang, Y.; Rink, D.E.; Todd, T.; Chen, H.; Shen, Y.; Yan, Y.; Xie, J. Protein-Adsorbed Magnetic-Nanoparticle-Mediated Assay for Rapid Detection of Bacterial Antibiotic Resistance. *Bioconjugate Chem.* **2017**, *28*, 890–896. [CrossRef] [PubMed]
14. Shi, X.; Kadiyala, U.; VanEpps, J.S.; Yau, S.-T. Culture-free bacterial detection and identification from blood with rapid, phenotypic, antibiotic susceptibility testing. *Sci. Rep.* **2018**, *8*, 3416. [CrossRef] [PubMed]
15. Kittel, M.; Findeisen, P.; Ghebremedhin, B.; Miethke, T.; Grundt, A.; Ahmad-Nejad, P.; Neumaier, M. Rapid susceptibility testing of multi-drug resistant *Escherichia coli* and *Klebsiella* by glucose metabolization monitoring. *Clin. Chem. Lab. Med.* **2019**, *57*, 1271–1279. [CrossRef] [PubMed]
16. Chung, C.-Y.; Wang, J.-C.; Chuang, H.-S. Rapid Bead-Based Antimicrobial Susceptibility Testing by Optical Diffusometry. *PLoS ONE* **2016**, *11*, e0148864. [CrossRef] [PubMed]
17. Iriya, R.; Syal, K.; Jing, W.; Mo, M.; Yu, H.; Haydel, S.E.; Wang, S.; Tao, N. Real-time detection of antibiotic activity by measuring nanometer-scale bacterial deformation. *J. Biomed. Opt.* **2017**, *22*, 126002–126009. [CrossRef] [PubMed]
18. Sparbier, K.; Schubert, S.; Kostrzewa, M. MBT-ASTRA: A suitable tool for fast antibiotic susceptibility testing? *Methods* **2016**, *104*, 48–54. [CrossRef] [PubMed]
19. Ibarlucea, B.; Rim, T.; Baek, C.K.; de Visser, J.A.G.M.; Baraban, L.; Cuniberti, G. Nanowire sensors monitor bacterial growth kinetics and response to antibiotics. *Lab Chip* **2017**, *17*, 4283–4293. [CrossRef] [PubMed]
20. Kuss, S.; Couto, R.A.D.S.; Evans, R.M.; Lavender, H.; Tang, C.C.; Compton, R.G. Versatile Electrochemical Sensing Platform for Bacteria. *Anal. Chem.* **2019**, *91*, 4317–4322. [CrossRef] [PubMed]
21. Idelevich, E.A.; Hoy, M.; Görlich, D.; Knaack, D.; Grünastel, B.; Peters, G.; Borowski, M.; Becker, K. Rapid Phenotypic Detection of Microbial Resistance in Gram-Positive Bacteria by a Real-Time Laser Scattering Method. *Front. Microbiol.* **2017**, *8*, 1064. [CrossRef]
22. e Silva, D.F.; Silva-Dias, A.; Gomes, R.; Martins-Oliveira, I.; Ramos, M.; Rodrigues, A.; Cantón, R.; Pina-Vaz, C. Evaluation of rapid colistin susceptibility directly from positive blood cultures using a flow cytometry assay. *Int. J. Antimicrob. Agents* **2019**, *54*, 820–823. [CrossRef]
23. Tannert, A.; Grohs, R.; Popp, J.; Neugebauer, U. Phenotypic antibiotic susceptibility testing of pathogenic bacteria using photonic readout methods: Recent achievements and impact. *Appl. Microbiol. Biotechnol.* **2018**, *103*, 549–566. [CrossRef] [PubMed]

24. Gao, J.; Li, H.; Torab, P.; Mach, K.E.; Craft, D.W.; Thomas, N.J.; Puleo, C.M.; Liao, J.C.; Wang, T.-H.; Wong, P.K. Nanotube assisted microwave electroporation for single cell pathogen identification and antimicrobial susceptibility testing. *Nanomed. Nanotechnol. Biol. Med.* **2019**, *17*, 246–253. [CrossRef]
25. Novelli-Rousseau, A.; Espagnon, I.; Filiputti, D.; Gal, O.; Douet, A.; Mallard, F.; Josso, Q. Culture-free Antibiotic-susceptibility Determination from Single-bacterium Raman Spectra. *Sci. Rep.* **2018**, *8*, 3957. [CrossRef] [PubMed]
26. Pitruzzello, G.; Thorpe, S.; Johnson, S.; Evans, A.; Gadêlha, H.; Krauss, T.F. Multiparameter antibiotic resistance detection based on hydrodynamic trapping of individual *E. coli*. *Lab Chip* **2019**, *19*, 1417–1426. [CrossRef] [PubMed]
27. Yu, H.; Jing, W.; Iriya, R.; Yang, Y.; Syal, K.; Mo, M.; Grys, T.E.; Haydel, S.E.; Wang, S.; Tao, N. Phenotypic Antimicrobial Susceptibility Testing with Deep Learning Video Microscopy. *Anal. Chem.* **2018**, *90*, 6314–6322. [CrossRef]
28. Correa-Martínez, C.L.; Idelevich, E.A.; Sparbier, K.; Kostrzewa, M.; Becker, K. Rapid Detection of Extended-Spectrum β-Lactamases (ESBL) and AmpC β-Lactamases in Enterobacterales: Development of a Screening Panel Using the MALDI-TOF MS-Based Direct-on-Target Microdroplet Growth Assay. *Front. Microbiol.* **2019**, *10*, 13. [CrossRef] [PubMed]
29. Köck, R.; Wüllenweber, J.; Horn, D.; Lanckohr, C.; Becker, K.; Idelevich, E.A. Implementation of short incubation MALDI-TOF MS identification from positive blood cultures in routine diagnostics and effects on empiric antimicrobial therapy. *Antimicrob. Resist. Infect. Control.* **2017**, *6*, 12. [CrossRef]
30. Kohler, A.; Venturelli, L.; Longo, G.; Dietler, G.; Kasas, S. Nanomotion detection based on atomic force microscopy cantilevers. *Cell Surf.* **2019**, *5*, 100021. [CrossRef]
31. Longo, G.; Alonso-Sarduy, L.; Rio, L.M.; Bizzini, A.; Trampuz, A.; Notz, J.; Dietler, G.; Kasas, S. Rapid detection of bacterial resistance to antibiotics using AFM cantilevers as nanomechanical sensors. *Nat. Nanotechnol.* **2013**, *8*, 522–526. [CrossRef]
32. Kasas, S.; Ruggeri, F.S.; Benadiba, C.; Maillard, C.; Stupar, P.; Tournu, H.; Dietler, G.; Longo, G. Detecting nanoscale vibrations as signature of life. *Proc. Natl. Acad. Sci. USA* **2014**, *112*, 378–381. [CrossRef]
33. Villalba, M.I.; Stupar, P.; Chomicki, W.; Bertacchi, M.; Dietler, G.; Arnal, L.; Vela, M.E.; Yantorno, O.; Kasas, S. Nanomotion Detection Method for Testing Antibiotic Resistance and Susceptibility of Slow-Growing Bacteria. *Small* **2017**, *14*. [CrossRef]
34. Venturelli, L.; Kohler, A.; Stupar, P.; Villalba, M.I.; Kalauzi, A.; Radotic, K.; Bertacchi, M.; Dinarelli, S.; Girasole, M.; Pešić, M.; et al. A perspective view on the nanomotion detection of living organisms and its features. *J. Mol. Recognit.* **2020**, *33*, e2849. [CrossRef]
35. Kasas, S.; Malovichko, A.; Villalba, M.; Vela, M.; Yantorno, O.; Willaert, R. Nanomotion Detection-Based Rapid Antibiotic Susceptibility Testing. *Antibiotics* **2021**, *10*, 287. [CrossRef]
36. Syal, K.; Iriya, R.; Yang, Y.; Yu, H.; Wang, S.; Haydel, S.E.; Chen, H.-Y.; Tao, N. Antimicrobial Susceptibility Test with Plasmonic Imaging and Tracking of Single Bacterial Motions on Nanometer Scale. *ACS Nano* **2015**, *10*, 845–852. [CrossRef] [PubMed]
37. Syal, K.; Shen, S.; Yang, Y.; Wang, S.; Haydel, S.E.; Tao, N. Rapid Antibiotic Susceptibility Testing of Uropathogenic *E. coli* by Tracking Submicron Scale Motion of Single Bacterial Cells. *ACS Sens.* **2017**, *2*, 1231–1239. [CrossRef] [PubMed]
38. Bermingham, C.R.; Murillo, I.; Payot, A.D.; Balram, K.C.; Kloucek, M.B.; Hanna, S.; Antognozzi, M. Imaging of sub-cellular fluctuations provides a rapid way to observe bacterial viability and response to antibiotics. *bioRxiv* **2018**, 460139. [CrossRef]
39. Leonard, H.; Halachmi, S.; Ben-Dov, N.; Nativ, O.; Segal, E. Unraveling Antimicrobial Susceptibility of Bacterial Networks on Micropillar Architectures Using Intrinsic Phase-Shift Spectroscopy. *ACS Nano* **2017**, *11*, 6167–6177. [CrossRef] [PubMed]
40. Willaert, R.G.; Boer, P.V.; Malovichko, A.; Alioscha-Perez, M.; Radotić, K.; Bartolić, D.; Kalauzi, A.; Villalba, M.I.; Sanglard, D.; Dietler, G.; et al. Single yeast cell nanomotions correlate with cellular activity. *Sci. Adv.* **2020**, *6*, eaba3139. [CrossRef] [PubMed]
41. Radonicic, V.; Yvanoff, C.; Villalba, M.I.; Kasas, S.; Willaert, R.G. The Dynamics of Single-Cell Nanomotion Behaviour of *Saccharomyces cerevisiae* in a Microfluidic Chip for Rapid Antifungal Susceptibility Testing. *Fermentation* **2022**, *8*, 195. [CrossRef]
42. Marchetti, O.; Majcherczyk, P.A.; Glauser, M.P.; Bille, J.; Moreillon, P.; Sanglard, D. Sensitive Bioassay for Determination of Fluconazole Concentrations in Plasma Using a Candida albicans Mutant Hypersusceptible to Azoles. *Antimicrob. Agents Chemother.* **2001**, *45*, 696–700. [CrossRef]
43. A Fonzi, W.; Irwin, M.Y. Isogenic strain construction and gene mapping in Candida albicans. *Genetics* **1993**, *134*, 717–728. [CrossRef]
44. Davis, R.H.; Hunt, T.P. Modeling and Measurement of Yeast Flocculation. *Biotechnol. Prog.* **1986**, *2*, 91–97. [CrossRef]
45. Fontana, A.; Bore, C.; Ghommidh, C.; Guiraud, J.P. Structure and Sucrose Hydrolysis Activity of Saccharorn yces cerevisiae Aggregates. *Biotechnol. Bioeng.* **1992**, *40*, 475–482. [CrossRef]
46. van Hamersveld, E.H.; van der Lans, R.G.J.M.; Luyben, K.C.A.M. *Quantification of Brewers' Yeast Flocculation in a Stirred Tank: Effect of Physical Parameters on Flocculation*; John Wiley & Sons, Inc.: Hoboken, NJ, USA, 1997.
47. Haddad, S.A.; Lindegren, C.C. A Method for Determining the Weight of an Individual Yeast Cell. *Appl. Environ. Microbiol.* **1953**, *1*, 153–156. [CrossRef] [PubMed]
48. Guizar-Sicairos, M.; Thurman, S.T.; Fienup, J.R. Efficient subpixel image registration algorithms. *Opt. Lett.* **2008**, *33*, 156–158. [CrossRef] [PubMed]
49. Kwolek-Mirek, M.; Zadrag-Tecza, R. Comparison of methods used for assessing the viability and vitality of yeast cells. *FEMS Yeast Res.* **2014**, *14*, 1068–1079. [CrossRef]
50. Suwunnakorn, S.; Wakabayashi, H.; Kordalewska, M.; Perlin, D.S.; Rustchenko, E. *FKS2* and *FKS3* Genes of Opportunistic Human Pathogen Candida albicans Influence Echinocandin Susceptibility. *Antimicrob. Agents Chemother.* **2018**, *62*, e02299-17. [CrossRef]
51. Sanglard, D.; Ischer, F.; Marchetti, O.; Entenza, J.; Bille, J. Calcineurin A of Candida albicans: Involvement in antifungal tolerance, cell morphogenesis and virulence. *Mol. Microbiol.* **2003**, *48*, 959–976. [CrossRef]

52. Cordeiro, R.A.; Teixeira, C.E.C.; Brilhante, R.S.N.; Castelo-Branco, D.S.C.M.; Paiva, M.A.N.; Leite, J.J.G.; Lima, D.T.; Monteiro, A.J.; Sidrim, J.J.C.; Rocha, M.F.G. Minimum inhibitory concentrations of amphotericin B, azoles and caspofungin against *Candida* species are reduced by farnesol. *Med. Mycol.* **2013**, *51*, 53–59. [CrossRef] [PubMed]
53. Marchetti, O.; Moreillon, P.; Entenza, J.M.; Vouillamoz, J.; Glauser, M.P.; Bille, J.; Sanglard, D. Fungicidal Synergism of Fluconazole and Cyclosporine in *Candida albicans* Is Not Dependent on Multidrug Efflux Transporters Encoded by the *CDR1*, *CDR2*, *CaMDR1*, and *FLU1* Genes. *Antimicrob. Agents Chemother.* **2003**, *47*, 1565–1570. [CrossRef] [PubMed]
54. Rettig, J.R.; Folch, A. Large-Scale Single-Cell Trapping and Imaging Using Microwell Arrays. *Anal. Chem.* **2005**, *77*, 5628–5634. [CrossRef] [PubMed]
55. Hoang, A. Caspofungin acetate: An antifungal agent. *Am. J. Health Pharm.* **2001**, *58*, 1206–1214. [CrossRef] [PubMed]
56. Lu, H.; Shrivastava, M.; Whiteway, M.; Jiang, Y. *Candida albicans* targets that potentially synergize with fluconazole. *Crit. Rev. Microbiol.* **2021**, *47*, 323–337. [CrossRef] [PubMed]
57. Pristov, K.E.; Ghannoum, M.A. Resistance of Candida to azoles and echinocandins worldwide. *Clin. Microbiol. Infect.* **2019**, *25*, 792–798. [CrossRef] [PubMed]
58. Berman, J.; Krysan, D.J. Drug resistance and tolerance in fungi. *Nat. Rev. Genet.* **2020**, *18*, 319–331. [CrossRef]
59. Willaert, R.G. Micro- and Nanoscale Approaches in Antifungal Drug Discovery. *Fermentation* **2018**, *4*, 43. [CrossRef]
60. Morschhäuser, J. The genetic basis of fluconazole resistance development in Candida albicans. *Biochim. Biophys. Acta Mol. Basis Dis.* **2002**, *1587*, 240–248. [CrossRef]
61. Pourakbari, B.; Teymuri, M.; Mahmoudi, S.; Valian, S.K.; Movahedi, Z.; Eshaghi, H.; Mamishi, S. Expression of Major Efflux Pumps in Fluconazole-Resistant Candida albicans. *Infect. Disord. Drug Targets* **2017**, *17*, 178–184. [CrossRef]
62. Calabrese, D.; Bille, J.; Sanglard, D. A novel multidrug efflux transporter gene of the major facilitator superfamily from Candida albicans (FLU1) conferring resistance to fluconazole. *Microbiology* **2000**, *146*, 2743–2754. [CrossRef]
63. Villalba, M.I.; Rossetti, E.; Bonvallat, A.; Yvanoff, C.; Radonicic, V. Simple optical nanomotion method for single-bacterium viability and antibiotic response testing. *Proc. Natl. Acad. Sci. USA* **2023**.
64. Roslon, I.; Japaridze, A.; Rodenhuis, S.; Hamoen, L.; Ghatkesar, M.; Steeneken, P.; Alijani, F. Microwell-enhanced optical detection of single bacteria. *bioRxiv* **2023**. [CrossRef]
65. Rosłoń, I.E.; Japaridze, A.; Steeneken, P.G.; Dekker, C.; Alijani, F. Probing nanomotion of single bacteria with graphene drums. *Nat. Nanotechnol.* **2022**, *17*, 637–642. [CrossRef] [PubMed]
66. Starodubtseva, M.N.; Chelnokova, I.A.; Shkliarava, N.M.; Villalba, M.I.; Tapalski, D.V.; Kasas, S.; Willaert, R.G. Modulation of the nanoscale motion rate of Candida albicans by X-rays. *Front. Microbiol.* **2023**, *14*, 833. [CrossRef]

Disclaimer/Publisher's Note: The statements, opinions and data contained in all publications are solely those of the individual author(s) and contributor(s) and not of MDPI and/or the editor(s). MDPI and/or the editor(s) disclaim responsibility for any injury to people or property resulting from any ideas, methods, instructions or products referred to in the content.

fermentation

MDPI

Article

Development of a Cost-Effective Process for the Heterologous Production of SARS-CoV-2 Spike Receptor Binding Domain Using *Pichia pastoris* in Stirred-Tank Bioreactor

Diego G. Noseda [1,2], Cecilia D'Alessio [1,3,4], Javier Santos [1,3,4], Tommy Idrovo-Hidalgo [3], Florencia Pignataro [1,3], Diana E. Wetzler [1,4,5], Hernán Gentili [3], Alejandro D. Nadra [1,3], Ernesto Roman [1,4], Carlos Paván [1,6] and Lucas A. M. Ruberto [7,8,9,*]

1 Consejo Nacional de Investigaciones Científicas y Técnicas (CONICET), Godoy Cruz 2290, Buenos Aires C1425FQB, Argentina; dnoseda@iib.unsam.edu.ar (D.G.N.)
2 Instituto de Investigaciones Biotecnológicas (IIBio), Universidad Nacional de San Martín-CONICET, Av. 25 de Mayo 1301-1367, Villa Lynch, Provincia de Buenos Aires 1650, Argentina
3 Facultad de Ciencias Exactas y Naturales, Departamento de Fisiología y Biología Molecular y Celular, Instituto de Biociencias, Biotecnología y Biología Traslacional (iB3), Ciudad Universitaria-Pabellón II, Buenos Aires 1428, Argentina
4 Departamento de Química Biológica, Facultad de Ciencias Exactas y Naturales, Universidad de Buenos Aires, Ciudad Universitaria-Pabellón II, Ciudad Autónoma de Buenos Aires 1428, Argentina
5 Instituto de Química Biológica de la Facultad de Ciencias Exactas y Naturales (IQUIBICEN), CONICET-Universidad de Buenos Aires, Ciudad Universitaria-Pabellón II, Ciudad Autónoma de Buenos Aires 1428, Argentina
6 Instituto de Química y Fisicoquímica Biológicas, LANAIS PROEM, Facultad de Farmacia y Bioquímica, Universidad de Buenos Aires, Ciudad Autónoma de Buenos Aires C1113AAD, Argentina
7 Departamento deMicrobiología, Inmunología, Biotecnología y Genética, Facultad de Farmacia y Bioquímica, Universidad de Buenos Aires, Junín 956 6to Piso, Ciudad Autónoma de Buenos Aires C1113AAD, Argentina
8 Facultad de Farmacia y Bioquímica, Instituto de Nanobiotecnología (NANOBIOTEC), CONICET-Universidad de Buenos Aires, Junín 956 6th Floor, Ciudad Autónoma de Buenos Aires C1113AAD, Argentina
9 Instituto Antártico Argentino, Ministerio de Relaciones Exteriores y Culto, Buenos Aires, Argentina. Av. 25 de Mayo 1147, Villa Lynch, Provincia de Buenos Aires 1650, Argentina
* Correspondence: luruberto@gmail.com

Abstract: SARS-CoV-2 was identified as the pathogenic agent causing the COVID-19 pandemic. Among the proteins codified by this virus, the Spike protein is one of the most-external and -exposed. A fragment of the Spike protein, named the receptor binding domain (RBD), interacts with the ACE2 receptors of human cells, allowing the entrance of the viruses. RBD has been proposed as an interesting protein for the development of diagnosis tools, treatment, and prevention of the disease. In this work, a method for recombinant RBD production using *Pichia pastoris* as a cell factory in a stirred-tank bioreactor (SRTB) up to 7 L was developed. Using a basal saline medium with glycerol, methanol, and compressed air in a four-stage procedure, around 500 mg/L of the raw RBD produced by yeasts (yRBD) and 206 mg/L of purified (>95%) RBD were obtained. Thereby, the proposed method represents a feasible, simple, scalable, and inexpensive procedure for the obtention of RBD for diagnosis kits and vaccines' formulation.

Keywords: SARS-CoV-2; RBD; antigen; STBR; *Pichia pastoris*

Citation: Noseda, D.G.; D'Alessio, C.; Santos, J.; Idrovo-Hidalgo, T.; Pignataro, F.; Wetzler, D.E.; Gentili, H.; Nadra, A.D.; Roman, E.; Paván, C.; et al. Development of a Cost-Effective Process for the Heterologous Production of SARS-CoV-2 Spike Receptor Binding Domain Using *Pichia pastoris* in Stirred-Tank Bioreactor. *Fermentation* **2023**, *9*, 497. https://doi.org/10.3390/fermentation9060497

Academic Editor: Ronnie Willaert

Received: 20 April 2023
Revised: 15 May 2023
Accepted: 16 May 2023
Published: 23 May 2023

Copyright: © 2023 by the authors. Licensee MDPI, Basel, Switzerland. This article is an open access article distributed under the terms and conditions of the Creative Commons Attribution (CC BY) license (https://creativecommons.org/licenses/by/4.0/).

1. Introduction

The 2020, the SARS-CoV-2 pandemic demanded the development of suitable tools to face and manage a widespread human infection. Although massive contagion seems to be currently a picture of the recent past, novel viral variants and global increases in the number of infected people, such as those taking place in China and the Northern Hemisphere during the second half of 2022, are issues of current concern related to SARS-CoV-2 [1].

Viral proteins or some of its domains would be useful molecules to detect, treat, and prevent viral disease and its consequences. SARS-CoV-2 belongs to the Coronaviridae family [2,3]. Coronaviruses are enveloped non-segmented positive-sense RNA viruses. SARS-CoV-2 virus presents a genome with four open reading frames (ORFs) for the structural proteins: Spike, Envelope, Membrane, and Nucleocapsid. The Spike complex (~150 kDa) mediates the viral and cellular membrane interaction and fusion by binding mainly to angiotensin-converting enzyme 2 (ACE2) through the receptor binding domain (RBD) [4–7]. Some regions of the Spike protein were suggested as suitable targets for drug development [8]. Considering this specific function, it is possible to assume that RBD heterologous expression would provide a useful tool for diagnosis purposes, as well as for immunization to obtain neutralizing antibodies or even a protein-based vaccine. It was reported that in human natural infection, a large fraction of the neutralizing antibodies target RBD [9–11]. What is more, Liu et al. (2020), working with nineteen potent neutralizing antibodies (in vitro) obtained from infected patients, found that almost half of them were directed against the RBD [10], highlighting its potential role as a vaccine antigen. For these reasons, SARS-CoV-2 RBD was selected for its heterologous expression aiming to obtain large amounts of such protein.

The heterologous production of several SARS-CoV-2 proteins was reported using different expression systems, the whole Spike protein and its RBD being the most-common ones [11–14]. For example, Li et al. (2020) [15] expressed the RBD, the S1 subunit, the WT S ectodomain, and a prefusion-trimer-stabilized form of S using Sf9 insect cells.

It is important to consider that the Spike protein presents 22 possible N-glycosylations and 4 O-glycosylation sites, some of them being on the RBD [16,17]. Additionally, the RBD presents 9 cysteines, 8 of them forming S-S bridges [8]. Glycosylation, as well as disulfide bridges are issues of special attention for heterologous protein production, since these kinds of modifications affect protein folding and, in some cases, biological activity [18]. This is particularly relevant when the recombinant proteins are produced for medical use in humans [19,20]. The selection of the expression system is usually strongly conditioned by the requirement of such post-translational modifications [21]. In the case of the RBD, glycosylations and S-S bridges' formation seem to be important for adequate protein folding, and thus, the expression host selection is a critical decision in the development of its production process [22].

Some yeasts such as *Saccharomyces cerevisiae*, *Kluyveromyces lactis*, *Yarrowia lipolytica*, and *Pichia pastoris* (*Komagataella phaffii*) are suitable and convenient hosts for recombinant protein production [23]. Singularly, *P. pastoris* is a methylotrophic non-conventional yeast considered as a biological model [24,25] and used for heterologous protein production, usually taking advantage of the strong alcohol oxidase 1 (AOX1) promoter and its ability to achieve high cellular density in bioreactors, for which values near 100 g dry cell weight (DCW)/L have been reported [26–28]. The AOX1 promoter strongly responds to methanol, while its activity is repressed by glucose and glycerol [29,30]. This yeast is also able to secrete large amounts of properly folded heterologous proteins with only a few other secreted proteins and is, for these reasons, widely used as an expression system. Additionally, this microorganism is able to perform some post-translational protein modifications (glycosylation, proteolytic processing, and disulfide bonds formation) usually observed in higher eukaryotes, a relevant feature when the production of proteins for medical purposes is involved [31–34] or when those modifications are required for proper protein folding [35–37]. Beyond heterologous protein production, *P. pastoris* has been recently used for the expression of other metabolic pathways, leading to the obtention of non-proteinous molecules [38]. It has also been evaluated as a key component in a probiotic preparation for poultry [39]. Finally, a complete reference genome of these microorganisms is available [40].

Optimum conditions for heterologous protein production using *P. pastoris* depend on several factors, such as the medium's composition, temperature, and culture strategy, among others [41,42]. For this reason, research must be performed to find the conditions that maximize protein production with the selected microbial construction [43]. Methanol feeding strategies, flows, and concentration are considered as relevant factors affecting cellular activity and protein production [43,44]. Staggered fed, exponentially fed, DO-stat, and methanol concentration feedback control are strategies commonly reported for *P. pastoris* cultivation [45]. Among them, DO-stat consists of looping the methanol feeding to the dissolved oxygen concentration. In such a way, methanol feeding is activated as pulses when the percent of dissolved oxygen rises beyond a setpoint, thus avoiding an excessive O_2 demand, heat production, as well as anaerobiosis or methanol accumulation [46–48].

Antigens such as recombinant RBD are considered useful for subunit vaccines' development, especially when they are produced using high-yield hosts such as *P. pastoris*, allowing the production of large amounts of antigen doses at a relatively low cost [12,13,49], for this reason and its scalability are being a suitable tool to face a pandemic.

The SARS-CoV-2 pandemic presented nucleic-acid-based vaccines as state-of-the-art tools for massive vaccination; however, some safety and logistical aspects of this recently implemented vaccination strategy for humans raised concerns among a great part of the population. The toxicity of synthetic raw materials used to conjugate lipids in mRNA vaccines, the possibility of nucleic acids persisting in vivo, and the risk of the theoretical integration of foreign DNA into the host chromosome [50], in addition to the high costs and strict cold chain requirements for some vaccines based on mRNA—having a difficult distribution in remote areas where ultra-low temperature freezers are unavailable—represent some of its disadvantages. Although this new technology is promising, protein subunit vaccines are also a functional and safe alternative. Due to their higher safety profile, subunit vaccines are primarily developed for use with elderly and infant patients [51,52], and vaccines based on the RBD alone effectively boost an immune response originally generated against a full-length Spike protein trimer, increasing interest in using RBD-based vaccine boosters to provide immunity against emerging variants [53]. Additionally, protein subunit vaccines do not require ultra-freezing conditions and can also be safely stored in a regular fridge or lyophilized for their distribution [54], making this type of vaccine useful for complementing vaccination campaigns all over the world.

In a previous work, we reported that recombinant SARS-CoV-2 RBD produced using *P. pastoris* as the expression host presented a similar and comparable conformation as the one produced using HEK293T mammalian cells. A bioreactor production procedure was used, yielding 45 mg/L of 90% pure protein [13]. This was a first attempt at the production of the RBD at a scale large enough for small-scale protein characterization and immunization assays.

In this work, we propose a new procedure that improves by five-times the production yield of the recombinant RBD antigen from the SARS-CoV-2 Spike protein. It consists of a four-step procedure that was optimized by comparing two culture strategies in a 7 L stirred-tank bioreactor. Furthermore, we report the scaling up of the procedure of RBD production to a 14 L stirred-tank bioreactor.

Goal: The goal of the project was to produce a low-cost antigen to be used in diagnosis (antibodies detection), therapies (generation of neutralizing antibodies), and prevention (vaccine antigen production).

2. Methods

2.1. Plasmid and Strain

The RBD sequence and plasmid construction, as well as the *Pichia pastoris* strain used in this work were the same as described previously [12,13]. Briefly, the sequence of amino acid residues 319–537 of the SARS-CoV2 Spike protein (RBD) was codon-optimized for expression in *P. pastoris*. Furthermore, the alpha-factor secretion signal (N-terminal) of *S. cerevisiae* was fused to direct the heterologous protein into the culture medium. In addition, a *Staphylococcus aureus* sortase A recognition sequence for covalent coupling and a His6 tag for purification were inserted into the C-terminal extreme. The entire sequence was synthesized and cloned into the pPICZalphaA vector under transcriptional control of the *P. pastoris* AOX1 promoter by GenScript (Piscataway, NJ, USA) and linearized and transformed into the *P. pastoris* X-33 strain.

2.2. Determination of Dry Cell Weight

The optical density of *P. pastoris* samples from the shake flask cultures and bioreactor fermentations was determined at 600 nm using a UV–Vis spectrophotometer and converted to dry cell weight (DCW, in g/L) using the following equation: DCW = 0.269 OD_{600nm} (R^2 = 0.99), corresponding to a calculated DCW to OD_{600nm} calibration curve.

2.3. Quantification of Total Proteins and RBD

Samples collected during the methanol induction phase of the flask cultures and bioreactor fermentations were centrifuged at 10,000× *g* for 20 min to obtain cell-free supernatants. The total protein concentration during cultures and fermentations was determined in the supernatants by the Bradford method [55,56] using a calibration curve of the BSA standard. Proteins in the supernatants were run in a 12% SDS-PAGE stained with Coomassie brilliant blue G-250 (Sigma-Aldrich; St. Louis, MO, USA) to visualize protein bands. Protein quantification was performed comparing densitometry with known standards of BSA. Chen et al. (2017) [57] reported that Coommasie Blue dye binds 1.33-fold less to RBD from SARS-CoV than to BSA. It is worth mentioning that, although in this work, we did not compare the binding capability of RBD from SARS-CoV2 to Coommasie Blue with that of BSA and, thus, there may be some underestimation of the yield of the process, this will be the same in all samples. The relative abundance of recombinant RBD, in both supernatants and purification fractions, was determined by band densitometry using the ImageJ software (http://rsb.info.nih.gov/ij, accessed on 10 December 2020) and considering the total protein concentration determined by the Bradford method.

2.4. Medium Composition for Flask Cultures and Bioreactor Fermentations

The cultivation of *P. pastoris* inoculums in Erlenmeyer flasks was performed either in BMGY medium (1% yeast extract, 2% bactopeptone, 1.34% YNB, 400 μg/L biotin, 100 mM potassium phosphate pH 6, and 1% glycerol) or in a low-salt medium (LSM) with 10 g/L glycerol at 30 °C and 250 rpm. The LSM medium contained: 4.55 g/L potassium sulfate, 3.73 g/L magnesium sulfate heptahydrate, 1.03 g/L potassium hydroxide, 0.23 g/L calcium sulfate anhydrous, and 10.9 mL/L phosphoric acid 85%. After the sterilization of the medium, 3.5 mL/L of filtered biotin solution (0.2 g/L) and 3.5 mL/L of filtered trace metal solution (PTM1) were added. PTM1 contained per liter: 6.0 g copper (II) sulfate pentahydrate, 0.08 g sodium iodide, 3.0 g manganese sulfate-monohydrate, 0.2 g sodium molybdate-dihydrate, 0.02 g boric acid, 0.5 g cobalt chloride, 20.0 g zinc chloride, 65.0 g ferrous sulfate-heptahydrate, 0.2 g biotin, and 5.0 mL sulfuric acid (98% *w*/*w*). The cultivations of *P. pastoris* in the bioreactors were performed in LSM supplemented with 40 g/L glycerol. The use of LSM prevents salt precipitation during the pH rise in the downstream process, as was previously reported [57].

2.5. Inoculum Preparation

To obtain the inoculum for flask cultures and bioreactor fermentations, a single colony of *P. pastoris* clone grown on a YPD agar plate was inoculated into a 250 mL flask containing 40 mL of LSM (supplemented with PTM1 and biotin) with 10 g/L glycerol or in 250 mL BMGY and cultured overnight at 30 ± 1 °C and 250 rpm in an orbital shaker. A volume of 400 mL of LSM (supplemented with PTM1 and biotin) containing 10 g/L glycerol in a 2 L Erlenmeyer flask was inoculated with the overnight culture and incubated under the same conditions until the culture reached an OD_{600} of ~14. This culture was used to simultaneously inoculate a set of Erlenmeyer flasks and a stirred-tank bioreactor with LSM at a ratio of $V_{seed} = V_0/10$, where V_{seed} is the volume of the inoculum and V_0 is the initial volume of the culture.

2.6. Cultivation in Erlenmeyer Flask

Cultures of *P. pastoris* expressing the RBD clone were started in Erlenmeyer flasks to evaluate the growth kinetics and recombinant RBD production. For this purpose, three 250 mL Erlenmeyer flasks containing 50 mL of LSM (supplemented with PTM1 and biotin) with 10 g/L glycerol were inoculated with 5 mL of the previously described culture and incubated at 30 $\pm$ 1 °C and 250 rpm in an orbital shaker. After 24 h, the induction conditions were established by adding pure methanol at a final concentration of 1% (*v*/*v*). This procedure was repeated every 24 h to maintain methanol induction for a period of 120 h. During induction, flasks were incubated on a shaker at 250 rpm and 25 $\pm$ 1 °C, as low culture temperatures increase the yield of soluble recombinant proteins in *P. pastoris* due to reduced extracellular proteolysis without affecting cell growth [58,59]. Culture samples for the determination of biomass, total proteins, and RBD concentration were collected every 24 h. Specific growth rates (μ) for each culture stage were calculated from the slope of the regression line of the growth curve.

2.7. Fermentations in Stirred-Tank Bioreactor

Fermentations were carried out in a stirred-tank bioreactor (BioFlo 115, New Brunswick Scientific; Edison, NJ, USA) using a four-stage procedure based on previous work with modifications [13,60–62]. The first stage consisted of a batch culture using LSM with unlimited glycerol (40 g/L) as the sole carbon and energy source, supplemented with 3.5 mL/L PTM1 and 3.5 mL/L biotin solution (0.2 g/L). After an abrupt peak in the percentage of dissolved oxygen (Spike) indicating carbon source depletion, the second phase—fed-batch with glycerol—was initiated. In this phase, the culture was fed with a solution containing 600 g/L of glycerol, 12.25 mL/L PTM1, and 12.25 mL/L biotin solution (0.2 g/L). To ensure glycerol limitation and, thus, gradually derepress the AOX1 promoter, feeding was automatically regulated according to the percentage of dissolved oxygen (%DO) in the culture, a strategy referred to as DO-stat [46,63]. Later, a short transition phase was performed to allow the adaptation of the culture to growth in the presence of methanol as the sole carbon source. For this, two strategies were compared: (1) feeding with a glycerol/methanol mixture (3:1) for 5 h and (2) feeding with a pulse of 4 g/L of methanol. Finally, the last induction phase was initiated by adding pure methanol, supplemented with 12.25 mL/L PTM1 and 12.25 mL/L biotin solution (0.2 g/L), as the sole carbon and energy source, applying a fed-batch procedure with a growth-limiting rate. The feeding of methanol at limiting concentration was also regulated with the level of DO in the culture (DO-stat). The stirred-tank bioreactor operated by the interface of the Biocommand Bioprocessing (New Brunswick Scientific) software for parameter control and data acquisition. The temperature was maintained at 30 $\pm$ 1 °C during the batch and glycerol-fed-batch phases and at 25 $\pm$ 1 °C during the transition and induction phases. The pH was kept at 5.0 in the first two phases and 5.3 in the last two phases by adding 42.5% (*v*/*v*) H_3PO_4 and 14% (*v*/*v*) NH_4OH, which also served as a nitrogen source. Dissolved oxygen was controlled by stirring (maximum 1000 rpm) and by filtered (0.22 µm) compressed air (1–1.5 VVM). The pH was measured using a pH electrode (Mettler-Toledo GmbH, Gießen, Hessen, Germany), and the dissolved oxygen content was determined using a polarographic probe (InPro6110/320, Mettler-Toledo GmbH). Foam formation was prevented by adding 3% (*v*/*v*) antifoam agent 204 (Sigma-Aldrich; St. Louis, MO, USA). Samples were taken during the different fermentation phases to determine the concentrations of the biomass, total protein, and recombinant RBD. Biomass evolution during the fermentation was expressed as DCW (g/L) = f(t). Fermentations were performed in vessels of 7 and 14 L to compare the RBD production process with two different vessel volumes. After fermentation, the biomass was removed from the culture by centrifugation at 18,600$\times$ *g* for 20 min at 4 °C in a Sorvall high-speed centrifuge (Lynx 4000 Thermo, Waltham, MA, USA) equipped with a F10 rotor, and the supernatant was used for RBD purification.

2.8. Purification of Recombinant RBD and Quality Control

The purification of recombinant RBD from cell-free supernatants, as well as RBD quality control was as already described [13]. Briefly, the purification was performed using a NTA-Ni^{2+} column, previously equilibrated with 20 mM Tris–HCl, 150 mM NaCl, and 20 mM imidazole, pH 7.4 (equilibration solution). The supernatants were adjusted to pH 7.4 with NaOH and to 20 mM Tris and 20 mM imidazole, centrifuged 20 min at 12,000× *g*, and loaded into the column. The column was washed with an equilibration solution. Finally, the recombinant RBD was eluted in Tris 20 mM, NaCl 150 mM, and 300 mM imidazole, pH 7.4. The purified protein was dialyzed twice in 20 mM Tris–HCl and 150 mM NaCl buffer, pH 7.4. Absorption spectra (240–340 nm range, using a 0.1 nm sampling interval) were acquired at 20°C with a JASCO V730 BIO spectrophotometer (Japan). The RBD concentration was determined using: ε = 33,850 M^{-1} cm^{-1} (Abs = 1.304 for a 1 mg mL^{-1} protein solution). After that, the RBD was stored at −80°C. The purity of the recombinant RBD was analyzed by reverse-phase HPLC using an analytical C18 column (Higgins Analytical, Inc., Mountain View, CA, USA) and a JASCO system (equipped with an autoinjector, an oven at 25 °C, and a UV detector). The elution was made at a 1.0 mL/min flow using an ACN gradient from 0 to 100% over 40 min (10–50 min of the run) with a mobile phase of 0.05% TFA.

2.9. Statistical Analysis

The data for the cell density, total protein concentration, and RBD concentration were taken in triplicate. The means and standard deviations were calculated. Statistical significance was evaluated by the Student's *t*-test using the Infostat V2020 Software. Differences were considered significant if $p < 0.05$.

3. Results

3.1. Growth Kinetics and RBD Expression at Flask Level

The *P. pastoris* clone expressing RBD [13] was cultured in Erlenmeyer flasks to evaluate the growth kinetics and recombinant RBD expression. The culture in LSM containing 10 g/L glycerol exhibited a maximum specific growth rate (μ_{max}) of 0.15 h^{-1} and a biomass yield coefficient based on consumed substrate ($Y_{X/S}$) of 0.43 g DCW/g glycerol (Figure 1A). At the end of the exponential phase, corresponding to an incubation period of 24 h, the culture displayed a cell concentration of 4.6 g DCW/L, while the glycerol was completely depleted. Moreover, during the expression induction phase, in which methanol was added in pulses, cells of the *P. pastoris* clone continued growing with a specific growth rate (μ) of 0.072 h^{-1} in the first 24 h of induction. The culture continued growing, showing a decrease in the growth rate, due to methanol limitation, to a final value of 0.002 h^{-1}. Hence, the average specific growth rate over the induction phase was 0.026 h^{-1}. After 120 h of methanol induction, the culture reached a maximum biomass level of 8.0 g DCW/L in the induction phase with a biomass yield coefficient ($Y_{X/S}$) of 0.085 g DCW/g methanol on average (Figure 1A). Figure 1B shows the decrease in the specific growth rate during the methanol induction phase.

As shown in Table 1, the total protein concentration in the culture supernatant increased from 5.2 mg/L at the beginning of the methanol induction phase to 72.3 mg/L after 120 h of induction. Moreover, the RBD concentration was 1.6 mg/L at the beginning of this phase and 21.7 mg/L at the end, representing a 14-fold increase. The percentage of the recombinant RBD in the supernatants with respect to the concentration of total proteins was 30%. The RBD yield based on biomass formation ($Y_{RBD/X}$) showed a value of 2.71 mg/g at 120 h methanol induction, corresponding to an eight-fold increase with respect to the initial value. Furthermore, the volumetric RBD productivity (vP) at the end of incubation was 0.15 mg/Lh, representing an increase of 100% compared to the initial induction time (Table 1). The specific RBD productivity (sP) was 18.8 μg RBD/g DCW h at the end, indicating that this parameter increased 1.4-fold with respect to the initial point (Table 1). Finally, the RBD yield based on methanol consumed ($Y_{RBD/S}$) reached

0.54 mg RBD/g methanol at the end of the whole process. It is worth mentioning that the results obtained with Erlenmeyer cultures were important for planning the scale-up of the recombinant RBD production in the stirred-tank bioreactor.

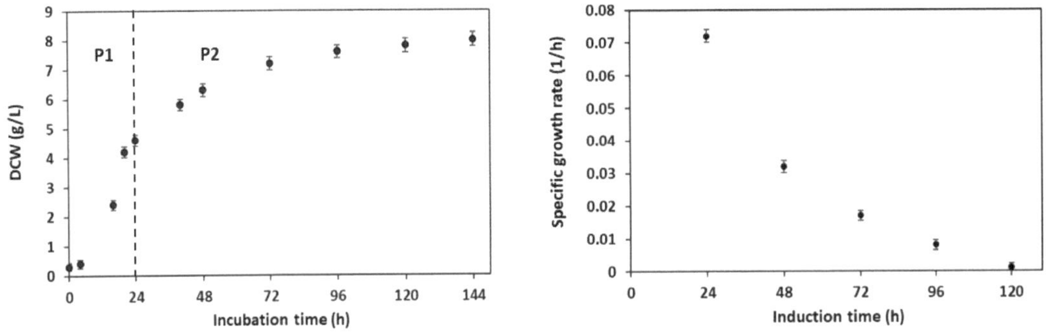

Figure 1. Growth of *P. pastoris* expressing RBD in flask cultures. (**A**) Biomass concentration (g DCW/L) evolution during flask cultivation. **P1**: exponential growth phase in LSM with 10 g/L glycerol, **P2**: Induction phase with methanol pulses at a final concentration of 1% (*v*/*v*), each 24 h incubation. Error bars indicate 2 SD. (**B**) Variation of the specific growth rate (μ) during the induction phase. During the induction of expression, the culture continued growing with a decrease in the growth rate due to the limitation of methanol.

Table 1. Parameters obtained from the culture of *P. pastoris* expressing RBD in Erlenmeyer flasks. Detailed parameters: Total protein concentration (mg/L), RBD concentration (mg/L), RBD concentration increase (fold), $Y_{RBD/X}$ (mg RBD/g DCW): RBD yield based on biomass formation; $_{V}P$ (mg RBD/L h): volumetric RBD productivity, $_{S}P$ (ug RBD/g DCW h): specific RBD productivity.

Induction Time (h)	Total Protein Concentration (mg/L)	RBD Concentration (mg/L)	RBD Increase (fold)	$Y_{RBD/x}$ (mg/g)	vP (mg/L h)	sP (μg/g h)
0	5.2	1.6	1.0	0.34	0.07	14.1
24	12.8	3.8	2.5	0.61	0.08	12.7
48	26.4	7.9	5.7	1.12	0.11	15.5
72	41.8	12.5	8.0	1.67	0.13	17.4
96	57.1	17.1	11.0	2.17	0.14	18.1
120	72.3	21.7	13.9	2.71	0.15	18.8

3.2. Production of Recombinant RBD in 7 L and 14 L Stirred Bioreactors

The fermentation of the *P. pastoris* clone for the production of the recombinant RBD was first performed in a 7 L stirred-tank bioreactor according to the four-phase procedure described above. Two fermentation strategies were compared, using an initial culture volume of 1.5 L in the bioreactor in both cases. The fermentation carried out with Strategy 1 showed a maximum biomass concentration of 16.7 g DCW/L at 18 h of the batch phase. During this stage, the *P. pastoris* culture displayed a maximum specific growth rate (μ_{max}) of 0.16 h^{-1} and a biomass yield coefficient ($Y_{X/S}$) of 0.42 DCW/g glycerol. After an increase in dissolved oxygen due to glycerol depletion (DO spike), the fed-batch phase was initiated, with glycerol feeding controlled by the percentage of dissolved oxygen (%DO), with a cut-off of 50% saturation. Glycerol feeding was maintained for 23 h, reaching a biomass level of 61.4 g DCW/L. Next, the transition phase was performed by feeding the culture with a glycerol:methanol (3:1) mixture for 5 h, allowing the cells to adapt to the methanol and reaching a cell concentration of 64.1 g DCW/L at the end of this phase. Then, the methanol-fed-batch phase was carried out to induce the expression of the recombinant

RBD, regulating the feeding of pure methanol in response to %DO with a saturation cut-off of 60%. After 52 h of methanol induction and a total fermentation process of 98 h, the culture reached a biomass level of 78.2 g DCW/L. At this fermentation time, the total protein concentration in the culture supernatant was 296.3 mg/L, while the RBD reached 98.4 mg/L, corresponding to 33% of the total proteins in the supernatant (Table 2). Since the volume of the supernatant was 2.6 L, a total amount of the RBD of 255.8 mg was obtained. The RBD yield based on biomass formation ($Y_{RBD/X}$) exhibited a value of 1.3 mg RBD/g DCW at the end of the induction. The volumetric RBD productivity ($_{V}P$) of the whole process reached 1.0 mg RBD/L h, and the total RBD productivity ($_{T}P$) was 2.6 mg RBD/h (Table 2). Moreover, the specific RBD productivity ($_{S}P$) of the whole process was 12.8 µg RBD/g DCW h, and the RBD yield based on methanol consumed ($Y_{RBD/S}$) reached 0.5 mg DCW/g methanol at the end of the fermentation.

Table 2. Production parameters of *P. pastoris* fermentations using Strategies 1 and 2 in 7 L stirred-tank bioreactor. Detailed production parameters: Final biomass level (g DCW/L); total protein concentration (mg/L); RBD concentration (mg/L); total RBD (mg); $Y_{RBD/biomass}$ (RBD yield based on biomass formation, mg/g); whole process volumetric RBD productivity ($_{V}P$, mg RBD/L h); whole process total RBD productivity ($_{T}P$, mg RBD/h); whole process specific RBD productivity ($_{S}P$, ug RBD/g DCW h); $Y_{RBD/Methanol}$ (final RBD yield based on methanol consumed, mg/g).

	Strategy 1	Strategy 2
Final biomass level (g DCW/L)	78.2	89.2
Total protein concentration (mg/L)	296.3	1378.5
RBD concentration (mg/L)	98.4	519.6
Total RBD (mg)	255.8	1402.9
YRBD/Biomass (mg/g)	1.3	5.8
Volumetric RBD productivity (mg/L h)	1.0	3.8
Total RBD productivity (mg/h)	2.6	10.3
Specific RBD productivity (µg/g DCW h)	12.8	42.2
YRBD/Methanol (mg/g)	0.5	1.8

The fermentation performed with Strategy 2 in the 7 L stirred-tank bioreactor presented a maximum biomass level of 17.2 g DCW/L at the end of the batch phase (18 h). At this stage, the culture showed similar values of the specific growth rate and biomass yield coefficient as in Strategy 1. In the fed-batch phase, feeding with glycerol was maintained for 8 h in response to %DO with a saturation cut-off of 50%, reaching a biomass level of 40.9 g DCW/L. Then, in the transition phase, feeding was carried out with a dose of 4 g/L methanol, which allowed culture adaptation to methanol and to reach a biomass concentration of 41.4 g DCW/L after 4 h. Complete consumption of methanol was indicated by an abrupt increase of %DO (DO spike). During the induction phase, methanol feeding was regulated by %DO with a saturation cut-off of 60%, reaching a final biomass level of 89.2 g DCW/L after 108 h of induction and 138 h of total process. At this time, the total protein and RBD concentrations in the culture reached 1378.5 mg/L and 519.6 mg/L, respectively, while the total recombinant RBD yielded 1402.9 mg, as the supernatant volume was 2.7 L (Table 2). Thereby, the RBD yield based on biomass ($Y_{RBD/X}$) at the end of the process showed a value of 5.8 mg RBD/g DCW, an increase of 4.5-fold compared to Strategy 1. The volumetric RBD productivity ($_{V}P$) of the whole process was 3.8 mg RBD/L h, and the total RBD productivity ($_{T}P$) reached 10.3 mg RBD/h (Table 2), which corresponds to an approximately 4-fold increase of both parameters compared to Strategy 1. Moreover, the specific RBD productivity ($_{S}P$) of the whole process was 42.2 µg RBD/g DCW h, and the RBD yield based on methanol consumed ($Y_{RBD/S}$) reached 1.8 mg RBD/g methanol at the end of the fermentation, corresponding to an approximately 3.5-fold increase of both parameters compared to Strategy 1. Figure 2 shows the parameter profile of the fermentation carried out with Strategy 2 and describes the variation of the stirring, temperature, pH, dissolved oxygen, and feeding throughout the fermentation process. This profile reports the decrease in dissolved oxygen content due to culture growth during the batch phase until the dissolved oxygen spike, the point at which glycerol feeding began, and also shows the variations in dissolved oxygen content by which glycerol or methanol feeding was regulated. SDS-PAGE analysis of the fermentation supernatants corresponding to Strategy 2 displayed the increase in the total protein and recombinant RBD during the expression induction phase with methanol. The three bands of around 30, 35, and 40 kDa correspond to different mannose content during Golgi glycosylation extensions of the RBD, as all bands merged to a single one of about 26 kDa after deglycosylation with endoglycosidase H, as previously reported by our group (Figure 3A). In addition, Figure 3B describes the increase in the total proteins and RBD concentration, as well as the percentage of the RBD in the total proteins every 12 h of the methanol induction phase corresponding to the fermentation performed with Strategy 2. Table 3 describes the evolution of the production parameters during the fermentation carried out with Strategy 2 in the 7 L stirred-tank bioreactor. It should be noted that the RBD concentration increased more than 20-fold during the induction, and the percentage of the RBD in the total proteins increased from 26.5% to 37.7% at the end of the induction phase. Under both strategies, a low RBD concentration was detected at the beginning of the induction (0 h). These amounts of recombinant protein were produced due to the RBD expression during the glycerol-fed-batch stage when this substrate was provided in a limited amount and during the adaptation phase when methanol was added. It is well known that both conditions allow the derepression of the AOX1 gene and, thus, the production of low levels of the recombinant protein [64–66].

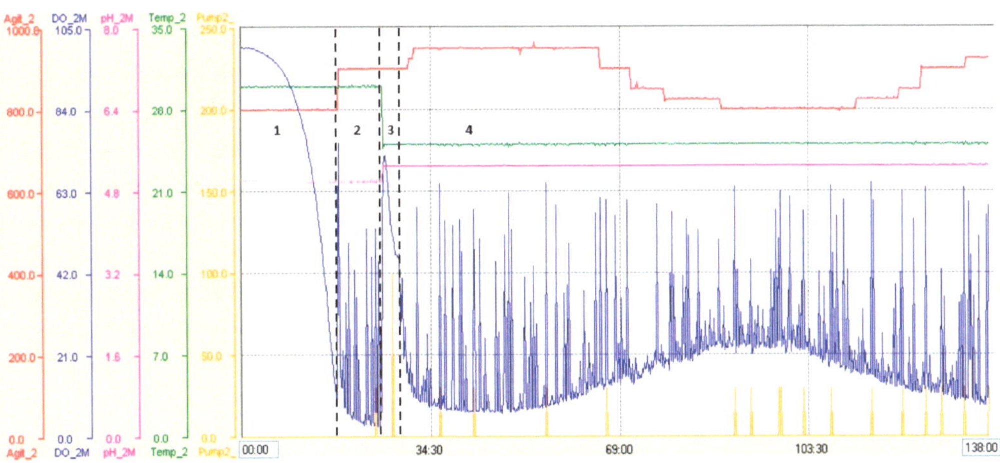

Figure 2. Parameter profile of *P. pastoris* fermentation carried out with Strategy 2 in 7-L stirred-tank bioreactor. Blue line: dissolved oxygen level (saturation percentage), green line: temperature (°C), pink line: pH, red line: stirring (RPM), yellow line: feeding (pumping percentage). 1: Batch phase, 2: glycerol-fed-batch phase, 3: transition phase, 4: induction phase.

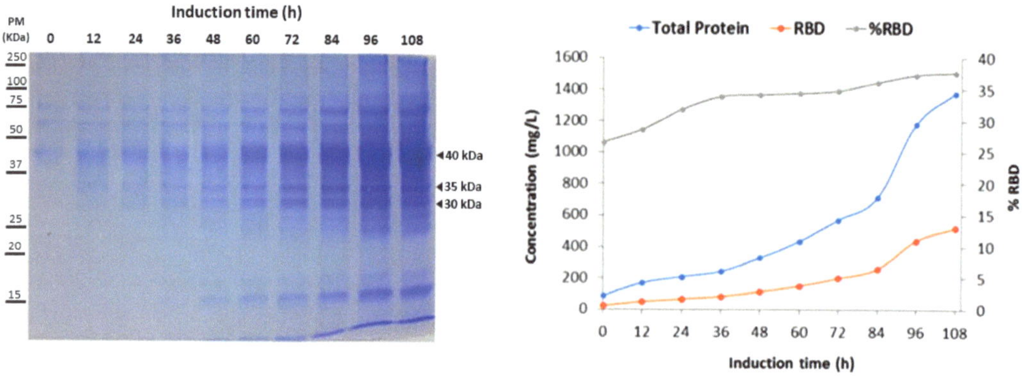

Figure 3. SDS-PAGE analysis and production parameters of samples from the induction phase of 7-L bioreactor fermentation. (**A**) SDS-PAGE protein profile of fermentation supernatants corresponding to methanol induction phase. Arrows indicate the molecular weights of the different glycosylated RBD variants. PM: Protein marker. (**B**) Variation in total protein concentration, RBD concentration, and RBD percentage in total protein during methanol induction.

Table 3. Evolution of production parameters during fermentation carried out with Strategy 2 in 7-L stirred-tank bioreactor: Detailed parameters every 12 h of methanol induction: total protein concentration (mg/L); RBD concentration (mg/L); RBD percentage with respect to total protein (%): RBD concentration increase (fold); $Y_{RBD/X}$: RBD yield based on biomass formation (mg/g); vP: volumetric RBD productivity (mg RBD/L h).

Induction Time (h)	Total Protein Concentration (mg/L)	RBD Concentration (mg/L)	RBD Percentage (%)	RBD Increase (fold)	Y RBD/x (mg/g) vP	vP (mg/L h)
0	85.0	22.6	26.5	1.0	0.5	0.8
12	168.7	48.2	28.6	2.1	0.8	1.2
24	207.7	66.0	31.8	2.9	1.0	1.2
36	241.0	81.7	33.9	3.6	1.1	1.3
48	328.5	112.3	34.2	5.0	1.4	1.5
60	435.0	149.6	34.4	6.6	1.9	1.7
72	570.6	198.6	34.8	8.8	2.4	2.0
84	711.4	256.8	36.1	11.4	2.9	2.3
96	1178.3	439.8	37.3	19.5	4.9	3.5
108	1378.5	519.6	37.7	23.0	5.8	3.8

After fermentation in a 7 L bioreactor, RBD production was scaled up to a 14 L stirred-tank bioreactor with an initial volume of 4 L and using Strategy 2 as a four-step procedure. In the batch phase, a biomass concentration of 17.6 g DCW/L was achieved after 18 h, reaching similar μ_{max} and $Y_{X/S}$ values to those previously obtained. Next, in the fed-batch phase, glycerol feeding was performed for 8 h, in response to %DO with a saturation cut-off of 60%, obtaining a biomass concentration of 41.7 g DCW/L. The transition phase was started by supplying the culture with a dose of 4 g/L methanol as the sole carbon and energy source. A biomass level of 42.5 g DCW/L was achieved after 4 h. During the subsequent fed-batch induction phase, feeding of pure methanol was performed using a DO-stat strategy, in which methanol feeding was controlled in response to %DO with a saturation cut-off of 60%. After 115 h of methanol induction and a total fermentation process of 145 h, the fermented culture reached a final biomass level of 90.3 g DCW/L. At this time, the total protein concentration in the fermented culture was 1032.7 mg/L, while the RBD concentration reached 533.4 mg/L, corresponding to 51.7% of the total proteins in the supernatant.

The total recombinant RBD was 2987 mg, as the volume of the supernatant was 5.6 L (Table 4). The RBD yield based on biomass ($Y_{RBD/X}$) at the end of fermentation was 5.9 mg RBD/g DCW; the volumetric RBD productivity ($_VP$) of the whole process reached 3.7 mg RBD/L h; the total RBD productivity ($_TP$) was 20.7 mg RBD/h (Table 4). Moreover, the specific RBD productivity ($_SP$) of the whole process reached 40.7 μg RBD/g DCW h, and the RBD yield based on methanol consumed ($Y_{RBD/S}$) was 1.9 mg RBD/g methanol at the end of the induction period. It is important to note that the values obtained for the RBD yield based on the biomass ($Y_{RBD/X}$), volumetric productivity ($_VP$), and specific productivity ($_SP$) of the whole process were similar to those obtained with the 7 L bioreactor applying Strategy 2, indicating that the scale-up was successful.

Table 4. Evolution of production parameters during fermentation carried out with Strategy 2 in 14 L stirred-tank bioreactor: Detailed parameters every 24 h of methanol induction: total protein concentration (mg/L); RBD concentration (mg/L); RBD percentage with respect to total protein (%); RBD concentration increase (fold); $Y_{RBD/X}$: RBD yield based on biomass formation (mg/g); $_VP$: volumetric RBD productivity (mg RBD/L h).

Induction Time (h)	Total Protein Concentration (mg/L)	RBD Concentration (mg/L)	RBD Percentage (%)	RBD Increase (fold)	Y RBD/x (mg/g) vP	vP (mg/L h)
0	120.3	37.3	31.0	1.0	1.0	1.2
24	169.7	65.2	38.4	1.7	1.1	1.2
48	237.0	101.2	42.7	2.7	1.4	1.3
72	415.3	188.1	45.3	5.0	2.4	1.8
96	669.2	321.8	48.1	8.6	3.9	2.6
115	1032.7	533.4	51.7	14.3	5.9	3.7

The parameter profile of the fermentation shown in Figure 4 specifically displays the variation of stirring, dissolved oxygen, and feeding during the process. It shows the decrease of the DO level in the batch phase until its abrupt increase (DO spike) and, later, the variations of dissolved oxygen that controlled the feeding of glycerol and methanol. In the DO-stat cultivation strategy, the input of the substrate was regulated by %DO. This means that, during the whole procedure, pulses of glycerol or methanol will be associated with fluctuations in %DO, which do not affect cell growth. The SDS-PAGE analysis revealed the increase in total protein and recombinant RBD during the induction phase with methanol (Figure 5A). It is worth mentioning that the glycosylated variants of the recombinant RBD corresponded to the diffuse bands of around 40, 35, and 30 kDa. Figure 5B shows the increase in total proteins and RBD concentration, as well as the percentage of RBD in total proteins every 24 h of the induction phase. The RBD concentration increased more than 14-fold, and the percentage of RBD in total proteins increased from 31% to 52% at the end of this phase.

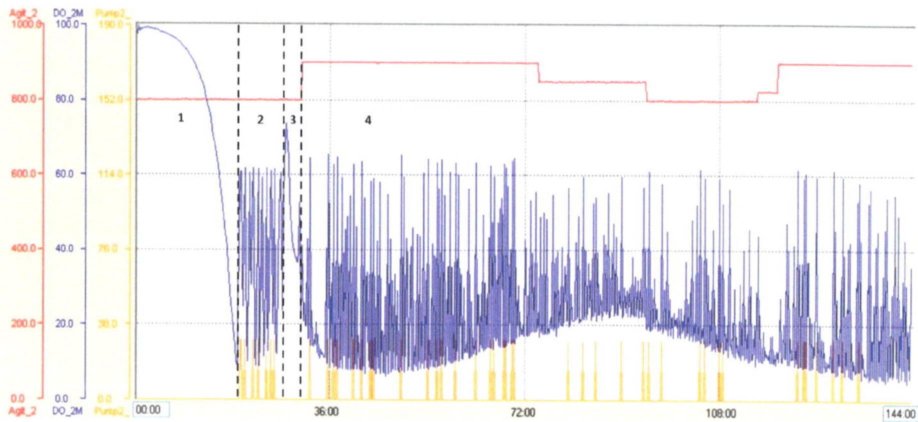

Figure 4. Parameter profile of fermentation carried out in 14 L stirred-tank bioreactor with Strategy 2. Blue line: dissolved oxygen level (saturation percentage), red line: stirring (RPM), yellow line: feeding (pumping percentage). 1: Batch phase, 2: glycerol-fed-batch phase, 3: transition phase, 4: induction phase.

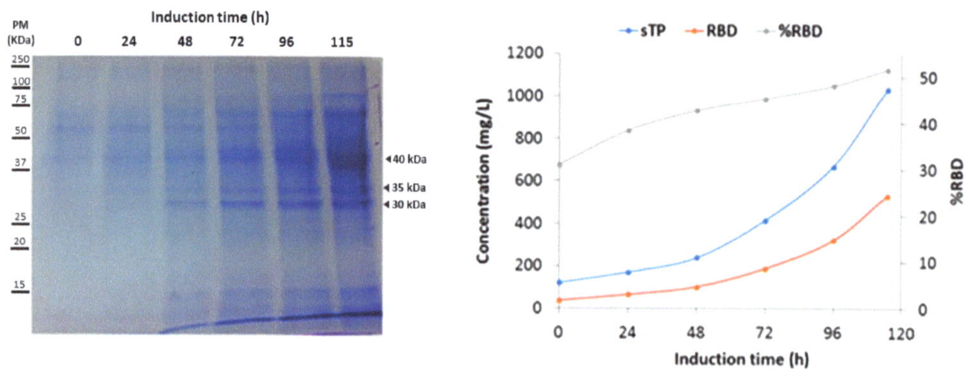

Figure 5. SDS-PAGE analysis and production parameters of the induction phase in a 14 L bioreactor fermentation performed with Strategy 2. (**A**) SDS-PAGE total protein profile of fermentation supernatants corresponding to induction phase. Arrows indicate the molecular weights of the different glycosylated RBD variants. PM: Protein marker. (**B**) Variation in total protein concentration, RBD concentration, and RBD percentage in total protein during methanol induction.

3.3. *Purification and Analysis of RBD*

The RBD was recovered from the fermentation supernatant as already described [13]. Briefly, 1–2 L fractions of the supernatant were purified in a single step using 20 mL of a Ni^{2+}-NTA affinity column. Pure RBD was eluted using imidazole as a competitor, and the protein was dialyzed with Tris 20 mM and NaCl 150 mM, pH 7.4. The protein analysis resulted in being identical to the one already characterized for the production at a smaller-scale fermentation, as the UV spectrum, mass spectrum, SDS-PAGE, and circular dichroism were indistinguishable from those previously reported by our group [13,14]. The protein batches of 7 L reported here were also obtained at more than 95% purity as determined by the HPLC analysis (Figure 6).

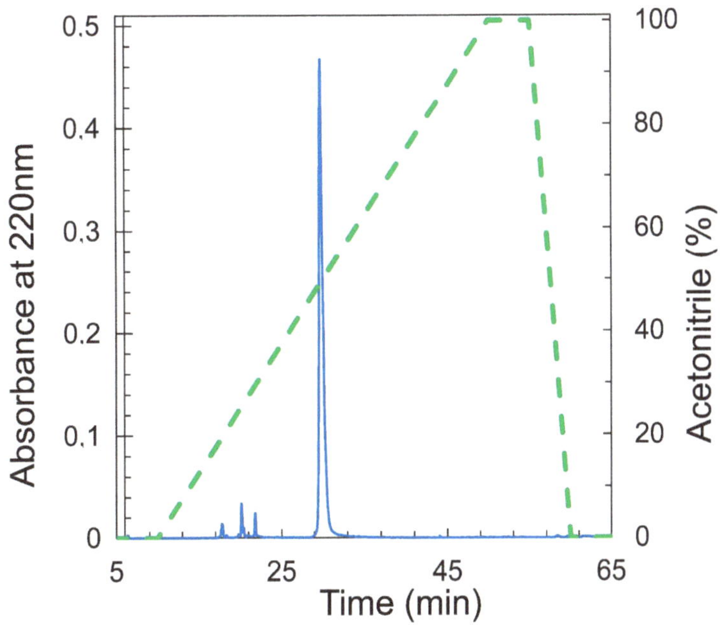

Figure 6. Analysis by RP-HPLC of RBD produced in *P. pastoris* and purified by NTA-Ni^{2+}. Analytical C18 reverse-phase HPLC chromatogram for RBD produced in *P. pastoris* (20 µg) was obtained upon an ACN gradient 0 to 100% over 40 min (10–50 min of the run, dashed line) and with a mobile phase of 0.05% TFA. The integration of the main peak was 95% (filled line).

4. Discussion

As a quick response to the SARS-CoV-2 pandemic, several research groups around the world started to develop biological tools to provide raw materials for the diagnosis, treatment, and prevention of the disease. In such a scenario, during 2020, we reported the production of SARS-CoV-2 RBD using *P. pastoris* as a cell factory applying a preliminary, non-optimized bioreactor culture procedure. We also compared its structural features with RBD produced in mammalian HEK293T cells to verify that they were similar [13]. As the next step, in this article, we describe a rational procedure for the improvement of SARS-CoV2 RBD production using *P. pastoris* in 7 and 14 L bioreactors. To do that, two different culture strategies were tested, Strategy 2, a combination of batch and fed-batch using a DO-stat feeding, showing the best performance by allowing the obtention of more than 500 mg/L of raw yRBD in the culture broth. This cultivation procedure represents a simple, robust, scalable, and low-cost method since it involves the use of a stirred-tank bioreactor (STBR), a defined basal salt medium with simple carbon sources (glycerol and methanol), and the provision of oxygen exclusively from compressed air, avoiding the use of pure O_2 and its associated risks [67,68]. This affirmation is especially valid when comparing RBD production with other expression hosts such as eukaryotic cells [69,70]. Additionally, the use of a defined saline medium, avoiding or minimizing the requirement of complex undefined ingredients, allows the monitoring of component concentrations throughout the cultivation period, which is considered as a valuable feature for industrial processes [71]. All these advantages turn RBB produced in *P. pastoris* following the procedure reported here into an attractive molecule to be used in diagnostic tools or for vaccine developments, especially considering its cost and easy production.

Based on the specific growth rate (µ) obtained during the first 24 h of induction (0.072 h^{-1}) of the Erlenmeyer flask cultivation, it can be surmised that this recombinant clone behaves as a Mut^+ strain, as expected. In this sense, Orman et al. [72] reported a maximum specific growth rate (μ_{max}) of 0.16 h^{-1} for a recombinant *P. pastoris* clone expressing hGH in a defined medium with methanol as the sole carbon and energy source. Pla et al. [73], working with Mut^+ and Mut^s clones expressing scFV, obtained a μ_{max} of 0.044 h^{-1} and 0.015 h^{-1}, respectively. A recent study proposed that the use of Mut^+ phenotype is convenient for high levels of heterologous protein production considering that pAOX1 is induced not only by methanol, but also by its metabolites: formaldehyde and formate [74]. Therefore, a high methanol utilization results in a stronger pAOX1 induction, increasing heterologous protein production compared to a lower methanol utilization metabolism. These authors reported that the Mut^+ clone showed a specific β-galactosidase (heterologous protein) expression rate 5- and 10-fold higher than the Mut^s and Mut^- ones. In our work, the observed Mut^+ phenotype could be one of the reasons supporting the high level of RBD expression.

It is well established that an adequate induction phase design is crucial for high heterologous protein titers in *P. pastoris* when the AOX1 promoter is used [75,76]. In our procedure, the transition phase from glycerol was performed as a pure methanol pulse to achieve a concentration of 4 g/L in the culture broth followed by the induction phase under a DO-stat feeding for 108 h. The criteria underlying this design are related to providing enough methanol for adaptation and induction, minimizing the risk of methanol accumulation, as well as keeping O_2 demand and heat production controlled. Some methanol feeding strategies keeping its concentration constant or within a defined range have been reported for heterologous protein production in *P. pastoris* and *P. methanolica* [77,78]. These strategies allow a strong and constant AOX1 induction, while avoiding methanol accumulation and the consequent cell intoxication. Several analytical methods to achieve this feeding profile have been developed [79,80]. Most of them require specific and expensive equipment. Considering such arguments, the strategy applied in this work emerges as an alternative procedure when permanent methanol concentration surveillance is not available. The induction strategy used in this work is based on and shares some features with those reported by Yamawaki et al. [44] for the production of an antibody fragment (scFv). For induction, these authors combined a stage in which the methanol concentration was kept at 15.7 g/L for 5 h (controlled by a methanol sensor feedback) followed by a DO-stat for 36 h (total induction time: 41 h). Under these conditions, 247 mg/L of recombinant protein was obtained. In the case of the process reported here, cells in the bioreactor were exposed to a significant, but non-toxic methanol level (4 g/L) adequate for AOX1 activation. The adaptation stage end was deducted from a O_2 spike, and indicator of methanol exhaustion and a signal to start the methanol-fed-batch under a DO-stat strategy. Therefore, we combined an adaptation stage with a methanol pulse and the subsequent induction of the RBD expression with methanol feeding in response to %DO in the culture broth as a production strategy.

Using this procedure, around 500 mg/L of RBD was obtained, solubilized in the culture broth. Further purification steps resulted in the obtention of 206.4 mg/L of pure protein (95%). This amount of recombinant protein was significantly higher than those obtained and reported previously (96.1 mg/L of total RBD and 45 mg/L of 95% pure protein), representing an ~5-fold improvement [13].

Other authors [57] reported the production of 400 mg/L of a recombinant SARS-CoV RBD219-N1 using *P. pastoris* X-33. For this purpose, these authors developed a multistage process including several feed flows and gradients. In that case, the induction stage took ~70 h. During the process, %DO was maintained above 30%, adjusting the gas provision and agitation. These authors did not report the gas provision source (air or O_2).

In a recent article [81], the production of a modified SARS-CoV-2 RBD using *P. pastoris* X-33 at a 50 L scale was reported. The production process was based on a saline medium containing yeast extract and other elements (histidine, biotin, myo-innositol, calcium pantothenate, pyridoxal hydrochloride, thiamine di-hydrochloride, and nicotinic acid). The process involved the use of constant feedings rates and yielded a dry cell weight of 58.15 g/L and 68.38 mg/L of the RBD. Downstream consisted of an IMAC (Cu^{2+}) followed by a semi-preparative RP-HPLC. After that, the recombinant RBD showed a purity equal to or higher than 98%. The bioreactor culture lasted for 38 to 48 h. Both the DCW and RBD concentration were lower than those obtained under the proposed procedure (90 g/L vs. 58 g/L and 500 mg/L vs. 68 mg/L). However, the fermentation time was shorter in the procedure reported by Limonta-Fernandez et al. (2022) [81], providing a potential advantage for industrial production.

Methanol metabolism by alcohol oxidase is a process that requires high levels of oxygen and releases a large amount of energy as heat. *P. pastoris* high-density cultures using this substrate are usually carried out by providing pure oxygen. The DO-stat strategy applied for both the glycerol-fed-batch and methanol induction resulted in a moderate O_2 consumption and a gradual heat production.

As was stated by Cardoso et al. (2020) [82], the costs associated with heterologous protein production using microorganisms are driven by medium composition and cooling. The cultivation method proposed in this work is coherent with this affirmation, considering that it involves the use of a simple and relatively inexpensive medium and a DO stat strategy, thus diminished oxygenation and cooling requirements.

In the case of prokaryotic hosts, the RBD is usually obtained as a non-glycosylated, non-folded protein. For this reason, re-folding is needed during downstream processing. Related to that, the production of the RBD was reported using *E. coli* as an expression system. He et al. (2021) [83] developed a production method using a BL21 strain obtaining the most the RDB in inclusion bodies. After solubilization and renaturalization, they recovered 65.2% of the produced RBD by Nickel affinity chromatography, reporting a production yield of 13.3 mg/L. Meena et al. [84] also informed RBD production in bacterial cells as inclusion bodies. These authors cultivated an *E. coli* strain carrying the RBD gene in n Erlenmeyer flask containing 1 L of a complex undefined media (a modification of the Luria Bertani broth), ampicillin, and IPTG induction. After cultivation, they obtained 62.10 mg of raw RBD as inclusion bodies from 1 L of culture. After that, several downstream steps were needed, including solubilization at pH 12.5 and 3, refolding, and DEAE chromatography.

The RBD obtained from this process is a glycosylated protein showing three main variants (30, 35, and 40 kDa). These variants present the same primary structure differing only in the glycosylation pattern [13]. Beyond its crucial function in protein folding, glycosylation represents an issue of special concern for RBD industrial production, since the intra- and inter-batch heterogeneity could be considered as a drawback for standardization and regulation compliance [85]. Even though the glycosylation status was heterogeneous, the protein behaves identically to the RBD expressed in mammalian HEK-293T cells, in its conformation, stability, and immunogenic properties. Moreover, antibodies raised when using our *P. pastoris* produced the RBD cross-reacted with the RBD produced in mammalian cells and vice versa [14]. Further research is needed to clarify the impact of this post-translational modification on biological activity and to improve protein homogeneity. We are currently working on several factors to improve protein homogeneity. Regarding the addition of a Hisx6 tag, it is worth mentioning that, even though the purification methods of a non-tagged RBD were reported [57,86], our strategy is based on the addition of a tag that allows purification to a high purity in a single step with the capability to be easily removed, prioritizing procedure simplicity and robustness. Our construction bears a Sortase A sequence at the N-terminal of the Hisx6 tag, which allows its removal in a transpeptidation reaction to a multimer [13] or Gly3 trisaccharide or even water (manuscript in preparation). The strategy provides the versatility to—depending on the use of the protein: diagnosis, generation of neutralizing antibodies, or vaccine development—remove or leave the tag in a single step.

In summary, in this work, we presented an efficient and reliable scaling up method for the production of a low-cost RBD antigen using the methylotrophic yeast *P. pastoris*, with multiple advantages. One possible drawback of this procedure is the extension of the induction stage. This extension was partially caused by the DO-stat strategy, which involves a kind of "dead time" among methanol feeding pulses. These periods are minutes in which no or very low levels of methanol are available in the culture broth, probably resulting in a waste of time for the metabolic machinery of *P. pastoris*. As an alternative, a strategy in which the methanol concentration is kept constant could be more effective. However, this strategy would require a method for real-time methanol concentration evaluation, through a GC-associated method or a specific detector, which requires specific equipment. As Strategy 2 resulted in the obtention of a well-folded, immunologically active RBD, we propose the method provided here to scale up the production of RBD.

5. Conclusions

Industrial production of recombinant proteins, especially those for pharmaceutical products, draw upon different living organisms as hosts, harnessing their advantages while dealing with their disadvantages [87,88]. There is no universal host for this task, and each protein can be produced "conveniently" in a process meeting a swarm of factors, the protein features (structure and post-translational modifications), production performance and costs, protein application, and legal regulations being some of the most-relevant. Here, we proposed such a convenient process for RBD production.

Author Contributions: D.G.N.: methodology, investigation, writing-original draft, visualization. C.D.: conceptualization, investigation, methodology, resources, writing—review and editing, funding acquisition, project administration, J.S.: funding acquisition, writing—review and editing, project administration, T.I.-H.: methodology, investigation. F.P.: methodology, investigation. D.E.W.: methodology, resources. H.G.: methodology. A.D.N.: funding acquisition, conceptualization, project administration, writing—review and editing. E.R.: methodology and validation. C.P.: methodology. L.A.M.R.: conceptualization, investigation, methodology, writing—original draft, project administration. All authors have read and agreed to the published version of the manuscript.

Funding: This work was supported partially by the grants IP COVID-19–234 and PICTO 2021-0007 provided by the ANPCYT (Argentina).

Institutional Review Board Statement: Not applicable.

Informed Consent Statement: Not applicable.

Data Availability Statement: Data will be made available upon request.

Acknowledgments: We would like to thank the Consorcio Argentino AntiCovid for helpful discussions and collaborative work during the pandemic.

Conflicts of Interest: The authors declare that they have no known competing financial interest or personal relationships that could have appeared to influence the work reported in this paper. The strain and some steps of the processes presented are included in the pending Patents PCT/US22/43578 and AR P210102579.

References

1. Dennehy, J.J.; Gupta, R.K.; Hanage, W.P.; Johnson, M.C.; Peacock, T.P. Where Is the next SARS-CoV-2 Variant of Concern? *Lancet* **2022**, *399*, 1938–1939. [CrossRef] [PubMed]
2. Forster, P.; Forster, L.; Renfrew, C.; Forster, M. Phylogenetic Network Analysis of SARS-CoV-2 Genomes. *Proc. Natl. Acad. Sci. USA* **2020**, *117*, 9241–9243. [CrossRef]
3. Wang, X.-M.; Mannan, R.; Xiao, L.; Abdulfatah, E.; Qiao, Y.; Farver, C.; Myers, J.L.; Zelenka-Wang, S.; McMurry, L.; Su, F.; et al. Characterization of SARS-CoV-2 and Host Entry Factors Distribution in a COVID-19 Autopsy Series. *Commun. Med.* **2021**, *1*, 24. [CrossRef] [PubMed]
4. Arimori, T.; Takagi, J. Structure of SARS-CoV-2 Spike Receptor-Binding Domain Complexed with High Affinity ACE2 Mutant 3N39 2020. *Nat. Commun.* **2021**, *12*, 3802.
5. Xu, Z.P.; Liu, K.F.; Han, P.; Qi, J.X. Structure of SARS-CoV-2 Spike Receptor-Binding Domain Complexed with Its Receptor Equine ACE2 2022. *Structure* **2022**, *30*, 1432–1442.e4.
6. Wang, X.; Lan, J.; Ge, J.; Yu, J.; Shan, S. Crystal Structure of SARS-CoV-2 Spike Receptor-Binding Domain Bound with ACE2 Structure summary MMDB-NCBI 2020. Available online: https://www.ncbi.nlm.nih.gov/Structure/pdb/6M0J (accessed on 8 January 2022).
7. Lan, J.; Ge, J.; Yu, J.; Shan, S.; Zhou, H.; Fan, S.; Zhang, Q.; Shi, X.; Wang, Q.; Zhang, L.; et al. Structure of the SARS-CoV-2 Spike Receptor-Binding Domain Bound to the ACE2 Receptor. *Nature* **2020**, *581*, 215–220. [CrossRef]

8. Shamsi, A.; Mohammad, T.; Anwar, S.; Amani, S.; Khan, M.S.; Husain, F.M.; Rehman, M.T.; Islam, A.; Hassan, M.I. Potential Drug Targets of SARS-CoV-2: From Genomics to Therapeutics. *Int. J. Biol. Macromol.* **2021**, *177*, 1–9. [CrossRef]
9. Ju, B.; Zhang, Q.; Ge, J.; Wang, R.; Sun, J.; Ge, X.; Yu, J.; Shan, S.; Zhou, B.; Song, S.; et al. Human Neutralizing Antibodies Elicited by SARS-CoV-2 Infection. *Nature* **2020**, *584*, 115–119. [CrossRef]
10. Liu, L.; Wang, P.; Nair, M.S.; Yu, J.; Rapp, M.; Wang, Q.; Luo, Y.; Chan, J.F.-W.; Sahi, V.; Figueroa, A.; et al. Potent Neutralizing Antibodies against Multiple Epitopes on SARS-CoV-2 Spike. *Nature* **2020**, *584*, 450–456. [CrossRef]
11. Fujita, R.; Hino, M.; Ebihara, T.; Nagasato, T.; Masuda, A.; Lee, J.M.; Fujii, T.; Mon, H.; Kakino, K.; Nagai, R.; et al. Efficient Production of Recombinant SARS-CoV-2 Spike Protein Using the Baculovirus-Silkworm System. *Biochem. Biophys. Res. Commun.* **2020**, *529*, 257–262. [CrossRef]
12. Argentinian AntiCovid Consortium. Covalent Coupling of Spike's Receptor Binding Domain to a Multimeric Carrier Produces a High Immune Response against SARS-CoV-2. *Sci. Rep.* **2022**, *12*, 692. [CrossRef] [PubMed]
13. Argentinian AntiCovid Consortium. Structural and Functional Comparison of SARS-CoV-2-Spike Receptor Binding Domain Produced in Pichia Pastoris and Mammalian Cells. *Sci. Rep.* **2020**, *10*, 21779. [CrossRef] [PubMed]
14. Smith, I.; Mc Callum, G.J.; Sabljic, A.V.; Marfía, J.I.; Bombicino, S.S.; Trabucchi, A.; Iacono, R.F.; Birenbaum, J.M.; Vázquez, S.C.; Minoia, J.M.; et al. Rapid and Cost-Effective Process Based on Insect Larvae for Scale-up Production of SARS-CoV-2 Spike Protein for Serological COVID-19 Testing. *Biotechnol. Bioeng.* **2021**, *118*, 4129–4137. [CrossRef]
15. Li, T.; Zheng, Q.; Yu, H.; Wu, D.; Xue, W.; Xiong, H.; Huang, X.; Nie, M.; Yue, M.; Rong, R.; et al. SARS-CoV-2 Spike Produced in Insect Cells Elicits High Neutralization Titres in Non-Human Primates. *Emerg. Microbes Infect.* **2020**, *9*, 2076–2090. [CrossRef]
16. Shajahan, A.; Supekar, N.T.; Gleinich, A.S.; Azadi, P. Deducing the N- and O-Glycosylation Profile of the Spike Protein of Novel Coronavirus SARS-CoV-2. *Glycobiology* **2020**, *30*, 981–988. [CrossRef] [PubMed]
17. Azad, T.; Singaravelu, R.; Taha, Z.; Jamieson, T.R.; Boulton, S.; Crupi, M.J.F.; Martin, N.T.; Brown, E.E.F.; Poutou, J.; Ghahremani, M.; et al. Nanoluciferase Complementation-Based Bioreporter Reveals the Importance of N-Linked Glycosylation of SARS-CoV-2 S for Viral Entry. *Mol. Ther.* **2021**, *29*, 1984–2000. [CrossRef]
18. Jenkins, N.; Curling, E.M. Glycosylation of Recombinant Proteins: Problems and Prospects. *Enzym. Microb. Technol.* **1994**, *16*, 354–364. [CrossRef]
19. Brooks, S.A. Appropriate Glycosylation of Recombinant Proteins for Human Use. *Mol. Biotechnol.* **2004**, *28*, 241–255. [CrossRef]
20. Lingg, N.; Zhang, P.; Song, Z.; Bardor, M. The Sweet Tooth of Biopharmaceuticals: Importance of Recombinant Protein Glycosylation Analysis. *Biotechnol. J.* **2012**, *7*, 1462–1472. [CrossRef]
21. Goh, J.B.; Ng, S.K. Impact of Host Cell Line Choice on Glycan Profile. *Crit. Rev. Biotechnol.* **2018**, *38*, 851–867. [CrossRef]
22. Casalino, L.; Gaieb, Z.; Goldsmith, J.A.; Hjorth, C.K.; Dommer, A.C.; Harbison, A.M.; Fogarty, C.A.; Barros, E.P.; Taylor, B.C.; McLellan, J.S.; et al. Beyond Shielding: The Roles of Glycans in the SARS-CoV-2 Spike Protein. *ACS Cent. Sci.* **2020**, *6*, 1722–1734. [CrossRef] [PubMed]
23. Vieira Gomes, A.M.; Souza Carmo, T.; Silva Carvalho, L.; Mendonça Bahia, F.; Parachin, N.S. Comparison of Yeasts as Hosts for Recombinant Protein Production. *Microorganisms* **2018**, *6*, 38. [CrossRef] [PubMed]
24. Bernauer, L.; Radkohl, A.; Lehmayer, L.G.K.; Emmerstorfer-Augustin, A. Komagataella Phaffii as Emerging Model Organism in Fundamental Research. *Front. Microbiol.* **2020**, *11*, 607028. [CrossRef]
25. Ata, Ö.; Ergün, B.G.; Fickers, P.; Heistinger, L.; Mattanovich, D.; Rebnegger, C.; Gasser, B. What Makes Komagataella Phaffii Non-Conventional? *FEMS Yeast Res.* **2021**, *21*, foab059. [CrossRef]
26. Wegner, G.H. Emerging Applications of the Methylotrophic Yeasts. *FEMS Microbiol. Rev.* **1990**, *7*, 279–283. [CrossRef] [PubMed]
27. Daly, R.; Hearn, M.T.W. Expression of Heterologous Proteins in Pichia Pastoris: A Useful Experimental Tool in Protein Engineering and Production. *J. Mol. Recognit.* **2005**, *18*, 119–138. [CrossRef] [PubMed]
28. Ergün, B.G.; Berrios, J.; Binay, B.; Fickers, P. Recombinant Protein Production in Pichia Pastoris: From Transcriptionally Redesigned Strains to Bioprocess Optimization and Metabolic Modelling. *FEMS Yeast Res.* **2021**, *21*, foab057. [CrossRef] [PubMed]
29. Hartner, F.S.; Glieder, A. Regulation of Methanol Utilisation Pathway Genes in Yeasts. *Microb. Cell Fact.* **2006**, *5*, 39. [CrossRef]
30. Hartner, F.S.; Ruth, C.; Langenegger, D.; Johnson, S.N.; Hyka, P.; Lin-Cereghino, G.P.; Lin-Cereghino, J.; Kovar, K.; Cregg, J.M.; Glieder, A. Promoter Library Designed for Fine-Tuned Gene Expression in Pichia Pastoris. *Nucleic Acids Res.* **2008**, *36*, e76. [CrossRef]
31. Lin Cereghino, G.P.; Sunga, A.J.; Cereghino, J.L.; Cregg, J.M. Expression of Foreign Genes in the Yeast Pichia Pastoris. *Genet. Eng. Princ. Methods* **2001**, *23*, 157–169.
32. Cregg, J.M.; Cereghino, J.L.; Shi, J.; Higgins, D.R. Recombinant Protein Expression in Pichia Pastoris. *Mol. Biotechnol.* **2000**, *16*, 23–52. [CrossRef] [PubMed]
33. Cereghino, G.P.L.; Cereghino, J.L.; Ilgen, C.; Cregg, J.M. Production of Recombinant Proteins in Fermenter Cultures of the Yeast Pichia Pastoris. *Curr. Opin. Biotechnol.* **2002**, *13*, 329–332. [CrossRef] [PubMed]
34. Cereghino, J.L.; Cregg, J.M. Heterologous Protein Expression in the Methylotrophic Yeast Pichia Pastoris. *FEMS Microbiol. Rev.* **2000**, *24*, 45–66. [CrossRef] [PubMed]
35. Shental-Bechor, D.; Levy, Y. Effect of Glycosylation on Protein Folding: A Close Look at Thermodynamic Stabilization. *Proc. Natl. Acad. Sci. USA* **2008**, *105*, 8256–8261. [CrossRef]
36. Jayaprakash, N.G.; Surolia, A. Role of Glycosylation in Nucleating Protein Folding and Stability. *Biochem. J.* **2017**, *474*, 2333–2347. [CrossRef]

37. Roth, J.; Zuber, C.; Park, S.; Jang, I.; Lee, Y.; Kysela, K.G. Protein N-Glycosylation, Protein Folding, and Protein Quality Control. *Molecules* **2010**, *30*, 497–506. [CrossRef]
38. Gao, J.; Jiang, L.; Lian, J. Development of Synthetic Biology Tools to Engineer Pichia Pastoris as a Chassis for the Production of Natural Products. *Synth. Syst. Biotechnol.* **2021**, *6*, 110–119. [CrossRef]
39. Gaboardi, G.C.; Alves, D.; Gil de Los Santos, D.; Xavier, E.; Nunes, A.P.; Finger, P.; Griep, E.; Roll, V.; Oliveira, P.; Silva, A.; et al. Influence of Pichia Pastoris X-33 Produced in Industrial Residues on Productive Performance, Egg Quality, Immunity, and Intestinal Morphometry in Quails. *Sci. Rep.* **2019**, *9*, 15372. [CrossRef]
40. Sturmberger, L.; Chappell, T.; Geier, M.; Krainer, F.; Day, K.J.; Vide, U.; Trstenjak, S.; Schiefer, A.; Richardson, T.; Soriaga, L.; et al. Refined Pichia Pastoris Reference Genome Sequence. *J. Biotechnol.* **2016**, *235*, 121–131. [CrossRef]
41. Sinha, J.; Plantz, B.A.; Zhang, W.; Gouthro, M.; Schlegel, V.; Liu, C.-P.; Meagher, M.M. Improved Production of Recombinant Ovine Interferon-Tau by Mut(+) Strain of Pichia Pastoris Using an Optimized Methanol Feed Profile. *Biotechnol. Prog.* **2003**, *19*, 794–802. [CrossRef]
42. Files, D.; Ogawa, M.; Scaman, C.H.; Baldwin, S.A. A Pichia Pastoris Fermentation Process for Producing High-Levels of Recombinant Human Cystatin-C. *Enzym. Microb. Technol.* **2001**, *29*, 335–340. [CrossRef]
43. Cos, O.; Ramón, R.; Montesinos, J.L.; Valero, F. Operational Strategies, Monitoring and Control of Heterologous Protein Production in the Methylotrophic Yeast Pichia Pastoris under Different Promoters: A Review. *Microb. Cell Fact.* **2006**, *5*, 17. [CrossRef] [PubMed]
44. Yamawaki, S.; Matsumoto, T.; Ohnishi, Y.; Kumada, Y.; Shiomi, N.; Katsuda, T.; Lee, E.K.; Katoh, S. Production of Single-Chain Variable Fragment Antibody (ScFv) in Fed-Batch and Continuous Culture of Pichia Pastoris by Two Different Methanol Feeding Methods. *J. Biosci. Bioeng.* **2007**, *104*, 403–407. [CrossRef] [PubMed]
45. Looser, V.; Bruhlmann, B.; Bumbak, F.; Stenger, C.; Costa, M.; Camattari, A.; Fotiadis, D.; Kovar, K. Cultivation Strategies to Enhance Productivity of Pichia Pastoris: A Review. *Biotechnol. Adv.* **2015**, *33*, 1177–1193. [CrossRef]
46. Ferreira, A.R.; Ataíde, F.; von Stosch, M.; Dias, J.M.L.; Clemente, J.J.; Cunha, A.E.; Oliveira, R. Application of Adaptive DO-Stat Feeding Control to Pichia Pastoris X33 Cultures Expressing a Single Chain Antibody Fragment (ScFv). *Bioprocess Biosyst. Eng.* **2012**, *35*, 1603–1614. [CrossRef] [PubMed]
47. Trinh, L.B.; Phue, J.N.; Shiloach, J. Effect of Methanol Feeding Strategies on Production and Yield of Recombinant Mouse Endostatin from Pichia Pastoris. *BioTechnology* **2003**, *82*, 438–444.
48. Liu, W.; Xiang, H.; Zhang, T.; Pang, X.; Su, J.; Liu, H. Development of a New High-Cell Density Fermentation Strategy for Enhanced Production of a Fungus β-Glucosidase in Pichia Pastoris. *Front. Microbiol.* **2020**, *11*, 1988. [CrossRef] [PubMed]
49. Dalvie, N.C.; Rodriguez-Aponte, S.A. Engineered SARS-CoV-2 Receptor Binding Domain Improves Manufacturability in Yeast and Immunogenicity in Mice. *Proc. Natl. Acad. Sci. USA* **2021**, *118*, e2106845118. [CrossRef]
50. Liu, Y.; Ye, Q. Nucleic Acid Vaccines against SARS-CoV-2. *Vaccines* **2022**, *10*, 1849. [CrossRef]
51. Jain, S.; Venkataraman, A.; Wechsler, M.E.; Peppas, N.A. Messenger RNA-Based Vaccines: Past, Present, and Future Directions in the Context of the COVID-19 Pandemic. *Adv. Drug Deliv. Rev.* **2021**, *179*, 114000. [CrossRef]
52. Heidary, M.; Kaviar, V.H.; Shirani, M.; Ghanavati, R.; Motahar, M.; Sholeh, M.; Ghahramanpour, H.; Khoshnood, S. A Comprehensive Review of the Protein Subunit Vaccines Against COVID-19. *Front. Microbiol.* **2022**, *13*, 927306. [CrossRef] [PubMed]
53. Dalvie, N.C.; Tostanoski, L.H.; Rodriguez-Aponte, S.A.; Kaur, K.; Bajoria, S.; Kumru, O.S.; Martinot, A.J.; Chandrashekar, A.; McMahan, K.; Mercado, N.B.; et al. SARS-CoV-2 Receptor Binding Domain Displayed on HBsAg Virus-like Particles Elicits Protective Immunity in Macaques. *Sci. Adv.* **2022**, *8*, eabl6015. [CrossRef] [PubMed]
54. Ghasemiyeh, P.; Mohammadi-Samani, S.; Firouzabadi, N.; Dehshahri, A.; Vazin, A. A Focused Review on Technologies, Mechanisms, Safety, and Efficacy of Available COVID-19 Vaccines. *Int. Immunopharmacol.* **2021**, *100*, 108162. [CrossRef] [PubMed]
55. Kruger, N.J. The Bradford Method for Protein Quantitation. In *Protein Protocols Handbook*; Humana: Louisville, KY, USA, 2009; pp. 15–22.
56. Kruger, N.J. The Bradford Method for Protein Quantitation. In *Springer Protocols Handbooks*; Humana Press: Totowa, NJ, USA, 2009; pp. 17–24. ISBN 9781603274746.
57. Chen, W.-H.; Chag, S.M.; Poongavanam, M.V.; Biter, A.B.; Ewere, E.A.; Rezende, W.; Seid, C.A.; Hudspeth, E.M.; Pollet, J.; McAtee, C.P.; et al. Optimization of the Production Process and Characterization of the Yeast-Expressed SARS-CoV Recombinant Receptor-Binding Domain (RBD219-N1), a SARS Vaccine Candidate. *J. Pharm. Sci.* **2017**, *106*, 1961–1970. [CrossRef] [PubMed]
58. Li, Z.; Xiong, F.; Lin, Q.; d'Anjou, M.; Daugulis, A.J.; Yang, D.S.; Hew, C.L. Low-Temperature Increases the Yield of Biologically Active Herring Antifreeze Protein in Pichia Pastoris. *Protein Expr. Purif.* **2001**, *21*, 438–445. [CrossRef] [PubMed]
59. Shi, X.; Karkut, T.; Chamankhah, M.; Alting-Mees, M.; Hemmingsen, S.M.; Hegedus, D. Optimal Conditions for the Expression of a Single-Chain Antibody (ScFv) Gene in Pichia Pastoris. *Protein Expr. Purif.* **2003**, *28*, 321–330. [CrossRef] [PubMed]
60. Celik, E.; Calık, P. Production of Recombinant Proteins by Yeast Cells. *Biotechnol. Adv.* **2012**, *30*, 1108–1118. [CrossRef]
61. Noseda, D.G.; Recúpero, M.N.; Blasco, M.; Ortiz, G.E.; Galvagno, M.A. Cloning, Expression and Optimized Production in a Bioreactor of Bovine Chymosin B in Pichia (Komagataella) Pastoris under AOX1 Promoter. *Protein Expr. Purif.* **2013**, *92*, 235–244. [CrossRef]
62. Noseda, D.G.; Recúpero, M.; Blasco, M.; Bozzo, J.; Galvagno, M.Á. Production in Stirred-Tank Bioreactor of Recombinant Bovine Chymosin B by a High-Level Expression Transformant Clone of Pichia Pastoris. *Protein Expr. Purif.* **2016**, *123*, 112–121. [CrossRef]

63. Picotto, L.D.; Sguazza, G.H.; Tizzano, M.A.; Galosi, C.M.; Cavalitto, S.F.; Pecoraro, M.R. An Effective and Simplified DO-Stat Control Strategy for Production of Rabies Glycoprotein in Pichia Pastoris. *Protein Expr. Purif.* **2017**, *132*, 124–130. [CrossRef]
64. Kim, S.; Warburton, S.; Boldogh, I.; Svensson, C.; Pon, L.; d'Anjou, M.; Stadheim, T.A.; Choi, B.-K. Regulation of Alcohol Oxidase 1 (AOX1) Promoter and Peroxisome Biogenesis in Different Fermentation Processes in Pichia Pastoris. *J. Biotechnol.* **2013**, *166*, 174–181. [CrossRef] [PubMed]
65. Looser, V.; Lüthy, D.; Straumann, M.; Hecht, K.; Melzoch, K.; Kovar, K. Effects of Glycerol Supply and Specific Growth Rate on Methanol-Free Production of CALB by P. Pastoris: Functional Characterisation of a Novel Promoter. *Appl. Microbiol. Biotechnol.* **2017**, *101*, 3163–3176. [CrossRef] [PubMed]
66. Capone, S.; Horvat, J.; Herwig, C.; Spadiut, O. Development of a Mixed Feed Strategy for a Recombinant Pichia Pastoris Strain Producing with a De-Repression Promoter. *Microb. Cell Fact.* **2015**, *14*, 101. [CrossRef] [PubMed]
67. Westwood, M.-M.; Rieley, W. Medical Gases, Their Storage and Delivery. *Anaesth. Intensive Care Med.* **2012**, *13*, 533–538. [CrossRef]
68. Malayaman, S.N.; Mychaskiw, G.I.I.; Berry, J.M.; Ehrenwerth, J. Medical Gases: Storage and Supply. In *Anesthesia Equipment*; Elsevier: Amsterdam, The Netherlands, 2021; pp. 3–24.
69. Farnós, O.; Venereo-Sánchez, A.; Xu, X.; Chan, C.; Dash, S.; Chaabane, H.; Sauvageau, J.; Brahimi, F.; Saragovi, U.; Leclerc, D.; et al. Rapid High-Yield Production of Functional SARS-CoV-2 Receptor Binding Domain by Viral and Non-Viral Transient Expression for Pre-Clinical Evaluation. *Vaccines* **2020**, *8*, 654. [CrossRef]
70. Li, W.; Fan, Z.; Lin, Y.; Wang, T.-Y. Serum-Free Medium for Recombinant Protein Expression in Chinese Hamster Ovary Cells. *Front. Bioeng. Biotechnol.* **2021**, *9*, 646363. [CrossRef]
71. Zhang, J.; Greasham, R. Chemically Defined Media for Commercial Fermentations. *Appl. Microbiol. Biotechnol.* **1999**, *51*, 407–421. [CrossRef]
72. Orman, M.A.; Çalık, P.; Özdamar, T.H. The Influence of Carbon Sources on Recombinant-human-growth-hormone Production by Pichia Pastoris Is Dependent on Phenotype: A Comparison of Muts and Mut+ strains. *Biotechnol. Appl. Biochem.* **2009**, *52*, 245–255. [CrossRef]
73. Pla, I.A.; Damasceno, L.M.; Vannelli, T.; Ritter, G.; Batt, C.A.; Shuler, M.L. Evaluation of Mut+ and MutS Pichia Pastoris Phenotypes for High Level Extracellular ScFv Expression under Feedback Control of the Methanol Concentration. *Biotechnol. Prog.* **2006**, *22*, 881–888. [CrossRef]
74. Singh, A.; Narang, A. The Mut+ Strain of Komagataella Phaffii (Pichia Pastoris) Expresses PAOX1 5 and 10 Times Faster than Muts and Mut− Strains: Evidence That Formaldehyde or/and Formate Are True Inducers of PAOX1. *Appl. Microbiol. Biotechnol.* **2020**, *104*, 7801–7814. [CrossRef]
75. Barrigón, J.M.; Montesinos, J.L.; Valero, F. Searching the Best Operational Strategies for Rhizopus Oryzae Lipase Production in Pichia Pastoris Mut+ Phenotype: Methanol Limited or Methanol Non-Limited fed-batch cultures? *Biochem. Eng. J.* **2013**, *75*, 47–54. [CrossRef]
76. Garrigós-Martínez, J.; Nieto-Taype, M.A. Specific Growth Rate Governs AOX1 Gene Expression, Affecting the Production Kinetics of Pichia Pastoris (Komagataella Phaffii) PAOX1-Driven Recombinant producer strains with different target gene dosage. *Microb. Cell Fact.* **2019**, *18*, 187. [CrossRef] [PubMed]
77. Mayson, B.E.; Kilburn, D.G.; Zamost, B.L.; Raymond, C.K.; Lesnicki, G.J. Effects of Methanol Concentration on Expression Levels of Recombinant Protein in Fed-Batch Cultures of Pichia Methanolica. *Biotechnol. Bioeng.* **2003**, *81*, 291–298. [CrossRef]
78. Katakura, Y.; Zhang, W.; Zhuang, G.; Omasa, T. Effect of Methanol Concentration on the Production of Human B2-Glycoprotein I Domain V by a Recombinant Pichia Pastoris: A Simple System for the Control of Methanol concentration using a semiconductor gas sensor. *J. Ferment. Bioeng.* **1998**, *86*, 482–487. [CrossRef]
79. Guarna, M.M.; Lesnicki, G.J.; Tam, B.M.; Robinson, J.; Radziminski, C.Z.; Hasenwinkle, D.; Boraston, A.; Jervis, E.; MacGillivray, R.T.; Turner, R.F.; et al. On-Line Monitoring and Control of Methanol Concentration in Shake-Flask Cultures of Pichia Pastoris. *Biotechnol. Bioeng.* **1997**, *56*, 279–286. [CrossRef]
80. Schenk, J.; Marison, I.W.; von Stockar, U. A Simple Method to Monitor and Control Methanol Feeding of Pichia Pastoris Fermentations Using Mid-IR Spectroscopy. *J. Biotechnol.* **2007**, *128*, 344–353. [CrossRef]
81. Limonta-Fernández, M.; Chinea-Santiago, G.; Martín-Dunn, A.M.; Gonzalez-Roche, D.; Bequet-Romero, M.; Marquez-Perera, G.; González-Moya, I.; Canaan-Haden-Ayala, C.; Cabrales-Rico, A.; Espinosa-Rodríguez, L.A.; et al. An Engineered SARS-CoV-2 Receptor-Binding Domain Produced in Pichia Pastoris as a Candidate Vaccine Antigen. *New Biotechnol.* **2022**, *72*, 11–21. [CrossRef]
82. Cardoso, V.M.; Campani, G.; Santos, M.P.; Silva, G.G. Cost Analysis Based on Bioreactor Cultivation Conditions: Production of a Soluble Recombinant Protein Using Escherichia Coli BL21 (DE3). *Biotechnology* **2020**, *26*, e00441. [CrossRef]
83. He, Y.; Qi, J.; Xiao, L.; Shen, L.; Yu, W.; Hu, T. Purification and Characterization of the Receptor-Binding Domain of SARS-CoV-2 Spike Protein from Escherichia Coli. *Eng. Life Sci.* **2021**, *21*, 453–460. [CrossRef]
84. Meena, J.; Singhvi, P.; Srichandan, S.; Dandotiya, J.; Verma, J.; Singh, M.; Ahuja, R.; Panwar, N.; Wani, T.Q.; Khatri, R.; et al. RBD Decorated PLA Nanoparticle Admixture with Aluminum Hydroxide Elicit Robust and Long Lasting Immune Response against SARS-CoV-2. *Eur. J. Pharm. Biopharm.* **2022**, *176*, 43–53. [CrossRef]
85. Hossler, P.; Khattak, S.F.; Li, Z.J. Optimal and Consistent Protein Glycosylation in Mammalian Cell Culture. *Glycobiology* **2009**, *19*, 936–949. [CrossRef] [PubMed]

86. Kalyoncu, S.; Yilmaz, S.; Kuyucu, A.Z.; Sayili, D.; Mert, O.; Soyturk, H.; Gullu, S.; Akinturk, H.; Citak, E.; Arslan, M.; et al. Process Development for an Effective COVID-19 Vaccine Candidate Harboring Recombinant SARS-CoV-2 Delta plus Receptor Binding Domain Produced by Pichia Pastoris. *Sci. Rep.* **2023**, *13*, 5224. [CrossRef] [PubMed]
87. Dumont, J.; Euwart, D.; Mei, B.; Estes, S.; Kshirsagar, R. Human Cell Lines for Biopharmaceutical Manufacturing: History, Status, and Future Perspectives. *Crit. Rev. Biotechnol.* **2016**, *36*, 1110–1122. [CrossRef] [PubMed]
88. Kiss, B.; Gottschalk, U.; Pohlscheidt, M. *New Bioprocessing Strategies: Development and Manufacturing of Recombinant Antibodies and Proteins*; Springer: Berlin/Heidelberg, Germany, 2018; ISBN 9783319971100.

Disclaimer/Publisher's Note: The statements, opinions and data contained in all publications are solely those of the individual author(s) and contributor(s) and not of MDPI and/or the editor(s). MDPI and/or the editor(s) disclaim responsibility for any injury to people or property resulting from any ideas, methods, instructions or products referred to in the content.

fermentation

Article

Exploring the Potential of Non-Conventional Yeasts in Wine Fermentation with a Focus on *Saccharomycopsis fermentans*

Madina Akan [1,2], Andreas Gudiksen [1], Yasemin Baran [1], Heike Semmler [1], Silvia Brezina [1], Stefanie Fritsch [1], Doris Rauhut [1] and Jürgen Wendland [1,2,3,*]

1 Department of Microbiology and Biochemistry, Hochschule Geisenheim University, Von-Lade-Strasse 1, D-65366 Geisenheim, Germany; madik001@gmail.com (M.A.); angrgu@gmail.com (A.G.); syaseminbaran@gmail.com (Y.B.); heike.semmler@hs-gm.de (H.S.); stefanie.fritsch@hs-gm.de (S.F.); doris.rauhut@hs-gm.de (D.R.)

2 Research Group of Microbiology (MICR)—Functional Yeast Genomics, Vrije Universiteit Brussel, Pleinlaan 2, BE-1050 Brussels, Belgium

3 Geisenheim Yeast Breeding Center (GYBC), Hochschule Geisenheim University, Von-Lade-Strasse 1, D-65366 Geisenheim, Germany

* Correspondence: juergen.wendland@hs-gm.de; Tel.: +49-6722-502-332

Abstract: Despite the increasing number of publications on non-conventional yeasts (NCYs), many areas in this field remain poorly understood, making the examination of these strains important for determining their potential in wine fermentations. The amino acid metabolic pathways involved, particularly the catabolic Ehrlich pathway but also anabolic pathways such as the leucine biosynthesis pathway, are crucial for producing high-value aroma compounds that contribute to the final flavour of wine. We examined the potential use of *Saccharomycopsis fermentans* in wine fermentations. We selected mutant strains resistant to the toxic compound trifluoro-leucine (TFL), verified mutations in the SfLEU4 gene, and characterized the ability of the resulting strains to contribute to fermentation bouquets. Resistance to TFL relieves feedback inhibition in the leucine biosynthesis pathway and resulted in increased leucine biosynthesis. Concomitantly, the *S. fermentans* TFL-resistant mutants generated increased amounts of isoamyl alcohol and isovalerate during wine fermentation. Selection of TFL-resistant strains thus provides a generally applicable strategy for the improvement in NCYs and their utilization in co-fermentation processes for different grape must varieties.

Keywords: Ehrlich pathway; grape must; trifluoro-leucine-resistant mutants; *LEU4*

Citation: Akan, M.; Gudiksen, A.; Baran, Y.; Semmler, H.; Brezina, S.; Fritsch, S.; Rauhut, D.; Wendland, J. Exploring the Potential of Non-Conventional Yeasts in Wine Fermentation with a Focus on *Saccharomycopsis fermentans*. *Fermentation* **2023**, *9*, 786. https://doi.org/10.3390/fermentation9090786

Academic Editor: Ronnie Willaert

Received: 28 July 2023
Revised: 17 August 2023
Accepted: 22 August 2023
Published: 25 August 2023

Copyright: © 2023 by the authors. Licensee MDPI, Basel, Switzerland. This article is an open access article distributed under the terms and conditions of the Creative Commons Attribution (CC BY) license (https://creativecommons.org/licenses/by/4.0/).

1. Introduction

Wine fermentations involve complex biochemical processes to achieve the conversion of sugars to ethanol, carbon dioxide, and other volatile compounds by employing yeasts, specifically by *Saccharomyces cerevisiae*. Research has explored the roles of wine yeast strains and crucial parameters such as temperature, pH, oxygen levels, and the addition of nitrogen sources in the fermentation process and their impact on the final product [1,2]. Among the various secondary metabolites produced by yeasts during wine fermentations, active volatile flavour and aroma compounds crucially contribute to the wine bouquet [3]. Higher alcohols and esters, in particular, have a significant impact on the sensory attributes of wine [4].

During the early stages of fermentation, the presence of non conventional yeasts (NCYs) can also contribute to the sensory characteristics of wines. Yeasts belonging to the genera of *Hanseniaspora*, *Pichia*, *Torulaspora*, *Kluyveromyces*, and *Metschnikowia* were found to dominate early stages of spontaneous wine fermentations due to their dominance in the vineyard. Crabtree-positive yeasts like *S. cerevisiae* take over only several days later [5,6]. *S. cerevisiae* proliferates under winemaking conditions, particularly due to its ability to grow anaerobically and its tolerance to high levels of sugar, sulphur dioxide, and heat stress [7].

The use of NCYs in wine fermentations may result in lower-alcohol-content wines with increased complexity due to the formation of a variety of higher alcohols and aromatic esters [8].

The Ehrlich pathway comprises a set of reactions resulting in the degradation of several amino acids into higher alcohols and further into their respective esters or acids. The pathway involves the transamination of an amino acid to its corresponding keto acid followed by decarboxylation to produce an aldehyde, which is then reduced to an alcohol. These alcohols can then be reduced into acetate esters or oxidized into corresponding acids [9]. In this way, leucine degradation leads to isoamyl alcohol; its ester, isoamyl acetate; or its acid, isovalerate. The production of isoamyl alcohol during fermentation is an important factor in determining the final flavour and aroma of a beverage [10,11]. The leucine biosynthetic pathway provides the starting compound for the derived aroma products. This pathway and its involved genes in *Saccharomyces cerevisiae* are shown in Figure 1. Studies have shown that the overexpression of the *LEU4* gene in yeast strains can significantly increase the production of isoamyl alcohol during wine fermentation, providing a potential strategy for the improvement in wine flavour [12]. Previous studies have demonstrated that resistance to 5,5,5-trifluoro-DL-leucine (TFL) is attributed to a mutation in the *LEU4* gene, specifically the mutant allele *LEU4fbr*, which is correlated with the resistance of α-isopropyl malate (a-IPM) synthase to leucine feedback inhibition, as reported by Baichwal et al. [13]. The *LEU4* genes of different *S. cerevisiae* mutants exhibiting resistance to TFL have been extensively characterized in previous studies by Cavalieri et al. [14]. These mutations primarily affect the R-region, which is responsible for leucine feedback inhibition.

NCYs are less studied. They may contribute to more complex aromas. Therefore, the use of non-conventional yeast, such as *S. fermentans*, may offer alternative options for the production of unique flavour profiles in wine through manipulation of the Ehrlich pathway and especially the leucine biosynthesis pathway [15]. In co-fermentation cultures, different inoculation approaches are utilized, with sequential inoculation being the prevailing method. In this approach, the NCY starter culture is initially introduced, allowing the NCY to exert its effects independently of *S. cerevisiae*. The addition of *S. cerevisiae* culture typically occurs after one to seven days or once a certain level of sugar consumption has been reached, depending on the NCY strain type. Alternatively, both starter cultures can be added simultaneously at the onset of fermentations. Another influential factor in the outcome of the wine is the ratio of *S. cerevisiae* to NCY strains [16]. Comparative studies of these inoculation modalities have revealed notable differences in the final quality of various alcohol beverages, emphasizing the intricate and unpredictable nature of the interaction between these distinct yeast strains [17,18].

To the best of our knowledge, there has been no investigation into the utilization of *Saccharomycopsis* yeasts in grape winemaking, particularly with *Saccharomycopsis fermentans* and its TFL-resistant mutants. This study focused on exploring the capacity of *Saccharomycopsis fermentans* to metabolize sugar with reduced ethanol production, as well as its impact on the aromatic characteristics of grape wine. Several other *Saccharomycopsis* species have been studied and have shown potential in producing interesting aroma products in different alcohol beverages [19–24]. One of the most popular NCY strains from *Saccharomycopsis* species is *Saccharomycopsis fibuligera*, which is present in various types of fermentation starters and is recognized for its ability to secrete α-amylase, β-glucosidase, and acid protease with high efficiency, which contributes to desirable flavours in the wine [25,26]. Despite the increasing number of publications on NCYs, many areas in this field remain poorly understood, especially regarding *S. fermentans*, requiring detailed examinations of this strain to determine its potential in wine fermentation. In previous studies of *S. fibuligera*, it was evident that varying sugar utilization patterns occurred in brewer's wort, enabling the production of both non-alcoholic and alcoholic beers [27–29]. In our preliminary single fermentations of *S. fermentans* in grape must, we also established that it generates very low levels of alcohol and still consumes some sugar from the must. This has led to the conception of co-fermentation to reduce the alcohol level in the final must while preserving full flavour.

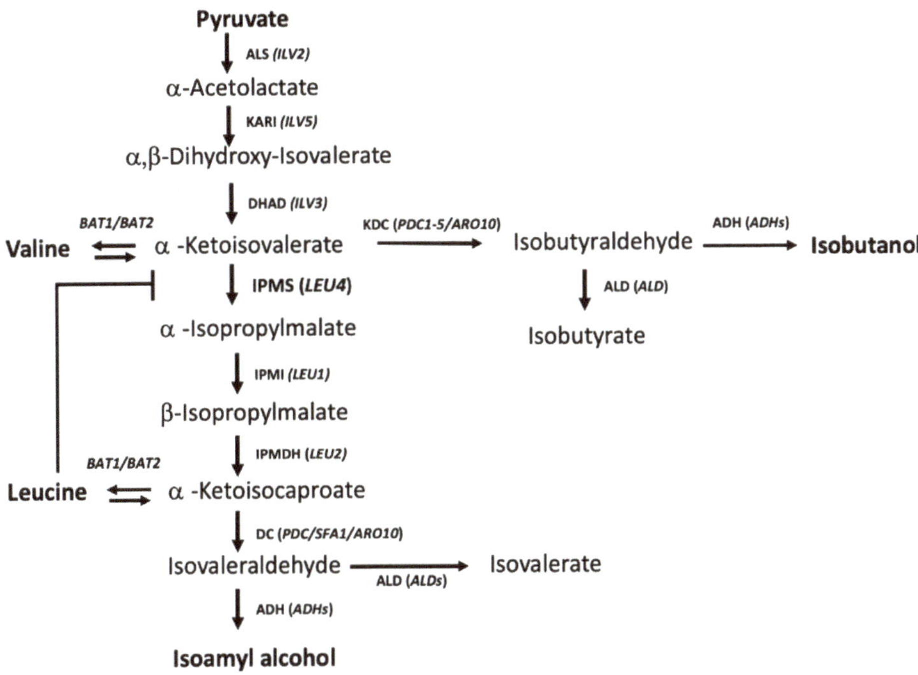

Figure 1. The process of isoamyl alcohol production and the regulation of the expression of the genes involved. Following the conversion of glucose into pyruvate and subsequently into α-ketoisovalerate, the enzyme α-isopropyl malate synthase (IPMS) becomes essential in producing both isoamyl alcohol and leucine. This is because IPMS is responsible for the feedback inhibition that results from leucine. Meanwhile, α-ketoisovalerate is also responsible for isobutanol production, with the help of alcohol dehydrogenase (ADH1). Isovaleraldehyde has another product in the form of isovalerate through aldehyde dehydrogenase (ALDH1, ALDH2).

2. Materials and Methods

2.1. Strains and Media

The yeast strains used in this study are shown in (Table 1). The *Saccharomyces cerevisiae* wine strain EC1118 was used for single fermentations and sequential co-fermentations. Rich medium (YPD, 20 g/L Bacto-peptone, 10 g/L yeast extract, 20 g/L glucose) was used for the propagation of yeast cells prior to fermentations.

Table 1. Strains used in this study.

Code Name	Strain Name	Description	Nucleotide Exchange	Amino Acid Residue in *S. cerevisiae*
EC1118	*Saccharomyces cerevisiae*	Wild-type wine strain	—	—
SFE	*Saccharomycopsis fermentans*	Wild type	—	—
G058, TFL1	*Saccharomycopsis fermentans*	Thr514Lys	ACA–AAA	Val522
G059, TFL2	*Saccharomycopsis fermentans*	His534Pro and Ala545Thr	GCT–ACT, CAC–CCC	His541, Ala551
G236, TFL3	*Saccharomycopsis fermentans*	Ser541Tyr	TCC–TAC	Ser547
G060, TFL4	*Saccharomycopsis fermentans*	Ser511Tyr	TCT–TAT	Ser519

The EC1118 strain is commercially available, SFE is wild-type CBS 7830, and TFL strains were obtained in this study. The *S. cerevisiae* residues correspond to ScLeu4p. Mutated positions are highlighted.

2.2. Trifluoro-Leucine (TFL)-Resistant Strain Selection

Wild-type *S. fermentans* was used for the selection of 5,5,5-trifluoro-DL-leucine (TFL)-resistant colonies. *S. fermentans* was inoculated in a Complete Supplement Mixture minus leucine (CSM-LEU) medium containing 20 g/L glucose, 1.7 g/L Yeast Nitrogen Base (YNB) without ammonium sulphate, 0.69 g/L CSM-LEU, 1 g/L asparagine, and 20 g/L agar. For the selection of TFL-resistant mutants, 150 µg/L TFL (5,5,5-trifluoro-DL-leucine, Sigma, Taufkirchen, Germany) was used according to Takagi et al. [30] on CSM-LEU plates. Twenty fast-growing colonies were picked and re-streaked on TFL selection spot assay plates with 150–300 µg/L TFL. The four fastest-growing colonies were chosen for further experiments.

2.3. PCR Amplification and Sequencing of LEU4 Amplicons

Publicly available sequences of the *LEU4* gene of *S. cerevisiae* and *S. fermentans* were aligned and compared in the lab before the PCR amplification. Over 70% identification was observed. TFL-resistant colonies were analysed for potential mutations in the *S. fermentans LEU4* gene. To this end, the *LEU4* ORF was amplified with two sets of primers (LEU4_U1–LEU4_M2 and LEU4_M1–LEU4_D1). The forward primer for the first part was 5′-GATTCAAGATTTTTGAAGAGAT-3′ and the reverse primer was 5′-CTTGCAGAAAACCGTTCATAAG-3′; as for the second part, the forward primer was 5′-AAACCTTTTCACGTTCAGTTAT-3′ and the reverse primer was 5′-AATCGAATATTTCT GCTCAAAT-3′. Primers were deduced according to the published draft genome sequence that is available at GenBank under accession number JNFW00000000 [21]. PCR products were sequenced (Starseq, Mainz, Germany). DNA sequence analyses were carried out using Geneious (Prime version 2019.2.1) software (Biomatters Ltd., NZL-1010 Auckland, New Zealand). Amino acids and nucleotide exchange sites were identified in each TFL-resistant mutant (see Table 1).

2.4. Fermentation Conditions

Lab-scale fermentations were carried out in triplicate with a standard pasteurized white wine Riesling, Chardonnay, and Müller-Thurgau musts of the indicated vintages derived from the university vineyards. These different types of musts were selected to potentially replicate and upscale the results of the lab-scale fermentations in the university winery setting. Riesling and Chardonnay fermentations were carried out in 250 mL Erlenmeyer flasks with 25 mL of must and Müller-Thurgau fermentation was carried out in 250 mL cylinder tubes with 50 mL of must. Musts were supplemented with the addition of 0.4 g/L Optimum-White (inactivated yeast product; according to the supplier's instructions; Lallemand, Vienna, Austria) and 0.3 g/L Vitaferm Ultra F3 (multi-nutrient complex containing nutrients such as amino acids, fatty acids, minerals, sterols, vitamins, etc.; according to the producer; Erbslöh Geisenheim GmbH, Geisenheim, Germany). *Saccharomycopsis fermentans* strains were inoculated at a density of 2×10^6 cells/mL, while the EC1118 strain was inoculated at a density of $OD_{600} = 0.5$. The single fermentation temperature was set to 18 °C, and cultures were incubated with constant stirring at 150 rpm for 11 days with only *S. fermentans* and TFL-resistant strains and the EC1118 strain as a control sample. Aliquots were then sampled and used for aroma profile analyses. EC1118 was added on the 12th day of the single fermentation. Sequential co-fermentations were run over a course of 23 days using the same conditions to allow EC1118 to finish fermentation. These sequential inoculation conditions were established through previous lab-scale fermentations trials. Daily measurements of weight loss were recorded, and at the end of the co-fermentation, residual sugars and ethanol content were measured via high-performance liquid chromatography (HPLC), and volatiles were analysed through gas chromatography and mass spectrometry (GC-MS).

2.5. Analytical Methods of Must Analysis

At the end of the fermentation process, high-performance liquid chromatography (HPLC) using an Agilent 1100 Series (Agilent Technologies, Waldbronn, Germany) coupled with an autosampler, a multi-wavelength (MWD), and a refractive index (RID) detector and a binary pump was utilized to analyse various compounds, such as fructose, glucose, ethanol, and organic acids. Quantitative analyses were conducted following the methodology outlined in Schneider et al. [31] and adapted as indicated in Scansani et al. [32] and are briefly described in the following: an Allure Organic Acids Column (length 250 mm, inner diameter 4.6 mm, and particle size 5 μm) from Restek (Bad Homburg v. d. Höhe, Germany) was used to separate the compounds. The organic acids were measured with the MWD (at a wavelength of 210 nm). The RID was applied for the detection of carbohydrates, organic acids, and ethanol. Purified water containing 0.5% ethanol acidified with 0.0139% concentrated sulphuric acid (95–97%) was used as the isocratic eluent. The flow rate was set at 0.6 mL/min using two column temperatures (29 °C and 46 °C). Chemstation software (Agilent, Germany) was used for analysis, integration, and determination of the concentrations of the individual analytes [32]. Headspace solid-phase microextraction (HS-SPME; multipurpose sampler MPS2 (Gerstel GmbH & Co. KG, Mülheim a. d. Ruhr, Germany)) alongside gas chromatography and mass spectrometry (GC-MS; 7890A and MS 5977B, Agilent Technologies, Waldbronn, Germany) was utilized to determine the presence of aroma compounds (higher alcohols, esters, fatty acids, etc.), following the analytical approach described in Tarasov et al. [33]. The process of sample preparation involved weighing 1.7 g of p.a. grade NaCl (Carl Roth GmbH & Co. KG, Karlsruhe, Germany) into a 20 mL brown, glass headspace vial. Subsequently, 5 mL of the sample was added to the vial, along with 10 μL of each internal standard solution (1-octanol at 600 mg/L and cumene at 52 mg/L). The vial was then securely sealed using a magnetic screw cap. SPME extraction was performed with a 1 cm fibre (65 μm polydimethylsiloxane/divinylbenzene (Supelco)) for 20 min (incubation temperature: 40 °C, incubation time: 10 min). The sample was transferred to the GC with a cooled injection system (CIS-4, Gerstel, Mülheim an der Ruhr, Germany). Calibration was carried out with a model wine (10% (*v*/*v*) solution of ethanol in water, 3 g/L tartaric acid, adjusted to pH 3). A gas chromatography column of 60 m (length) × 0.25 mm (internal diameter) × 1 μm (film thickness) (Rxi-5Sil, Restek, Bellefonte, PA, USA), together with particular GC-MS software and technical settings (temperature program: 30 °C (1 min), 12 °C/s to 240 °C (4 min); split ratio 1:10), was used. Agilent MassHunter Workstation software was used for instrumental control, data acquisition, and analysis of the qualitative and quantitative data.

The following chemicals were used for the mobile phase: Purified water was taken from the TKA Thermo Scientific GenPure (Dreieich, Germany). Ethanol (ROTISOLV HPLC Gradient Grade) from Roth (Karlsruhe, Germany). Sulphuric acid (95–97%) was obtained from Merck (Darmstadt, Germany).

As reference substances for the analysis, we used citric acid anhydrous (puriss p.a. > 99.5%) and L-(+)-tartaric acid (puriss. > 99.5%) from Fluka (Buchs, Switzerland); D-(−)-fructose (>99%), D-(+)-glucose (≥99.5%), L-(−)-malic acid (97%), and-L(+)-lactic acid lithium salt crystalline from BioChemica; shikimic acid (>99%) from Sigma Aldrich (Seelze, Germany); ethanol (ROTISOLV HPLC Gradient Grade) from Roth (Karlsruhe, Germany).

The following chemicals were used as reference compounds for the analysis and were purchased from Sigma Aldrich (Seelze, Germany): hexanoic acid (>99%), octanoic acid (≥99%), decanoic acid (99.9%), i-valeric acid (99.5%), i-butanol (99.8%) (isobutanol), 2-methylbutanol (≥99%) (active amyl alcohol), 3-methylbutanol (≥99%) (isoamyl alcohol), hexanol (99.9%), 2-phenylethanol (≥99%) (phenylethyl alcohol), acetic acid phenylethylester (99%), acetic acid ethylester (>99.8%), acetic acid 2-methylbutylester (≥95%), acetic acid 3-methylbutylester (99.7%), benzeneacetic acid ethylester (99.8%), propionic acid ethylester (99%), i-butyric acid ethylester (99.8%), butyric acid ethylester (99%), lactic acid ethylester (98%), hexanoic acid ethylester (≥99%), succinic acid diethylester (99.8%), octanoic acid ethylester (99.8%), decanoic acid ethylester (99.8%), 2-methylbutyric acid ethylester (99%), and 2-methylbutyric acid methylester (99%). For the internal standards, we used octanol (99.8%) and cumene (99.9%). Acetic acid hexylester (99%) was acquired from VWR (Darmstadt, Germany). 2-Hydroxy-4-methylvaleric acid ethylester (>98%) was obtained from TCI (Eschborn, Germany).

The amounts of residual amino acids present in the Riesling must single fermentations were measured prior to sequential inoculation via post-column derivatization with ninhydrin and detection at 440 nm and 570 nm using maintenance-free LED photometers with ARACUS amino acid analyser (membraPure GmbH, Hennigsdorf, Germany). Sample preparation and analysis were carried out following the methodology outlined in Krause et al. [34]. All measurable amino acids (L-alanine, L-arginine, L-asparagine, L-aspartic acid, L-cysteine, L-glutamine, L-glutamic acid, glycine, L-histidine, L-isoleucine, L-leucine, L-lysine, L-methionine, L-phenylalanine, L-proline, L-serine, L-threonine, L-tryptophan, L-tyrosine, L-valine) were purchased from Sykam (Eresing, Germany).

3. Results

3.1. Verification of Obtained TFL-Resistant Mutants

S. fermentans strains are haploid heterothallic yeast strains. These strains are characterized by having a single set of chromosomes (haploid) and being unable to mate and form diploid cells (heterothallic) without a compatible mating partner. This makes them useful for controlled genetic modifications. We were able to select mutants resistant to TFL, a known toxic leucine analogue, from the wild-strain *S. fermentans*. Initial selection of *S. fermentans* cells on TFL plates identified several colonies of spontaneously resistant cells. The sequencing of the *LEU4* gene site for the chosen *S. fermentans* TFL-resistant mutants was carried out and compared to the wild-type *S. fermentans*. The draft genome sequence of *S. fermentans* was obtained previously [21]. The *S. cerevisiae LEU4* gene is encoded by YNL104C. The *S. fermentans LEU4* homologue is located on scaffold 1 of the draft genome sequence [21]. *SfLEU4* is placed between the *S. cerevisiae* homologues of *YLR106C* and *YNL241C*. The SfLEU4 open reading frame is 1782 bp in length and encodes a protein of 593 amino acids. In *S. cerevisiae*, *YOR108W/LEU9* encodes a paralogue of *LEU4* as a result of its whole-genome duplication. The SfLEU4 gene bears 64.5% and 65% sequence identity with *ScLEU4* and *ScLEU9*, respectively. The SfLeu4 protein shares 62.8% sequence identity with ScLeu4p and 64.1% with ScLeu9p. Since ScLEU4 provides major a-isopropylmalate synthase activity and Leu9 is a minor isoenzyme, we annotated the *S. fermentans* gene as *SfLEU4*.

The alignment of DNA and the amino acid sequence was performed to identify sites of mutation in the TFL-resistant strains and resulting amino acid sequence changes thereof. Mutated positions are listed in Table 1 and compared with mutations identified in the *S. cerevisiae LEU4* gene of TFL-resistant strains [15,35,36].

3.2. Fermentation Performance

Three different types of grape musts were selected for fermentation trials: Riesling, Müller-Thurgau, and Chardonnay. The chosen varieties were selected because Chardonnay usually has a high nitrogen content, while Riesling and Müller-Thurgau have lower levels of yeast available nitrogen. These varietal distinctions were expected to yield variations in yeast fermentation kinetics and flavour output. Single fermentations were carried out over 11 days, and sequential co-fermentations were initiated on the 12th day with the additional inoculation of EC1118. These fermentations were continued for an additional 23 days. Summary fermentation curves were plotted based on mass loss by combining the single and co-fermentation phases (Figure 2). These curves demonstrated that the contribution of *S. fermentans* strains to overall sugar consumption was much lower than that of the EC1118 wine yeast in all single fermentations. After the addition of the EC1118 strain to these single fermentations to start co-fermentation, the sugar consumption of all samples increased drastically. In most single fermentations, the TFL2 sample outperformed all other *S. fermentans* strains. In the Riesling and Müller-Thurgau co-fermentations, the TFL2 sample showed the best performance, while in Chardonnay, the TFL1 sample was better. In general, the fermentation process of Chardonnay exhibited a slower rate of activity in comparison to the other two musts, despite the fact that the Chardonnay must had a higher nitrogen content, which was expected to enhance the fermentation process. Even EC1118 single fermentation demonstrated a slower rate in Chardonnay compared to Müller-Thurgau and Riesling. Overall, Müller-Thurgau co-fermentation showed more vigorous fermentation activity.

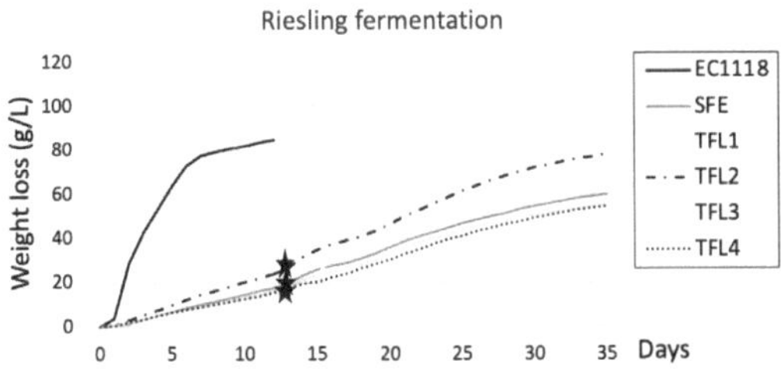

(A)

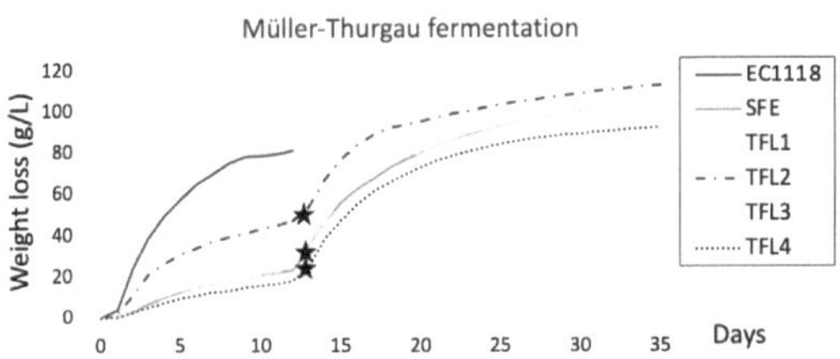

(B)

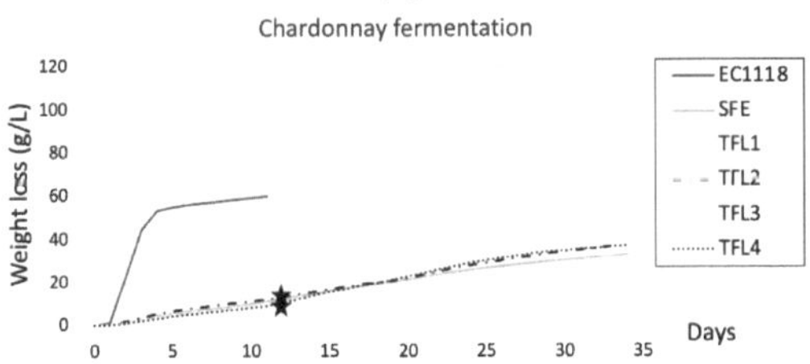

(C)

Figure 2. Fermentation curves based on CO_2 release (weight loss). Combined weight loss curves of single and co-fermentations as one sequential fermentation. Fermentation curves of three different musts, (**A**) Riesling, (**B**) Müller-Thurgau, and (**C**) Chardonnay, with (★) indicating when co-fermentation started on day 12 after the addition of EC1118.

3.3. Amino Acid Concentrations after Single Fermentation of Riesling Must

We analysed the amino acid composition of Riesling must before and after fermentations. The concentrations of amino acids involved in the Ehrlich pathway are listed in Table 2. As expected, most of the amino acids were metabolized by the yeast strains during the single fermentation, with EC1118 showing the highest degree of utilization overall. TFL resistance alleviates leucine's allosteric inhibition of *LEU4*, resulting in an excess of leucine production. This leucine overflow is catabolized via the Ehrlich pathway, as shown in the leucine pathway in Figure 1. Our amino acid measurement showed a higher concentration of leucine in the musts fermented with TFL-resistant strains compared to *S. fermentans*. As a consequence of *LEU4* inhibition, valine production is expected to decrease, which is reflected in the utilization of valine from must by three TFL strains (Figure 1, Table 2). Interestingly, the levels of several other amino acids also involved in the Ehrlich pathway, such as phenylalanine, isoleucine, tyrosine, and tryptophan, were significantly elevated in the TFL2 sample; in the case of phenylalanine and tryptophan, the levels present in must were well above normal concentrations. This suggests that TFL resistance also induces changes in other amino acid biosynthesis pathways.

Table 2. List of selected amino acid measurements.

	Valine	**Leucine**	**Isoleucine**	**Phenylalanine**	**Alanine**	**Methionine**	**Glycine**	**Glutamine**	**Threonine**	**Tyrosine**	**Tryptophan**	**Serine**	**Total**
EC1118	0.2	0.4	0.3	0.3	6.2	0.2	0.7	0.3	1.1	0.2	0.1	0.7	21.9
SFE	**1.6**	0.4	0.5	2.1	2.6	0.1	1.2	0.1	1.2	0.1	<0.1	0.3	126.1
TFL1	<0.1	**3.2**	0.6	1.3	5.4	0.4	1.9	0.7	1.7	0.1	<0.1	0.3	171.2
TFL2	0.2	**3.2**	**2.2**	**7.7**	2.6	0.3	**5.8**	1.1	**10.0**	**3.2**	**1.7**	0.6	196.9
TFL3	0.1	**2.8**	0.4	0.4	5.1	0.2	1.9	0.7	1.3	0.1	<0.1	0.7	172.3
TFL4	1.5	**2.9**	0.7	2.3	4.1	0.8	2.2	1.2	3.5	0.6	<0.1	0.8	201.2
Riesling Must	9.2	7.2	4.9	6.6	77.3	0.9	2.4	<0.1	30.3	3.9	<0.1	27.5	589.7

Residual amino acid measurements after the single fermentation, concentration in mg/L. Significant increases are highlighted (see also Table S2).

3.4. Aroma Analysis of Fermentations

We conducted standard analyses (HPLC and GC-MS) of aroma compounds, including for esters, higher alcohols, and acids. Amino acid degradation via the Ehrlich pathway generates higher alcohols such as isoamyl alcohol, active amyl alcohol, isobutanol, and phenethyl alcohol, which are important aroma compounds that contribute to the sensory characteristics of wine. We clearly observed the elevation of all these aroma compound levels in all co-fermentation samples compared to the single fermentations. The difference between wild-type *S. fermentans* and TFL-resistant mutants was also observed in the levels of isobutanol and isoamyl alcohol, which corresponds with the results of amino acid measurements, where valine and leucine levels are in accordance with the alleviation of the feedback inhibition of Leu4 (Figure 3). Although all strains generated various compounds, our findings show that EC1118, which is a prominent wine production strain, produced less isoamyl alcohol in comparison to the TFL2 sample in all single fermentation musts. Moreover, their joint effort in co-fermentation doubled isoamyl alcohol levels in all three musts.

The aroma profiles of single fermentations and co-fermentations showed significant differences. The most prominent difference could be observed in the levels of isovaleric acid, which differed significantly between the EC1118 strain and other strains, as well as across the different must types (Table 3). The increase in isovaleric acid levels in TFL-resistant mutants compared to wild-type *S. fermentans* can be attributed to the loss of feedback inhibition of *LEU4* and subsequent overproduction of leucine, as one of the products of the leucine biosynthetic pathway is isovalerate. Overall isovaleric acid is not one of the desired aroma compounds in wine.

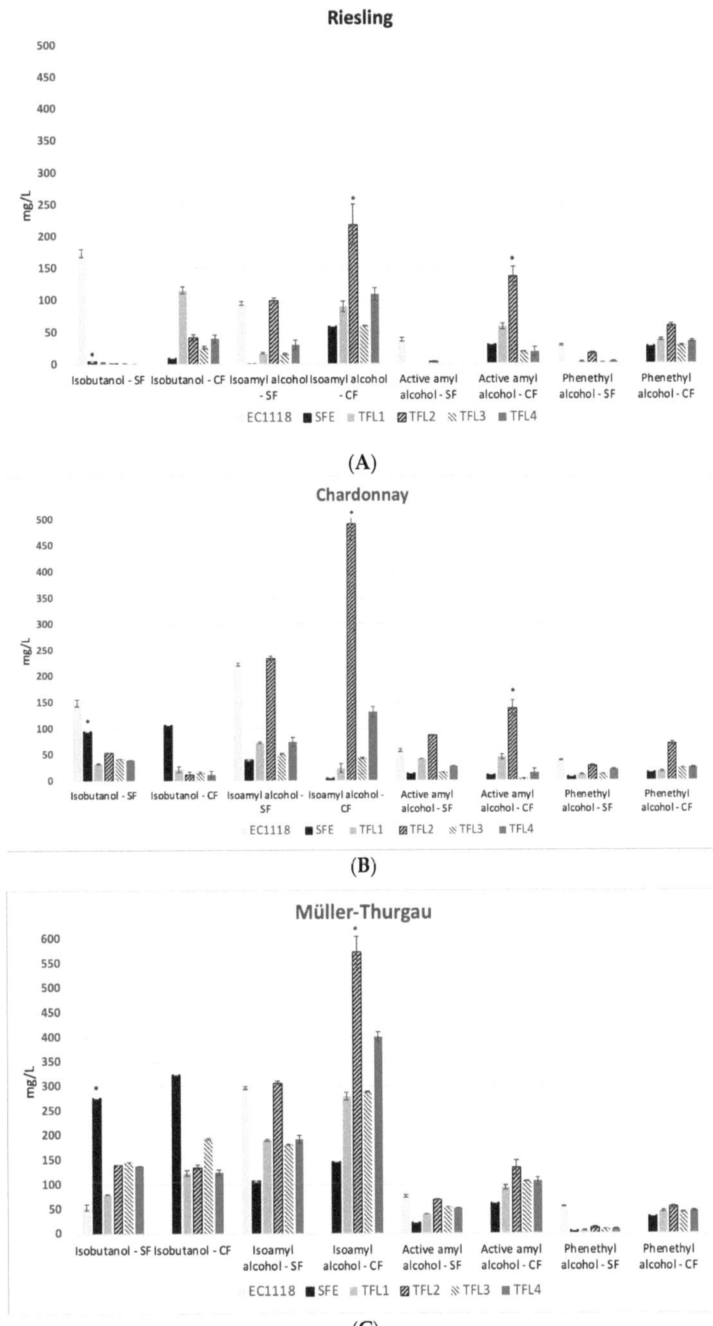

Figure 3. Bar charts with the amounts of selected higher alcohol compounds (in mg/L) in single fermentations (SFs) and co-fermentations (CFs) of different must types (**A**–**C**). Higher alcohol compounds were quantified via GC-MS at the end of each fermentation. (*) indicates a significant increase.

Table 3. Acids generated at the end of each fermentation *.

Must Type	Strain	Tartaric Acid (g/L)		Malic Acid (g/L)		Shikimic Acid (mg/L)		Lactic Acid (g/L)		Acetic Acid (g/L)		Citric Acid (g/L)		Isovaleric Acid ** (mg/L)	
		SF	CF	SF	CF	SF	CF	SF	CF	SF	CF	SF	CF	SF	CF
Riesling	EC1118	4.5		2.7		23.9		<0.1		5.4		<0.1		3	
	SFE	6.8	6.1	1.9	<0.1	28.6	23.3	<0.1	<0.1	<0.1	<0.1	<0.1	0.4	20	116
	TFL1	6.7	6.2	1.6	<0.1	29.9	23.3	<0.1	<0.1	<0.1	<0.1	0.1	0.3	76	302
	TFL2	6.8	6.2	<0.1	<0.1	29.4	21.9	<0.1	<0.1	<0.1	0.1	0.2	0.3	118	486
	TFL3	6.8	6.0	1.4	<0.1	29.0	22.5	<0.1	<0.1	<0.1	<0.1	0.1	0.3	80	419
	TFL4	5.4	5.6	1.9	<0.1	28.3	23.5	<0.1	<0.1	<0.1	<0.1	0.1	0.3	73	329
	Must	6.2	6.2	3.9	3.9	29	29.3	<0.1	<0.1	<0.1	<0.1	<0.1	<0.1	nd	nd
Chardonnay	EC1118	3.6		3.5		46		<0.1		6.9		<0.1		4	
	SFE	3.8	3.5	3.7	<0.1	55	50.8	0.17	<0.1	<0.1	0.3	<0.1	0.2	9	80
	TFL1	3.5	3.6	3.5	0.8	54	40.4	0.14	<0.1	<0.1	0.1	<0.1	0.2	34	71
	TFL2	3.5	3.3	3.1	<0.1	48	51.0	0.18	<0.1	<0.1	<0.1	<0.1	0.2	33	289
	TFL3	3.4	3.5	1.6	<0.1	55	49.1	<0.1	<0.1	<0.1	<0.1	<0.1	0.2	29	253
	TFL4	3.4	3.5	2.4	0.6	54	51.2	<0.1	<0.1	<0.1	0.2	<0.1	0.2	50	206
	Must	3.3	3.3	4.6	3.8	54	52.9	<0.1	<0.1	<0.1	<0.1	<0.1	0.2	nd	nd
Müller-Thurgau	EC1118	4.4		1.8		25		0.16		0.36		<0.1		5	
	SFE	5.4	5.1	1.6	1.5	16	13.4	0.48	0.3	<0.1	0.1	<0.1	0.1	3	9
	TFL1	5.5	5.2	1.3	1.2	26	15.1	0.44	0.3	<0.1	0.4	<0.1	0.1	11	19
	TFL2	5.4	5.1	1.1	1.2	25	6.7	<0.1	0.2	0.36	0.1	<0.1	<0.1	20	21
	TFL3	5.4	5.1	1.3	1.6	25	14.2	0.69	0.4	<0.1	0.2	<0.1	0.1	8	14
	TFL4	5.3	5.1	1.3	1.5	25	14.8	0.40	0.4	<0.1	0.1	<0.1	0.1	10	19
	Must	5.6	5.0	1.6	1.6	28	26.0	<0.1	<0.1	<0.1	<0.1	<0.1	<0.1	nd	nd

* This table contains the HPLC results for measured acid concentration generated during the fermentation of three grape varieties, Riesling, Chardonnay, and Müller-Thurgau, after both single fermentation (SF) and co-fermentation (CF) processes. ** Isovaleric acid was measured via GC-MS. nd: not detected. See also Table S3.

3.5. HPLC Results of Ethanol and Sugars

For more detailed analyses of the fermented wines, we determined the residual sugars and final ethanol content using HPLC. The alcohol produced at the end of the fermentations can be observed in Figure 4A, where it can be noted ethanol was typically not present in the pure must. The total sugar concentration of residual glucose + fructose is shown in Figure 4B, and pure must was used as an indicator of initial sugar levels compared to the consumed sugar levels after each fermentation. EC1118 completely consumed all sugars within the first 5–7 days. It was observed that *S. fermentans* strains did not produce ethanol in a single fermentation of Riesling must, but rather reduced sugars by an average of 10–18 g/L, with a maximum reduction of 42 g/L, as observed in the TFL2 sample. Similar results were obtained in Chardonnay single fermentations, where no significant ethanol production was observed in *Saccharomycopsis* strains, but there was a reduction of 10–15 g/L in sugar levels, with the TFL2 strain showing the highest reduction at 28 g/L, consistent with the earlier results for Riesling must. Interestingly, in Müller-Thurgau must single fermentations, there were low amounts of ethanol observed in the *S. fermentans* strains, with an overall reduction in sugars of around 40–50 g/L, and the TFL2 strain again showed the highest reduction at 105 g/L.

In all co-fermentations, EC1118 managed to carry the fermentation to completion and utilized all remaining sugars. EC1118 finished co-fermentations with an average ethanol production of 6–7% in Riesling must. Co-fermentation with *S. fermentans* strains reduced ethanol levels by 1–2% in Riesling must. As a consequence of the high performance of TFL2, alcohol reduction in this sample was minimal compared to other *S. fermentans* strains. The reduction in ethanol in the Chardonnay fermentation was slightly different. Low reduction was observed not only in TFL2, but also in the TFL3 and TFL4 samples, while *S. fermentans* and TFL1 showed higher ethanol reductions at 4.5% and 5.5%, respectively. The co-fermentation of Müller-Thurgau showed high ethanol production in all samples, including EC1118, which had the highest ethanol production among the controls. As a result, there was a reduction in ethanol of 1–2% in *Saccharomycopsis* strains.

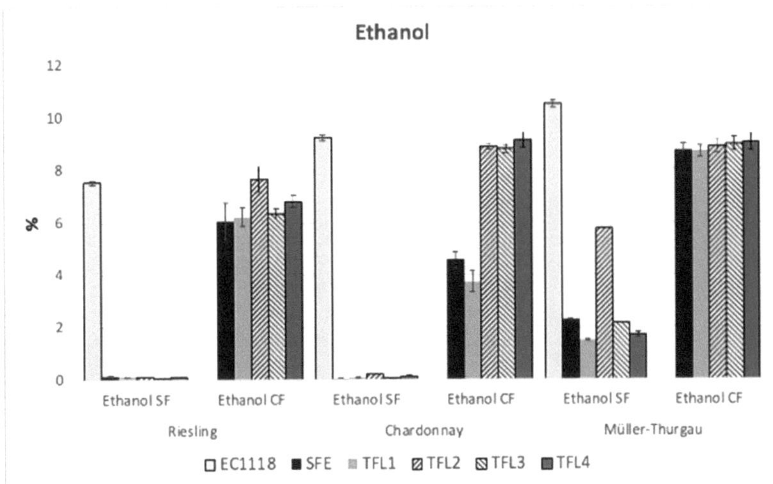

(A)

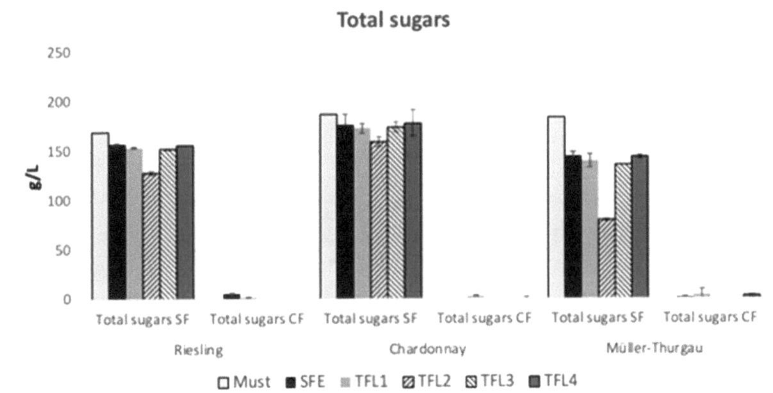

(B)

Figure 4. Bar charts with the HPLC measurements of (**A**) ethanol at the end of fermentation and (**B**) total sugar concentrations of glucose + fructose at the end of single fermentations (SFs) and co-fermentations (CFs) in different must types. Pure must was used as a control indicator for initial sugar amounts to compare the consumption after each fermentation. EC1118 utilized all sugars within the first week of fermentation (**B**).

4. Discussion

Saccharomycopsis fermentans is a non-conventional yeast which, to the best of our knowledge, has not previously been used in grape must fermentations. In our study, we focused on *S. fermentans*, which is a haploid yeast, making it easier to modify this strain through non-GMO methods [21]. This has been demonstrated in previous studies which focused on TFL-resistant mutants in *S. cerevisiae* [30]. Interestingly, the mutation sites observed in our four *S. fermentans* TFL-resistant mutants displayed a close similarity to those reported in previous studies on *S. cerevisiae* [35,37]. Our results also reveal mutations accumulated within the R-region of the IPMS enzyme (Leu4). IPMS catalyses the condensation of acetyl-CoA and α-ketoisovalerate, leading to the formation of 2-isopropylmalate, which is further converted to leucine [10]. Specifically, the amino acid exchange observed in TFL2 and TFL4 at the His534Pro and Ser511Tyr sites, respectively, correspond to the mutations previously reported in *S. cerevisiae*, namely His541Pro and Ser519Thr (see Table 1).

To further explore the implications of our findings, TFL-resistant mutants and wild-type *S. fermentans* were compared with respect to isoamyl alcohol production during single fermentations of various must types. The results show that TFL-resistant mutants produced significantly more isoamyl alcohol, phenethyl alcohol, and active amyl alcohol in single fermentations compared to the wild-type *S. fermentans* across all must types [38]. This is likely due to the leucine biosynthetic pathway with a negative feedback loop that was disrupted by mutation, leading to leucine accumulation. Interestingly, the increase in phenethyl alcohol was observable in all TFL2 single fermentations in all musts compared to wild-type *S. fermentans*. This may be due to the efficient mutation and overall high metabolic activity demonstrated by the TFL2 strain, which helps to elevate other higher alcohol productions in the Ehrlich pathway [39]. In a study by Vuralhan et al. [40], yeast cultures were grown with phenylalanine, leucine, or methionine as a nitrogen source, resulting in high levels of the corresponding fusel alcohols and organic acids, indicating utilization of the Ehrlich pathway and suggesting the involvement of *ARO10*-dependent common enzyme activity. These results correspond to our elevated IPMS activity, where α-ketoisovalerate is highly activated for leucine production and isobutanol side production is limited, but *ARO10* is utilized [41]. Additionally, we observed that wild-type *S. fermentans* produced more isobutanol than all TFL-resistant mutants. This was also due to the leucine biosynthetic pathway, as the undisturbed negative feedback loop in wild-type *S. fermentans* should accumulate valine, which is the precursor of isobutanol [42,43].

The *S. fermentans* strains studied produced lower levels of esters, which might have had a negative impact on wine quality due to the oxidative, rusty, and sweaty notes that are associated with it [42]. Volatile acid production was also high for the *Saccharomycopsis* yeasts, with isovaleric acid values being much higher compared to the reported acceptable odour thresholds [31], which can be explained by the high metabolic activity in all TFL strains. Oxygen is crucial in the Ehrlich pathway and can control the generation of desired fusel alcohols and fusel acids [36]. Our results suggest that oxygen exposure led to a shift in the Ehrlich pathway towards producing more fusel acids in all *Saccharomycopsis* yeast fermentations. The Riesling and Chardonnay fermentations especially exhibited significantly higher levels of isovaleric acid compared to the Müller-Thurgau fermentation. This difference could potentially be attributed to the utilization of cylindrical tubes during the Müller-Thurgau fermentation process, which provides less head space and, therefore, limits the availability of oxygen intake during fermentation.

In parallel to the laboratory-scale fermentation, upscale fermentations were conducted using Riesling must in a winery setting, and the resulting wine was subjected to sensory evaluation by a panel of experts. The results from this evaluation indicate that with certain refinements, this type of wine could be interesting for consumers (Table S1 in the Supplementary Materials). This suggests that better adjustment of fermentation properties and optimization can reduce alcohol concentration in wine, due to demand from consumers [44].

5. Conclusions

To the best of our knowledge, this is the first study on *S. fermentans* and grape must fermentation. In this study, we demonstrated the feasibility of utilizing TFL selection in a non-conventional yeast. Single fermentations and co-fermentations of *Saccharomycopsis* TFL-resistant mutants with a wine yeast provided insight into the behaviour of *S. fermentans* in small-scale grape must fermentations. Such studies may also be useful for the generation of flavour-adapted beverages with other non-conventional yeasts enhanced through TFL selection.

Supplementary Materials: The following supporting information can be downloaded at: https://www.mdpi.com/article/10.3390/fermentation9090786/s1, Table S1: results of simple descriptive test of winery fermentation of Riesling must (upscale), Table S2: amino acid measurements, Table S3: complete dataset of the measured acids with statistical values.

Author Contributions: Conceptualization, J.W. and M.A.; methodology, J.W., M.A. and D.R.; validation, J.W., M.A., A.G. and Y.B.; formal analysis, J.W., M.A., A.G., Y.B., H.S., S.B. and S.F.; investigation, J.W., M.A., A.G., Y.B., H.S., S.B. and S.F.; resources, J.W., data curation, J.W., M.A., H.S., S.B., S.F. and D.R.; writing—original draft preparation, M.A.; writing—review and editing, J.W., M.A. and D.R.; visualization, J.W. and M.A.; supervision, J.W.; project administration, J.W.; funding acquisition, J.W. All authors have read and agreed to the published version of the manuscript.

Funding: This research was funded in part by the European Union Marie Curie Innovative Training Network Aromagenesis, grant no. 764364.

Acknowledgments: We thank Bettina Mattner for her excellent technical service.

Conflicts of Interest: The authors declare no conflict of interest.

References

1. Pretorius, I.S.; Curtin, C.; Chambers, P. The Winemaker's Bug. *Bioengineered* **2012**, *3*, 149–158. [CrossRef] [PubMed]
2. Maicas, S. The Role of Yeasts in Fermentation Processes. *Microorganisms* **2020**, *8*, 1142. [CrossRef] [PubMed]
3. Lambrechts, M.G.; Pretorius, I.S. Yeast and Its Importance to Wine Aroma—A Review. *S. Afr. J. Enol. Vitic.* **2019**, *21*, 97–129. [CrossRef]
4. Goldner, M.C.; Zamora, M.C.; Di Leo Lira, P.; Gianninoto, H.; Bandoni, A.L. Effect of ethanol level in the perception of aroma attributes and the detection of volatile compounds in red wine. *J. Sens. Stud.* **2009**, *24*, 243–257. [CrossRef]
5. Pinto, C.; Pinho, D.; Cardoso, R.; Custódio, V.; Fernandes, J.O.; Sousa, S.; Pinheiro, M.; Egas, C.; Gomes, A.M.P. Wine Fermentation Microbiome: A Landscape from Different Portuguese Wine Appellations. *Front. Microbiol.* **2015**, *6*, 905. [CrossRef] [PubMed]
6. Jolly, N.P.; Varela, C.; Pretorius, I.S. Not Your Ordinary Yeast: Non-Saccharomyces Yeasts in Wine Production Uncovered. *FEMS Yeast Res.* **2013**, *14*, 215–237. [CrossRef] [PubMed]
7. Hansen, E.; Nissen, P.; Sommer, P.; Nielsen, J.; Arneborg, N. The Effect of Oxygen on the Survival of Non-Saccharomyces Yeasts during Mixed Culture Fermentations of Grape Juice with *Saccharomyces cerevisiae*. *J. Appl. Microbiol.* **2001**, *91*, 541–547. [CrossRef] [PubMed]
8. Fleet, G.H. Yeast Interactions and Wine Flavour. *Int. J. Food Microbiol.* **2003**, *86*, 11–22. [CrossRef]
9. Hazelwood, L.A.; Daran, J.-M.; Van Maris, A.J.A.; Pronk, J.T.; Dickinson, J.R. The Ehrlich Pathway for Fusel Alcohol Production: A Century of Research on *Saccharomyces cerevisiae* Metabolism. *Appl. Environ. Microbiol.* **2008**, *74*, 2259–2266. [CrossRef]
10. Kohlhaw, G.B. Leucine Biosynthesis in Fungi: Entering Metabolism through the Back Door. *Microbiol. Mol. Biol. Rev.* **2003**, *67*, 1–15. [CrossRef]
11. Takagi, H.; Yamamoto, K.; Matsuo, Y.; Furuie, M.; Kasayuki, Y.; Ohtani, R.; Shiotani, M.; Hasegawa, T.; Ohnishi, T.; Ohashi, M.; et al. Influence of Mutation in the Regulatory Domain of α-Isopropylmalate Synthase from *Saccharomyces cerevisiae* on Its Activity and Feedback Inhibition. *Biosci. Biotechnol. Biochem.* **2022**, *86*, 755–762. [CrossRef] [PubMed]
12. Hirata, D.; Aoki, S.; Watanabe, K.; Tsukioka, M.; Suzuki, T. Stable Overproduction of Isoamyl Alcohol by *Saccharomyces cerevisiae* with Chromosome-Integrated Multicopy *LEU4* Genes. *Biosci. Biotechnol. Biochem.* **1992**, *56*, 1682–1683. [CrossRef]
13. Baichwal, V.; Cunningham, T.J.; Gatzek, P.R.; Kohlhaw, G.B. Leucine Biosynthesis in Yeast. *Curr. Genet.* **1983**, *7*, 369–377. [CrossRef] [PubMed]
14. Cavalieri, D.J.; Casalone, E.; Bendoni, B.; Fia, G.; Polsinelli, M.; Barberio, C. Trifluoroleucine Resistance and Regulation of α-Isopropyl Malate Synthase in *Saccharomyces cerevisiae*. *Mol. Genet. Genom.* **1999**, *261*, 152–160. [CrossRef]
15. Oba, T.; Yamamoto, Y.; Nomiyama, S.; Suenaga, H.; Muta, S.; Tashiro, K.; Kuhara, S. Properties of a Trifluoroleucine-Resistant Mutant Of *Saccharomyces cerevisiae*. *Biosci. Biotechnol. Biochem.* **2006**, *70*, 1776–1779. [CrossRef] [PubMed]
16. Canonico, L.; Galli, E.; Ciani, E.; Comitini, F.; Ciani, M. Exploitation of Three Non-Conventional Yeast Species in the Brewing Process. *Microorganisms* **2019**, *7*, 11. [CrossRef]
17. Van Wyk, N.; Grossmann, M.; Wendland, J.; Von Wallbrunn, C.; Pretorius, I.S. The Whiff of Wine Yeast Innovation: Strategies for Enhancing Aroma Production by Yeast during Wine Fermentation. *J. Agric. Food Chem.* **2019**, *67*, 13496–13505. [CrossRef]
18. Badura, J.; Kiene, F.; Brezina, S.; Fritsch, S.; Semmler, H.; Rauhut, D.; Pretorius, I.S.; Von Wallbrunn, C.; van Wyk, N. Aroma Profiles of Vitis Vinifera L. Cv. Gewürztraminer Must Fermented with Co-Cultures of *Saccharomyces cerevisiae* and Seven *Hanseniaspora* Spp. *Fermentation* **2023**, *9*, 109. [CrossRef]

19. Lachance, M.-A.; Pupovac-Velikonja, A.; Natarajan, S.; Schlag-Edler, B. Nutrition and Phylogeny of Predacious Yeasts. *Can. J. Microbiol.* **2000**, *46*, 495–505. [CrossRef]
20. Kurtzman, C.P.; Smith, M.T. Saccharomycopsis Schiönning (1903). In *The Yeasts*; Elsevier: Amsterdam, The Netherlands, 2011; pp. 751–763.
21. Hesselbart, A.; Junker, K.; Wendland, J. Draft Genome Sequence of *Saccharomycopsis fermentans* CBS 7830, a Predacious Yeast Belonging to the *Saccharomycetales*. *Genome Announc.* **2018**, *6*, e01445-17. [CrossRef]
22. Kayacan, Y.; Griffiths, A.; Wendland, J. A Script for Initiating Molecular Biology Studies with Non-Conventional Yeasts Based on *Saccharomycopsis schoenii*. *Microbiol. Res.* **2019**, *229*, 126342. [CrossRef] [PubMed]
23. Dutra-Silva, L.; Pereira, G.E.; Batista, L.R.; Matteoli, F.P. Fungal Diversity and Occurrence of Mycotoxin Producing Fungi in Tropical Vineyards. *World J. Microbiol. Biotechnol.* **2021**, *37*, 112. [CrossRef] [PubMed]
24. Park, E.Y.; Kim, M.-D. Draft Genome Sequence Data of the Starch-Utilizing Yeast *Saccharomycopsis fibuligera* MBY1320 Isolated from Nuruk. *Data Brief* **2021**, *35*, 106888. [CrossRef] [PubMed]
25. Xie, Z.-B.; Zhang, K.; Kang, Z.; Yang, J. *Saccharomycopsis fibuligera* in Liquor Production: A Review. *Eur. Food Res. Technol.* **2021**, *247*, 1569–1577. [CrossRef]
26. Rensburg, P.; Stidwell, T.; Lambrechts, M.; Cordero-Otero, R.; Pretorius, I. Development and assessment of a recombinant *Saccharomyces cerevisiae* wine yeast producing two aroma-enhancing β-glucosidase encoded by the *Saccharomycopsis fibuligera* BGL1 and BGL2 genes. *Ann. Microbiol.* **2005**, *55*, 33–42.
27. Methner, Y.; Magalhães, F.; Raihofer, L.; Zarnkow, M.; Jacob, F.; Hutzler, M. Beer Fermentation Performance and Sugar Uptake of *Saccharomycopsis fibuligera*–A Novel Option for Low-Alcohol Beer. *Front. Microbiol.* **2022**, *13*, 1011155. [CrossRef]
28. Methner, Y.; Hutzler, M.; Matoulková, D.; Jacob, F.; Michel, M. Screening for the Brewing Ability of Different Non-Saccharomyces Yeasts. *Fermentation* **2019**, *5*, 101. [CrossRef]
29. Methner, Y.; Hutzler, M.; Zarnkow, M.; Prowald, A.; Endres, F.; Jacob, F. Investigation of non-Saccharomyces yeast strains for their suitability for the production of non-alcoholic beers with novel flavor profiles. *J. Am. Soc. Brew. Chem.* **2022**, *80*, 341–355. [CrossRef]
30. Takagi, H.; Hashida, K.; Watanabe, D.; Nasuno, R.; Ohashi, M.; Iha, T.; Nezuo, M.; Tsukahara, M. Isolation and Characterization of Awamori Yeast Mutants with L-Leucine Accumulation That Overproduce Isoamyl Alcohol. *J. Biosci. Bioeng.* **2015**, *119*, 140–147. [CrossRef]
31. Schneider, A.; Gerbi, V.; Redoglia, M. A Rapid HPLC Method for Separation and Determination of Major Organic Acids in Grape Musts and Wines. *Am. J. Enol. Vitic.* **1987**, *38*, 151–155. [CrossRef]
32. Scansani, S.; van Wyk, N.; Bou Nader, K.; Beisert, B.; Brezina, S.; Fritsch, S.; Semmler, H.; Pasch, L.; Pretorius, I.S.; von Wallbrunn, C.; et al. The film-forming *Pichia* spp. in a winemaker's toolbox: A simple isolation procedure and their performance in a mixed-culture fermentation of *Vitis vinifera* L. cv. Gewürztraminer must. *Int. J. Food Microbiol.* **2022**, *365*, 109549. [CrossRef] [PubMed]
33. Tarasov, A.; Garzelli, F.; Schuessler, C.; Fritsch, S.; Loisel, C.; Pons, A.; Patz, C.-D.; Rauhut, D.; Jung, R. Wine Storage at Cellar vs. Room Conditions: Changes in the Aroma Composition of Riesling Wine. *Molecules* **2021**, *26*, 6256. [CrossRef]
34. Krause, B.; Löhnertz, O. Saccharopin und Pipecolinsäure—Diagnostische Biomarker in der klassischen Aminosäureanalytik. *MTA Dialog* **2017**, *18*, 316–321. [CrossRef]
35. Oba, T.; Nomiyama, S.; Hirakawa, H.; Tashiro, K.; Kuhara, S. Asp578 in LEU4p Is One of the Key Residues for Leucine Feedback Inhibition Release in Sake Yeast. *Biosci. Biotechnol. Biochem.* **2005**, *69*, 1270–1273. [CrossRef]
36. Abe, T.; Toyokawa, Y.; Sugimoto, Y.; Azuma, H.; Tsukahara, K.; Nasuno, R.; Watanabe, D.; Tsukahara, M.; Takagi, H. Characterization of a New *Saccharomyces cerevisiae* Isolated From Hibiscus Flower and Its Mutant With L-Leucine Accumulation for Awamori Brewing. *Front. Genet.* **2019**, *10*, 490. [CrossRef] [PubMed]
37. Link, A.J.; Mock, M.L.; Tirrell, D.A. Non-Canonical Amino Acids in Protein Engineering. *Curr. Opin. Biotechnol.* **2003**, *14*, 603–609. [CrossRef]
38. Shen, L.; Nishimura, Y.; Matsuda, F.; Ishii, J.; Kondo, A. Overexpressing Enzymes of the Ehrlich Pathway and Deleting Genes of the Competing Pathway in *Saccharomyces cerevisiae* for Increasing 2-Phenylethanol Production from Glucose. *J. Biosci. Bioeng.* **2016**, *122*, 34–39. [CrossRef]
39. Atsumi, S.; Hanai, T.; Liao, J.K. Non-Fermentative Pathways for Synthesis of Branched-Chain Higher Alcohols as Biofuels. *Nature* **2008**, *451*, 86–89. [CrossRef]
40. Vuralhan, Z.; Luttik, M.a.H.; Tai, S.L.; Boer, V.M.; De Morais, M.A.; Schipper, D.; Almering, M.J.H.; Kötter, P.; Dickinson, J.R.; Daran, J.-M.; et al. Physiological Characterization of the *ARO10*-Dependent, Broad-Substrate-Specificity 2-Oxo Acid Decarboxylase Activity of *Saccharomyces cerevisiae*. *Appl. Environ. Microbiol.* **2005**, *71*, 3276–3284. [CrossRef]
41. Yuan, J.; Mishra, P.; Ching, C.B. Engineering the Leucine Biosynthetic Pathway for Isoamyl Alcohol Overproduction in *Saccharomyces cerevisiae*. *J. Ind. Microbiol. Biotechnol.* **2017**, *44*, 107–117. [CrossRef]
42. Herschorn, S.; Kim, S.-K.; Lee, Y.-G.; Park, K.W.; Seo, J.-H. Elimination of Biosynthetic Pathways for L-Valine and l-Isoleucine in Mitochondria Enhances Isobutanol Production in Engineered *Saccharomyces cerevisiae*. *Bioresour. Technol.* **2018**, *268*, 271–277. [CrossRef]

43. Wess, J.; Brinek, M.; Boles, E. Improving Isobutanol Production with the Yeast *Saccharomyces cerevisiae* by Successively Blocking Competing Metabolic Pathways as Well as Ethanol and Glycerol Formation. *Biotechnol. Biofuels* **2019**, *12*, 1–15. [CrossRef] [PubMed]
44. Saliba, A.; Ovington, L.; Moran, C.C. Consumer Demand for Low-Alcohol Wine in an Australian Sample. *Int. J. Wine Res.* **2013**, *1*, 1–8. [CrossRef]

Disclaimer/Publisher's Note: The statements, opinions and data contained in all publications are solely those of the individual author(s) and contributor(s) and not of MDPI and/or the editor(s). MDPI and/or the editor(s) disclaim responsibility for any injury to people or property resulting from any ideas, methods, instructions or products referred to in the content.

fermentation

MDPI

Communication

Candida albicans Adhesion Measured by Optical Nanomotion Detection

Maria I. Villalba [1,2], Salomé LeibundGut-Landmann [3], Marie-Elisabeth Bougnoux [4,5], Christophe d'Enfert [4], Ronnie G. Willaert [2,6] and Sandor Kasas [1,2,7,*]

1 Laboratory of Biological Electron Microscopy, Swiss Federal Institute of Technology Lausanne (École Polytechnique Fédérale de Lausanne), 1015 Lausanne, Switzerland; ines.villalba@epfl.ch
2 International Joint Research Group VUB-EPFL NanoBiotechnology & NanoMedicine (NANO), Vrije Universiteit Brussel, 1050 Brussels, Belgium; Swiss Federal Institute of Technology Lausanne (Ecole Polytechnique Fédérale de Lausanne), 1015 Lausanne, Switzerland; ronnie.willaert@vub.be
3 Section of Immunology, Vetsuisse Faculty, and Institute of Experimental Immunology, University of Zurich, 8006 Zurich, Switzerland; salome.leibundgut-landmann@uzh.ch
4 Unité Biologie et Pathogénicité Fongiques, Institut Pasteur, Université Paris Cité, INRAE USC 2019, 75015 Paris, France; marie-elisabeth.bougnoux@aphp.fr (M.-E.B.); christophe.denfert@pasteur.fr (C.d.)
5 Assistance Publique des Hôpitaux de Paris (APHP), Unité de Parasitologie-Mycologie, Service de Microbiologie Clinique, Hôpital Necker-Enfants-Malades, 75015 Paris, France
6 Research Group Structural Biology Brussels (SBB), Alliance Research Group VUB-UGent NanoMicrobiology (NAMI), Vrije Universiteit Brussel, 1050 Brussels, Belgium
7 Centre Universitaire Romand de Médecine Légale, UFAM, Université de Lausanne, 1015 Lausanne, Switzerland
* Correspondence: sandor.kasas@epfl.ch

Abstract: Cellular adhesion plays an important role in numerous fundamental physiological and pathological processes. Its measurement is relatively complex, requires sophisticated equipment, and, in most cases, cannot be carried out without breaking the links between the studied cell and its target. In this contribution, we propose a novel, nanomotion-based, technique that overcomes these drawbacks. The applied force is generated by the studied cell itself (nanomotion), whereas cellular movements are detected by traditional optical microscopy and dedicated software. The measurement is non-destructive, single-cell sensitive, and permits following the evolution of the adhesion as a function of time. We applied the technique on different strains of the fungal pathogen *Candida albicans* on a fibronectin-coated surface. We demonstrated that this novel approach can significantly simplify, accelerate, and make more affordable living cells–substrate adhesion measurements.

Keywords: *Candida albicans*; cellular nanomotion; optical nanomotion detection; yeast adhesion; fibronectin

Citation: Villalba, M.I.; LeibundGut-Landmann, S.; Bougnoux, M.-E.; d'Enfert, C.; Willaert, R.G.; Kasas, S. *Candida albicans* Adhesion Measured by Optical Nanomotion Detection. *Fermentation* **2023**, *9*, 991. https://doi.org/10.3390/fermentation9110991

Academic Editor: Patrizia Romano

Received: 10 October 2023
Revised: 3 November 2023
Accepted: 13 November 2023
Published: 20 November 2023

Copyright: © 2023 by the authors. Licensee MDPI, Basel, Switzerland. This article is an open access article distributed under the terms and conditions of the Creative Commons Attribution (CC BY) license (https://creativecommons.org/licenses/by/4.0/).

1. Introduction

Adhesion is a fundamental property of living cells that permits their attachment to other organisms or various organic and inorganic substrates. It plays a fundamental role in numerous physiological and pathological processes such as cell growth, migration, immune response, pathogen-host interaction, and tumor cell growth and spreading [1–5]. Despite the existence of different techniques to measure adhesion, its quantification is still a challenge [6–9]. A comprehensive review of cellular adhesion's importance in physiological and pathological processes, as well as measurement techniques, can be found in [10]. Most of the existing techniques rely on the detachment of the cells upon an applied force. Liquid flow (shear stress) [11–14], centrifugal acceleration [15,16], micropipette manipulation [17,18], optical tweezers [19,20], atomic force microscopy [6,21,22] or FRET force sensors [23] can generate this force. However, applying it with a force in the range of nano-newton on single cells is not a trivial task, and, therefore, the setup of such measurements is relatively complex and relies on expensive equipment. In addition, detachment-based measurements do not permit obtaining information about the adhesion process as a function of time at a single-cell level. An alternative to the force-applying devices is the use of planar optical waveguides that monitor the contact surface between the living organism and the substrate [24]. This last approach requires specially treated surfaces and only informs about the surface of contact between the cell and its substrate.

In recent years, it was highlighted that all living organisms oscillate at a nanometric scale as long as they are alive. These oscillations, referred to as nanomotion, since the displacements are in the nanometer range, stop as soon as the organism is dead. Nanomotion exists in virtually all living organisms on earth [25–30]. Its monitoring is nowadays applied to rapid antibiotic sensitivity tests; for example, if a bacterium is sensitive to a given drug, its nanomotion rapidly stops or decreases upon antibiotic exposure [30–33]. Nanomotion was initially highlighted by atomic force microscopy (AFM), but later it appeared that classical optical microscopes equipped with a camera also detect nanomotion [34]. The technique consists of recording a movie of the living organism and processing it with motion detection dedicated software that allows tracking bacterial or fungal displacements with a sub-pixel resolution; therefore, the method is referred to as optical nanomotion detection (ONMD). This last development dramatically increased the availability of the technique and its accessibility in terms of cost and complexity of use. The origin of cellular nanomotion is still under debate; the opening and closing of ion channels, rearrangements of the cytoskeleton, or conformational changes of membrane proteins certainly participate in its generation [35].

In the present study, we introduced a novel nanomotion-based method to measure single-cell adhesion of the human fungal pathogen *Candida albicans* on flat surfaces. The technique relies on the detection of cellular nanomotion, and is therefore essentially limited to living organisms. We focused in this manuscript to the adhesion to fibronectin since it is a physiologically relevant molecule that is present in patients' basal lamina [36] and plays a role in *Candida* adhesion in immunocompromised patients. It has been shown that fibronectin is an important protein ligand of the host extracellular matrix (ECM) that plays an essential role in *C. albicans* adhesion [37–40]. Furthermore, targeting fibronectin has shown to alter *C. albicans* biofilm formation [41]. The nanomotion-based adhesion measurement relied on the position of the studied cell (a *C. albicans* cell, in this study) as a function of time. Depending on the attachment force and the length of the link between the cell and the substrate, the displacements of the cell will be more or less constrained, i.e., the displacement envelope of a cell is inversely proportional to the adhesive force and the length of the anchoring linker. A strongly attached organism will be constrained to a smaller area than a loosely attached one. To confirm this hypothesis, we examined the nanomotion amplitude of different *C. albicans* strains deposited on an optical quality Petri dish coated with fibronectin. We measured the amplitude of nanomotion of the different strains and compared it to the values obtained by classical adhesion tests. It appeared that our hypothesis was confirmed, and strongly attached cells had a lower nanomotion amplitude than those that were loosely attached.

2. Materials and Methods

2.1. Yeast Cell Culture

In this study, *C. albicans* strains CEC3672 [42], CEC3609 [43], CEC3678 [42], CEC3621 [44], SC5314 [45], 101 [46] and CEC3675 [42] were used. The yeasts were plated from cryo stocks on yeast-extract peptone–dextrose (YPD) agar and cultured for 24 h at 30 °C. YPD was composed of d-glucose (Gibco™ 15023021, Grand Island, NY, USA) 20 g/L, peptone (Millipore 82303, Burlington, MA, USA) 20 g/L, yeast extract 10 g/L, and agar (plates only) 20 g/L, which was diluted in distilled water. Single colonies from the agar plates were then grown in 4 mL YPD liquid medium at 30 °C, shaking at 160 rpm overnight.

2.2. Adhesion Assay

Two different methods were used to measure *C. albicans* adhesion, as illustrated in Figure 1. The first method [46–48] consisted of incubating freely floating yeast cells in wells coated with fibronectin (Figure 1A). A 6-well culture plate (Thermo Scientific™-Nunc™, 140675, Waltham, MA, USA) was incubated for an hour with fibronectin (50 μg/mL) (fibronectin human plasma, lyophilized powder, Sigma-Aldrich, F2006, St. Louis, MO, USA). A PBS wash was performed following the incubation with fibronectin. The liquid was gently removed, and 250 cells suspended in 500 μL of YPD medium were added to each well. After an incubation at 30 °C without shaking during 25 min, the supernatant containing the freely floating cells was removed and inoculated in a new YPD agar Petri dish (Figure 1B); attached cells were immersed in Sabourand dextrose (SDA) agar, (Figure 1C), by adding melted agar at 40 °C. SDA was composed of d-glucose (Gibco™ 15023021, Grand Island, NY, USA) 40 g/L, peptone (Millipore 82303, Burlington, MA, USA) 10 g/L, and agar 15 g/L, which was diluted in distilled water. After an incubation of 48 h the yeast cells in both Petri dishes formed visible colonies. The % adhesion was calculated by dividing the number of grown cells in SDA agar by the total number of grown cells (in both YPD and SDA agar) for each strain.

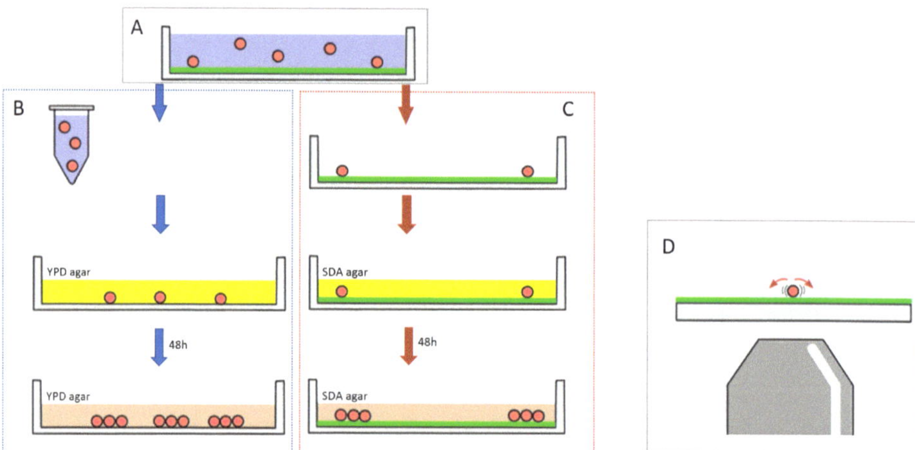

Figure 1. Two different techniques are used to measure adhesion: (**A–C**) Classical technique based on the ratio between floating and adhering cells and (**D**) nanomotion-based measurement. The classical method (**A–C**) consists of depositing yeast cells onto a fibronectin-coat Petri dish (**A**). The supernatant is than removed (**B**), then deposited into a new Petri dish with YPD agar. The single cells eventually divide and grow into colonies that enable the estimation their number. The cells that remained attached to the fibronectin covered Petri dish (**C**) were also covered with SDA agar, and developed CFUs, too. The nanomotion-based measurements (**D**) were performed by recording movies of 10 s.

2.3. Optical Nanomotion Adhesion Method

ONMD was performed by recording 10 s long movies with an inverted optical microscope (Zeiss Observer Z.1) of the attachment of *C. albicans* cells (Figure 1D). We used a 63× oil immersion objective, and a PCO Edge 5.5 camera. Petri dishes (μ-Dish 35 mm, low, uncoated, 80131 IBIDI) were incubated for an hour with fibronectin (50 μg/mL) (Sigma-Aldrich, F2006, St. Louis, MO, USA), followed by washing with PBS, according to the protocol recommended at www.ibidi.com. The liquid was removed, and 1.5 mL of the YPD liquid medium with 20 μL of overnight culture was added to the Petri dish with the fibronectin layer. Two minutes after placing the plates under the microscope, the first video (time 0) was recorded, followed by videos at 5, 10, 20, 30, and 40 min. All the measurements were carried out at room temperature without phase contrast nor fluorescent staining.

The camera typically recorded 10 s long videos (AVI format) at a frame rate of 30 frames per second (fps). The movies were processed with a custom-made tracking algorithm [34] implemented in Matlab, which tracked individual cells in 2D along 100–200 frames. Only individual cells (those that were not in contact with other cells) were selected. The algorithm reached a sub-pixel resolution. The Matlab program is freely available upon request. The total displacements of the cells were calculated and the average value represented the optical nanomotion signal [34].

2.4. Constrained Random Walk Simulation

A fixed step length random walk [49] was simulated with a Matlab (R2023a) computer program. We simulated a massless particle submitted to a random walk during 50,000 time cycles. We carried out two simulation rounds: one with a "rope" of 80 arbitrary length units, which restrains the distance to which the particle can diffuse, and a second round that simulates a free, "ropeless" particle diffusion. The random walk simulation consisted of randomly choosing a displacement direction at every simulation step and displacing the particle in that direction with an arbitrary unit length. In the "particle attached to a rope" simulation, the software checked the distance between the spot from which the simulation started (in our case x = 0 and y = 0) and the particle's actual position. If it exceeds that of the rope, another random displacement direction is tested and "refused", unless it decreases the distance between the particle and its starting position.

2.5. Statistical Analysis

Statistical analysis was performed using OriginPro, version 2021. We used the *t*-test to analyze the significance of the results (* $p < 0.05$). Each experiment was replicated at least three times. Optical nanomotion was analyzed for each case using more than 30 individual cells. For the adhesion assay, more than eight replicates were conducted.

3. Results

To confirm our hypothesis, according to which a low adhesion force corresponds to a higher displacement freedom, we carried out constrained and non-constrained random walk simulations. If assuming that the grid lines of the graphs delimit single pixels, Figure 2 clearly shows that a constrained particle "visits" (gray color squares) much less pixels than a free one. In the present simulation, only four pixels were visited by the attached particle, whereas 13 were visited by the free one.

We then compared seven different *C. albicans* strains by both the classical cell attachment assays and ONMD. Figure 3 displays the results obtained by the two methods after a 40 min-long attachment period. These results confirm our working hypothesis that strains whose cells have a low attachment capacity (located below the 50% adhesion line) have cells that display more freedom in their displacements (located above the blue dashed horizontal line). Inversely, *C. albicans* strains that show strong cell binding (located above the brown dashed line) display a smaller ONMD (are situated below the blue dashed line of the upper graph).

To highlight a putative correlation between classical adhesion tests and ONMD measurements, we displayed both data sets onto the same graph and fitted it with a first order polynomial as depicted in Figure 4. The R-square of the linear fit is 0.77, which indicates a high correlation.

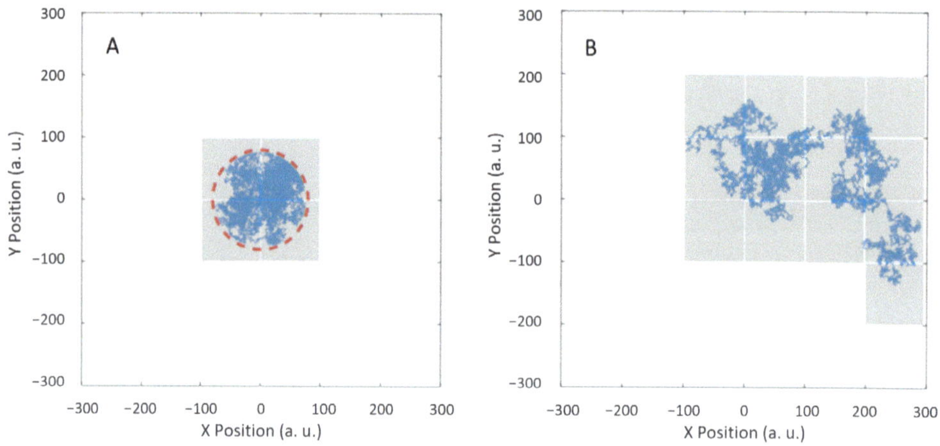

Figure 2. Random walk simulations of a constrained (**A**) and non-constrained (**B**) particle during 50,000 time steps. The "rope" that constrained the particle in frame A. had a length of 80 a.u. and is attached at the center of the graph. The red circle represents the maximum distance to which the particle can move if constrained by the rope. The number of pixels (gray squares) visited by the particle is higher in the non-constrained case (n = 13) than in the constrained one (n = 4).

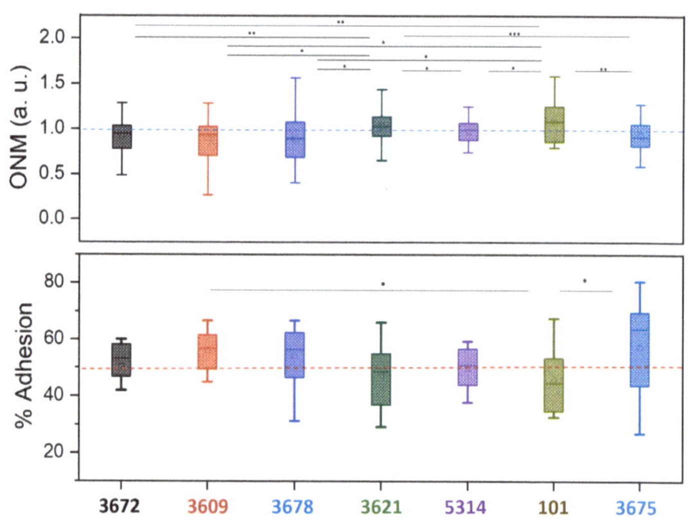

Figure 3. Adhesion results for 7 different *C. albicans* strains. The lower panel represents traditional adhesion test results, whereas the upper panel shows the corresponding ONMD data. The upper horizontal blue dashed line indicates the average displacement value of the cells at the beginning of the ONMD experiments (t = 0 min). The lower red dashed horizontal line indicates that 50% of the cells are attached to the fibronectin-coated surface in the adhesion tests. * $p < 0.05$; ** $p < 0.01$; *** $p < 0.001$. Number of cells per strain measured in ONM experiments: 3672 = 37, 3609 = 28, 3678 = 28, 3621 = 117, 5314 = 52, 101 = 28 and 3675 = 60. Number of replicates per strain measured in adhesion experiments: 3672 = 10, 3609 = 12, 3678 = 11, 3621 = 7, 5314 = 9, 101 = 11 and 3675 = 19.

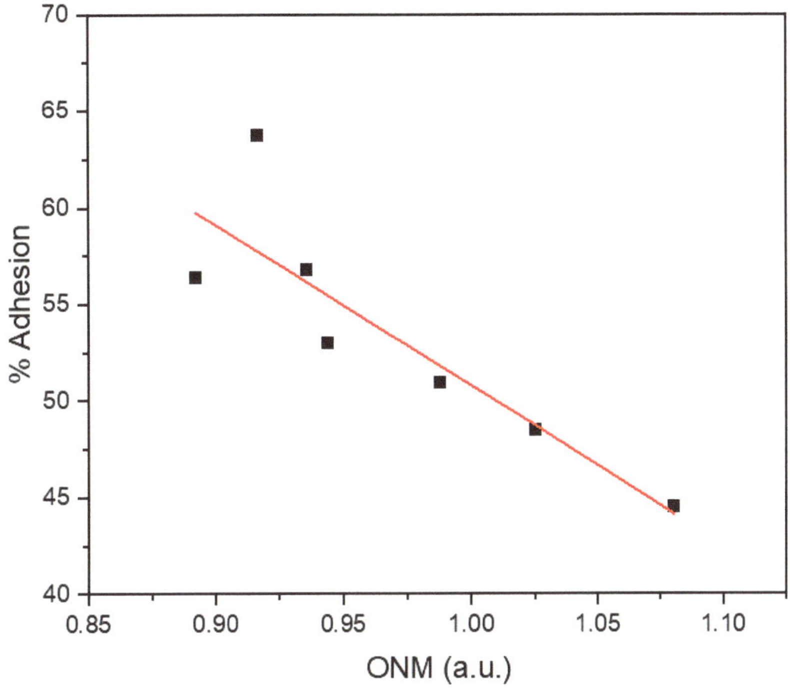

Figure 4. Linear fit of the traditional (percentage of adhesion) vs. the ONMD adhesion measurements results. The fitting equation has a slope-intercept form (y = a + bx), with intercept (a): 133.72 ± 19.88 and slope (b): −82.90 ± 20.47.

As mentioned previously, ONMD permits to monitor the adhesion force as a function of time. Figure 5 shows the evolution of the nanomotion of different *C. albicans* strains during 40 min. As visible on this graph, the different strains behave differently upon exposure to fibronectin-coated surfaces. Certain strains increase their adhesion (reduction of the optical nanomotion signal) as a function of time, such as strains CEC3672, CEC3609, and CEC3678, whereas others, such as CEC101, reduce adhesion, and some, CEC3621, CEC5314, and CEC3675, show hybrid behavior during the measurement period. This type of information is relatively easily accessible by ONMD. It provides valuable information about yeast attachment dynamics and subtle differences in the attachment processes of the different strains.

The free and constrained random walk simulations demonstrated that free cells cover a larger distance than attached ones. To confirm this behavior with attached living cells, we displayed the displacement trajectories of a strongly (3672) and a poorly attached (101) cell. Figure 6 displays the trajectories of the two cells during 200 frames (i.e., about 6.7 s).

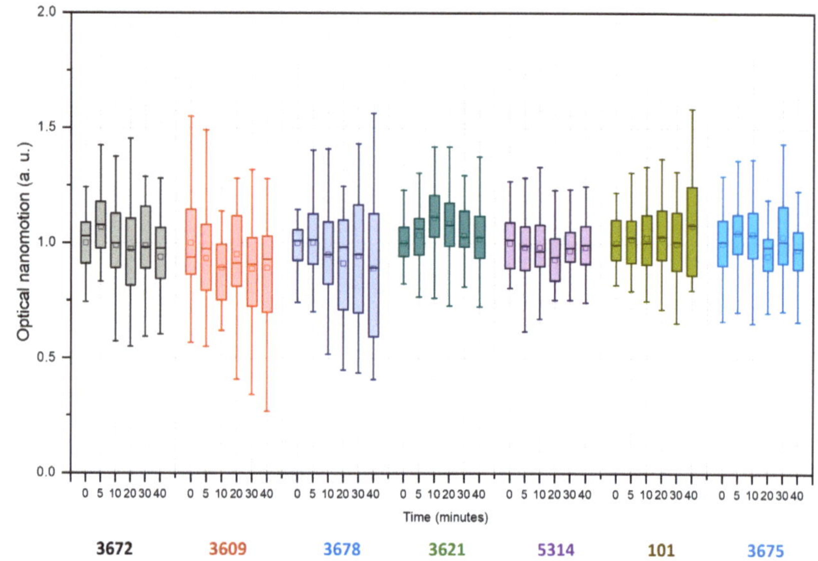

Figure 5. Evolution of the nanomotion of different *C. albicans* strains as a function of time. As shown, the different strains behave differently upon exposure to fibronectin-coated surfaces. The adhesion values for each strain are normalized to time 0. Number of cells: 3672: 0 min = 27, 5 min = 32, 10 min = 31, 20 min = 35, 30 min = 31 and 40 min = 32; 3609: 0 min = 35, 5 min = 37, 10 min = 37, 20 min = 33, 30 min = 27 and 40 min = 27; 3678: 0 min = 24, 5 min = 27, 10 min = 23, 20 min = 25, 30 min = 24 and 40 min = 24; 3621: 0 min = 54, 5 min = 59, 10 min = 69, 20 min = 74, 30 min = 71 and 40 min = 71; 5314: 0 min = 42, 5 min = 60, 10 min = 61, 20 min = 36, 30 min = 25 and 40 min = 35; 101: 0 min = 45, 5 min = 52, 10 min = 50, 20 min = 46, 30 min = 31 and 40 min = 28; 3675: 0 min = 41, 5 min = 42, 10 min = 48, 20 min = 37, 30 min = 35 and 40 min = 25.

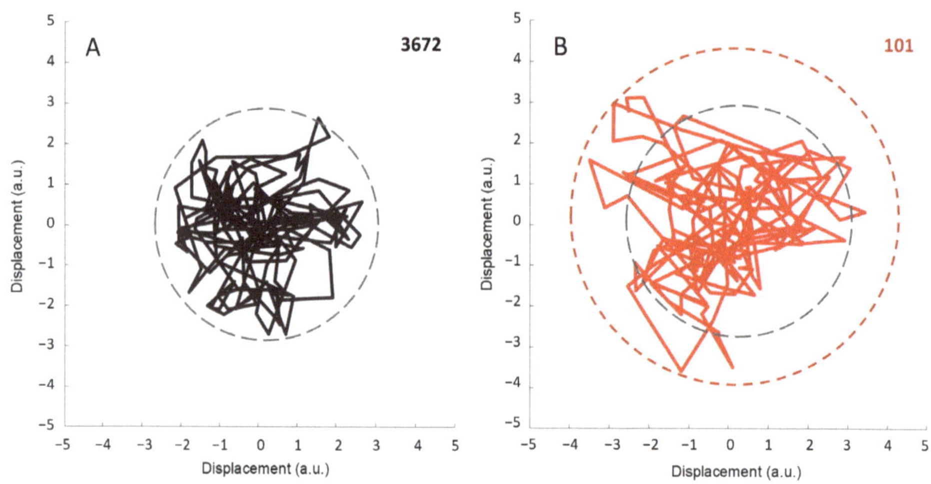

Figure 6. Typical path of single (**A**) CEC3672 and (**B**) 101 *C. albicans* cells. As depicted in this figure, the lower adherent cell (101- shown in red color) moves to larger distances than the higher adherent one (CEC3672- shown in gray color). The dashed circles correspond to the maximal distance reached by both cell types. In (**B**), both embedding circles are displayed for comparison.

4. Discussion

According to our experiments, ONMD can be used to determine the adhesion of yeast cells onto flat surfaces in a very rapid, cost-effective, and experimentally simple way. Recording by classical optical microscopy the nanomotion of living cells and processing the movies with a dedicated software permits not only to determine the adhesion, but also follow its modifications as a function of time. The technique is very straightforward and does not require any force-applying devices. The living cells' spontaneous nanomotion serves as a force generator, and the cellular attachment to the substrate constrains cellular displacements. Importantly, the measurement does not break the cell–substrate bond, and therefore enables measuring the adhesion as a function of time or as a function of various chemicals added in the analysis chamber at a given experimental time point. The evolution of the adhesion as a function of time is a poorly known parameter that is relatively difficult to measure with classical adhesion tests. It most likely reflects the speed at which different adhesive molecules that are present on the cell wall bind to the substrate, and could be a pertinent parameter to identify a given strain or inform about the invasive potential of the studied strain. These last two hypotheses would need additional experiments to be confirmed or invalidated.

The proposed technique has numerous advantages compared to traditional adhesion tests: it is extremely rapid, does not require any force-applying device, does not destroy the cell–substrate bond during the measurement, and can provide information on the adhesion evolution as a function of time. The experimental setup is also very simple and is limited to a traditional optical microscope equipped with a camera and dedicated image-processing software. Among the disadvantages, we can cite the difficulty, for the moment, of extracting absolute adhesion force values limiting the technique to relative measurements. Additional fluid dynamic studies, considering the geometry of the studied cell in addition to other physical parameters, could lead to some models that might correlate absolute adhesion values to nanomotion. A possible alternative could be calibration by an absolute adhesion measurement such as AFM, and its correlation with the nanomotion values.

Another limitation of the technique is the difficulty to separate the nanomotion itself from adhesion. Nanomotion measurements on cell-repellant surfaces could serve as a calibration technique to which the nanomotion on adherent surfaces could be normalized.

5. Conclusions

We introduced a new method to detect the dynamics of cellular adhesion. The new method is based on observing the cellular nanomotion of single cells by optical microscopy during the adhesion process. As a proof of concept, we successfully demonstrated the dynamics of *C. albicans'* adhesion to a fibronectin-coated surface. Among the advantages of the technique, we can mention its rapidity, the simplicity of the setup and the low cost of the required instrumentation. The technique is also cost effective, since the very same cell type can be employed in the very same experimental conditions to compare different adhesive surfaces. Importantly, the method offers the possibility of monitoring the evolution of the adhesion as a function of time without the need to destroy the link between the living organism and its substrate.

C. albicans is known to become a pathogen in patients having an impaired immune system. Its ability to adhere to catheter surfaces can be considered as a fundamental virulence feature [50]. The proposed technique could be very useful to rapidly evaluate *C. albicans'* adhesive properties on different medical devices. We are convinced that the technique can also be applied to bacterial adhesion too. In this last case, it would dramatically increase its application spectrum to fields such as medical material development, or antibacterial surface treatments.

Author Contributions: Conceptualization, S.K. and M.I.V.; methodology, S.K., M.I.V. and R.G.W.; software, S.K.; validation, M.I.V., S.L.-L., M.-E.B., C.d. and R.G.W.; investigation, S.K. and M.I.V.; resources, S.L.-L., M.-E.B., C.d., R.G.W. and S.K.; data curation, M.I.V.; writing—original draft preparation, S.K.; writing—review and editing, M.I.V., S.L.-L., M.-E.B., C.d., R.G.W. and S.K.; visualization, S.K.; supervision, S.K., R.G.W., S.L.-L. and C.d.; project administration, S.K.; funding acquisition, S.L.-L., M.-E.B., C.d., R.G.W. and S.K. All authors have read and agreed to the published version of the manuscript.

Funding: This research was funded by the Belgian Federal Science Policy Office (Belspo) and the European Space Agency, grant number PRODEX project Flumias Nanomotion; The Research Foundation—Flanders (FWO), grant numbers AUGE/13/19 and I002620; FWO-SNSF, grant number 310030L_197946, M.I.V., and S.K., S.L.-L. and C.d. were supported by Swiss National Science Foundation (SNSF) grant CRSII5_173863. Work in the laboratory of C.d. is supported by the Agence Nationale de Recherche (ANR-10-LABX-62-IBEID).

Institutional Review Board Statement: Not applicable.

Informed Consent Statement: Not applicable.

Data Availability Statement: The data presented in this study are available on request from the corresponding author.

Acknowledgments: The authors thank R. Horváth and I. Székács for highly constructive discussions.

Conflicts of Interest: The authors declare no conflict of interest.

References

1. Busscher, H.J.; van der Mei, H.C. Microbial Adhesion in Flow Displacement Systems. *Clin. Microbiol. Rev.* **2006**, *19*, 127–141. [CrossRef] [PubMed]
2. Achinas, S.; Charalampogiannis, N.; Euverink, G.J.W. A Brief Recap of Microbial Adhesion and Biofilms. *Appl. Sci.* **2019**, *9*, 2801. [CrossRef]
3. Di Martino, P. Bacterial Adherence: Much More than a Bond. *AIMS Microbiol.* **2018**, *4*, 563–566. [CrossRef] [PubMed]
4. Janiszewska, M.; Primi, M.C.; Izard, T. Cell Adhesion in Cancer: Beyond the Migration of Single Cells. *J. Biol. Chem.* **2020**, *295*, 2495–2505. [CrossRef]
5. Mackay, C.R.; Imhof, B.A. Cell Adhesion in the Immune System. *Immunol. Today* **1993**, *14*, 99–102. [CrossRef]
6. Gunaratnam, G.; Dudek, J.; Jung, P.; Becker, S.L.; Jacobs, K.; Bischoff, M.; Hannig, M. Quantification of the Adhesion Strength of Candida Albicans to Tooth Enamel. *Microorg.* **2021**, *9*, 2213. [CrossRef]
7. Hu, L.; Zhang, X.; Miller, P.; Ozkan, M.; Ozkan, C.; Wang, J. Cell Adhesion Measurement by Laser-Induced Stress Waves. *J. Appl. Phys.* **2006**, *100*, 084701. [CrossRef]
8. Ahmad Khalili, A.; Ahmad, M.R. A Review of Cell Adhesion Studies for Biomedical and Biological Applications. *Int. J. Mol. Sci.* **2015**, *16*, 18149–18184. [CrossRef]
9. Zhou, D.W.; García, A.J. Measurement Systems for Cell Adhesive Forces. *J. Biomech. Eng.* **2015**, *137*, 0209081–0209088. [CrossRef]
10. Ungai-Salanki, R.; Peter, B.; Gerecsei, T.; Orgovan, N.; Horvath, R.; Szabo, B. A Practical Review on the Measurement Tools for Cellular Adhesion Force. *Adv. Colloid Interface Sci.* **2019**, *269*, 309–333. [CrossRef]
11. Truskey, G.A.; Pirone, J.S. The Effect of Fluid Shear Stress upon Cell Adhesion to Fibronectin-Treated Surfaces. *J. Biomed. Mater. Res.* **1990**, *24*, 1333–1353. [CrossRef]
12. Kucik, D.F. Measurement of Adhesion Under Flow Conditions. *Curr. Protoc. Cell Biol.* **2009**, *43*, 9.6.1–9.6.10. [CrossRef] [PubMed]
13. Goldstein, A.S.; DiMilla, P.A. Effect of Adsorbed Fibronectin Concentration on Cell Adhesion and Deformation under Shear on Hydrophobic Surfaces. *J. Biomed. Mater. Res.* **2002**, *59*, 665–675. [CrossRef] [PubMed]
14. McCall, A.D.; Pathirana, R.U.; Prabhakar, A.; Cullen, P.J.; Edgerton, M. Candida Albicans Biofilm Development Is Governed by Cooperative Attachment and Adhesion Maintenance Proteins. *Npj Biofilms Microbiomes* **2019**, *5*, 21. [CrossRef] [PubMed]
15. Christ, K.V.; Turner, K.T. Methods to Measure the Strength of Cell Adhesion to Substrates. *J. Adhes. Sci. Technol.* **2010**, *24*, 2027–2058. [CrossRef]
16. Giacomello, E.; Neumayer, J.; Colombatti, A.; Perris, R. Centrifugal Assay for Fluorescence-Based Cell Adhesion Adapted to the Analysis of Ex Vivo Cells and Capable of Determining Relative Binding Strengths. *Biotech.* **1999**, *26*, 758–762, 764–766. [CrossRef]
17. Chu, Y.-S.; Thomas, W.A.; Eder, O.; Pincet, F.; Perez, E.; Thiery, J.P.; Dufour, S. Force Measurements in E-Cadherin–Mediated Cell Doublets Reveal Rapid Adhesion Strengthened by Actin Cytoskeleton Remodeling through Rac and Cdc42. *J. Cell Biol.* **2004**, *167*, 1183–1194. [CrossRef]
18. Daoudi, M.; Lavergne, E.; Garin, A.; Tarantino, N.; Debré, P.; Pincet, F.; Combadière, C.; Deterre, P. Enhanced Adhesive Capacities of the Naturally Occurring Ile249–Met280 Variant of the Chemokine Receptor CX3CR1. *J. Biol. Chem.* **2004**, *279*, 19649–19657. [CrossRef]
19. Thoumine, O.; Kocian, P.; Kottelat, A.; Meister, J.-J. Short-Term Binding of Fibroblasts to Fibronectin: Optical Tweezers Experiments and Probabilistic Analysis. *Eur. Biophys. J.* **2000**, *29*, 398–408. [CrossRef]
20. Castelain, M.; Rouxhet, P.G.; Pignon, F.; Magnin, A.; Piau, J.-M. Single-Cell Adhesion Probed in-Situ Using Optical Tweezers: A Case Study with Saccharomyces Cerevisiae. *J. Appl. Phys.* **2012**, *111*, 114701. [CrossRef]

21. Benoit, M.; Gabriel, D.; Gerisch, G.; Gaub, H.E. Discrete Interactions in Cell Adhesion Measured by Single-Molecule Force Spectroscopy. *Nat. Cell Biol.* **2000**, *2*, 313–317. [CrossRef]
22. Friedrichs, J.; Legate, K.R.; Schubert, R.; Bharadwaj, M.; Werner, C.; Müller, D.J.; Benoit, M. A Practical Guide to Quantify Cell Adhesion Using Single-Cell Force Spectroscopy. *Methods* **2013**, *60*, 169–178. [CrossRef]
23. Borghi, N.; Sorokina, M.; Shcherbakova, O.G.; Weis, W.I.; Pruitt, B.L.; Nelson, W.J.; Dunn, A.R. E-Cadherin Is under Constitutive Actomyosin-Generated Tension That Is Increased at Cell–Cell Contacts upon Externally Applied Stretch. *Proc. Natl. Acad. Sci. USA* **2012**, *109*, 12568–12573. [CrossRef]
24. Ramsden, J.J.; Horvath, R. Optical Biosensors for Cell Adhesion. *J. Recept. Signal Transduct.* **2009**, *29*, 211–223. [CrossRef]
25. Kasas, S.; Ruggeri, F.S.; Benadiba, C.; Maillard, C.; Stupar, P.; Tournu, H.; Dietler, G.; Longo, G. Detecting Nanoscale Vibrations as Signature of Life. *Proc. Natl. Acad. Sci. USA* **2015**, *112*, 378–381. [CrossRef]
26. Pleskova, S.N.; Lazarenko, E.V.; Bezrukov, N.A.; Bobyk, S.Z.; Boryakov, A.V.; Kriukov, R.N. Differences in Bacteria Nanomotion Profiles and Neutrophil Nanomotion during Phagocytosis. *Front. Microbiol.* **2023**, *14*, 1113353. [CrossRef]
27. Wu, S.; Zhang, Z.; Zhou, X.; Liu, H.; Xue, C.; Zhao, G.; Cao, Y.; Zhang, Q.; Wu, X. Nanomechanical Sensors for Direct and Rapid Characterization of Sperm Motility Based on Nanoscale Vibrations. *Nanoscale* **2017**, *9*, 18258–18267. [CrossRef]
28. Wu, S.; Liu, X.; Zhou, X.; Liang, X.M.; Gao, D.; Liu, H.; Zhao, G.; Zhang, Q.; Wu, X. Quantification of Cell Viability and Rapid Screening Anti-Cancer Drug Utilizing Nanomechanical Fluctuation. *Biosens. Bioelectron.* **2016**, *77*, 164–173. [CrossRef]
29. Girasole, M.; Dinarelli, S.; Longo, G. Correlating Nanoscale Motion and ATP Production in Healthy and Favism Erythrocytes: A Real-Time Nanomotion Sensor Study. *Front. Microbiol.* **2023**, *14*. [CrossRef]
30. Longo, G.; Alonso-Sarduy, L.; Rio, L.M.; Bizzini, A.; Trampuz, A.; Notz, J.; Dietler, G.; Kasas, S. Rapid Detection of Bacterial Resistance to Antibiotics Using AFM Cantilevers as Nanomechanical Sensors. *Nat. Nanotech* **2013**, *8*, 522–526. [CrossRef]
31. Al-Madani, H.; Du, H.; Yao, J.; Peng, H.; Yao, C.; Jiang, B.; Wu, A.; Yang, F. Living Sample Viability Measurement Methods from Traditional Assays to Nanomotion. *Biosens.* **2022**, *12*, 453. [CrossRef]
32. Vocat, A.; Sturm, A.; Jóźwiak, G.; Cathomen, G.; Świątkowski, M.; Buga, R.; Wielgoszewski, G.; Cichocka, D.; Greub, G.; Opota, O. Nanomotion Technology in Combination with Machine Learning: A New Approach for a Rapid Antibiotic Susceptibility Test for Mycobacterium Tuberculosis. *Microbes Infect.* **2023**, *25*, 105151. [CrossRef]
33. Rosłoń, I.E.; Japaridze, A.; Steeneken, P.G.; Dekker, C.; Alijani, F. Probing Nanomotion of Single Bacteria with Graphene Drums. *Nat. Nanotechnol.* **2022**, *17*, 637–642. [CrossRef]
34. Willaert, R.G.; Vanden Boer, P.; Malovichko, A.; Alioscha-Perez, M.; Radotić, K.; Bartolić, D.; Kalauzi, A.; Villalba, M.I.; Sanglard, D.; Dietler, G.; et al. Single Yeast Cell Nanomotions Correlate with Cellular Activity. *Sci. Adv.* **2020**, *6*, eaba3139. [CrossRef]
35. Venturelli, L.; Kohler, A.-C.; Stupar, P.; Villalba, M.; Kalauzi, A.; Radotic, K.; Bertacchi, M.; Dinarelli, S.; Girasole, M.; Pesic, M.; et al. A Perspective View on the Nanomotion Detection of Living Organisms and Its Features. *J. Mol. Recognit.* **2020**, *33*, e2849. [CrossRef]
36. Hostetter, M.K. Adhesins and Ligands Involved in the Interaction of Candida Spp. with Epithelial and Endothelial Surfaces. *Clin. Microbiol. Rev.* **1994**, *7*, 29–42. [CrossRef]
37. Calderone, R.A.; Scheld, W.M. Role of Fibronectin in the Pathogenesis of Candidal Infections. *Rev. Infect. Dis.* **1987**, *9* (Suppl. 4), S400–S403. [CrossRef]
38. Klotz, S.A. Plasma and Extracellular Matrix Proteins Mediate in the Fate of Candida Albicans in the Human Host. *Med. Hypotheses* **1994**, *42*, 328–334. [CrossRef]
39. Pendrak, M.L.; Klotz, S.A. Adherence of Candida Albicans to Host Cells. *FEMS Microbiol. Lett.* **1995**, *129*, 103–113. [CrossRef]
40. Kozik, A.; Karkowska-Kuleta, J.; Zajac, D.; Bochenska, O.; Kedracka-Krok, S.; Jankowska, U.; Rapala-Kozik, M. Fibronectin-, Vitronectin- and Laminin-Binding Proteins at the Cell Walls of Candida Parapsilosis and Candida Tropicalis Pathogenic Yeasts. *BMC Microbiol.* **2015**, *15*, 197. [CrossRef]
41. Nett, J.E.; Cabezas-Olcoz, J.; Marchillo, K.; Mosher, D.F.; Andes, D.R. Targeting Fibronectin To Disrupt In Vivo Candida Albicans Biofilms. *Antimicrob. Agents Chemother.* **2016**, *60*, 3152–3155. [CrossRef]
42. Bougnoux, M.-E.; Kac, G.; Aegerter, P.; d'Enfert, C.; Fagon, J.-Y.; Group, C.S. Candidemia and Candiduria in Critically Ill Patients Admitted to Intensive Care Units in France: Incidence, Molecular Diversity, Management and Outcome. *Intensive Care Med.* **2008**, *34*, 292–299. [CrossRef]
43. Sdoudi, K.; Bougnoux, M.-E.; El Hamoumi, R.; Diogo, D.; El Mdaghri, N.; Christophe, d. Aziza Razki Phylogeny and Diversity of Candida Albicans Vaginal Isolates from Three Continents. *Int. J. Curr. Microbiol. Appl. Sci.* **2014**, *3*, 471–480.
44. Ropars, J.; Maufrais, C.; Diogo, D.; Marcet-Houben, M.; Perin, A.; Sertour, N.; Mosca, K.; Permal, E.; Laval, G.; Bouchier, C.; et al. Gene Flow Contributes to Diversification of the Major Fungal Pathogen Candida Albicans. *Nat. Commun.* **2018**, *9*, 2253. [CrossRef]
45. Gillum, A.M.; Tsay, E.Y.; Kirsch, D.R. Isolation of the Candida Albicans Gene for Orotidine-5′-Phosphate Decarboxylase by Complementation of S. Cerevisiae Ura3 and E. Coli pyrF Mutations. *Mol. Gen. Genet.* **1984**, *198*, 179–182. [CrossRef]
46. Schönherr, F.A.; Sparber, F.; Kirchner, F.R.; Guiducci, E.; Trautwein-Weidner, K.; Gladiator, A.; Sertour, N.; Hetzel, U.; Le, G.T.T.; Pavelka, N.; et al. The Intraspecies Diversity of C. Albicans Triggers Qualitatively and Temporally Distinct Host Responses That Determine the Balance between Commensalism and Pathogenicity. *Mucosal Immunol.* **2017**, *10*, 1335–1350. [CrossRef]
47. Murciano, C.; Moyes, D.L.; Runglall, M.; Tobouti, P.; Islam, A.; Hoyer, L.L.; Naglik, J.R. Evaluation of the Role of Candida Albicans Agglutinin-like Sequence (Als) Proteins in Human Oral Epithelial Cell Interactions. *PLoS ONE* **2012**, *7*, e33362. [CrossRef]

48. Kohler, A.-C.; Venturelli, L.; Kannan, A.; Sanglard, D.; Dietler, G.; Willaert, R.; Kasas, S. Yeast Nanometric Scale Oscillations Highlights Fibronectin Induced Changes in C. Albicans. *Ferment.* **2020**, *6*, 28. [CrossRef]
49. Codling, E.A.; Plank, M.J.; Benhamou, S. Random Walk Models in Biology. *J. R. Soc. Interface* **2008**, *5*, 813–834. [CrossRef]
50. Jung, P.; Mischo, C.E.; Gunaratnam, G.; Spengler, C.; Becker, S.L.; Hube, B.; Jacobs, K.; Bischoff, M. Candida Albicans Adhesion to Central Venous Catheters: Impact of Blood Plasma-Driven Germ Tube Formation and Pathogen-Derived Adhesins. *Virulence* **2020**, *11*, 1453–1465. [CrossRef]

Disclaimer/Publisher's Note: The statements, opinions and data contained in all publications are solely those of the individual author(s) and contributor(s) and not of MDPI and/or the editor(s). MDPI and/or the editor(s) disclaim responsibility for any injury to people or property resulting from any ideas, methods, instructions or products referred to in the content.

MDPI AG
Grosspeteranlage 5
4052 Basel
Switzerland
Tel.: +41 61 683 77 34

Fermentation Editorial Office
E-mail: fermentation@mdpi.com
www.mdpi.com/journal/fermentation

Disclaimer/Publisher's Note: The statements, opinions and data contained in all publications are solely those of the individual author(s) and contributor(s) and not of MDPI and/or the editor(s). MDPI and/or the editor(s) disclaim responsibility for any injury to people or property resulting from any ideas, methods, instructions or products referred to in the content.

www.ingramcontent.com/pod-product-compliance
Lightning Source LLC
Chambersburg PA
CBHW041358210926
51850CB00013B/3114

9783725822935